TECHNOLOGY AND YOUTH: GROWING UP IN A DIGITAL WORLD

SOCIOLOGICAL STUDIES OF CHILDREN AND YOUTH

Series Editor: David A. Kinney (from 1999)

Series Editors: David A. Kinney and
Katherine Brown Rosier (2004–2010)

Series Editors: David A. Kinney and
Loretta E. Bass (from 2011)

Series Editor: Loretta E. Bass (from 2012)

Recent Volumes:

SOCIOLOGICAL STUDIES OF CHILDREN AND YOUTH
VOLUME 19

TECHNOLOGY AND YOUTH: GROWING UP IN A DIGITAL WORLD

EDITED BY

SAMPSON LEE BLAIR
State University of New York, Buffalo, NY, USA

PATRICIA NEFF CLASTER
Edinboro University of Pennsylvania, Edinboro, PA, USA

SAMUEL M. CLASTER
Edinboro University of Pennsylvania, Edinboro, PA, USA

SERIES EDITOR

LORETTA E. BASS
The University of Oklahoma, Norman, OK, USA

Emerald

United Kingdom – North America – Japan
India – Malaysia – China

Emerald Group Publishing Limited
Howard House, Wagon Lane, Bingley BD16 1WA, UK

First edition 2015

British Library Cataloguing in Publication Data
A catalogue record for this book is available from the British Library

ISBN: 978-1-78560-265-8
ISSN: 1537-4661 (Series)

Printed and bound by CPI Group (UK) Ltd, Croydon, CR0 4YY

CONTENTS

LIST OF CONTRIBUTORS

Åsa Andersson	University West, Trollhättan, Sweden
Troy E. Beckert	Utah State University, Logan, UT, USA
Margareta Bohlin	University West, Trollhättan, Sweden
Antonella Brighi	University of Bologna, Bologna, Italy
Ana Campos-Holland	Connecticut College, New London, CT, USA
Shelia R. Cotten	Michigan State University, East Lansing, MI, USA
Elmir de Almeida	*Universidade de São Paulo*, Ribeirão Preto, Brazil
Brooke Dinsmore	University of Virginia, New Ipswich, VA, USA
Sara L. Dodd	Texas Tech University, Lubbock, TX, USA
Harry T. Dyer	University of East Anglia, Norwich, UK
Holly E. Follmer-Reece	Texas Tech University, Lubbock, TX, USA
Giovanna Gianesini	University of Bologna, Bologna, Italy
Philip Gibson	University of Alabama at Birmingham, Birmingham, AL, USA
Gloria Gonzales	Texas Tech University, Lubbock, TX, USA
Timothy M. Hale	Harvard Medical School, Boston, MA, USA
Jodie L. Hertzog	Wichita State University, Wichita, KS, USA
Erin Kostina-Ritchey	Texas Tech University, Lubbock, TX, USA
Linda Lundin	University West, Trollhättan, Sweden

Yuping Mao	Erasmus University Rotterdam, Rotterdam, The Netherlands
Vanderlei Mariano	Centro Universitário Fundação Santo André, Santo André, Brazil
Christine McCauley Ohannessian	University of Connecticut School of Medicine, Farmington, CT, USA
Gregory J. Mills	University of Connecticut, Storrs, CT, USA
Marilena Nakano	Centro Universitário Fundação Santo André, Santo André, Brazil
Jessica Niblack	Hutchinson Community College, Sedgwick, KS, USA
LaToya O'Neal Coleman	Michigan State University, East Lansing, MI, USA
Gina Pol	Connecticut College, New London, CT, USA
Matthew H. Rafalow	University of California-Irvine, Irvine, CA, USA
Kayla Sherman	Texas Tech University, Lubbock, TX, USA
Lu Shi	Clemson University, Clemson, SC, USA
Brian Simpson	University of New England, Armidale, Australia
Jennifer E. Simpson	The Open University, Milton Keynes, UK
Radhamany Sooryamoorthy	University of KwaZulu-Natal, Durban, South Africa
Emma Sorbring	University West, Trollhättan, Sweden
Sarah Tulane	Utah State University, Logan, UT, USA
J. Mitchell Vaterlaus	Montana State University, Bozeman, MT, USA
Maria Elena Villar e Villar	Centro Universitário Fundação Santo André, Santo André, Brazil
Kevin Zevallos	Connecticut College, New London, CT, USA

EDITORIAL BOARD

FOREWORD

Over recent decades, the prominence of technology in the lives of children has grown dramatically. While previous generations of parents and researchers were concerned about the impact of television upon the development of children, the veritable leaps and bounds of technological innovation have made it difficult to keep track of the almost daily changes in the forms of technology in the lives of children. The widespread use of cellular technology has made it increasingly more common to see not only adolescents, but young children carrying cell phones. The ubiquitous nature of the internet, along with the growth of ever-changing forms of social media, has created a rather dichotomous existence for many youth – one in which they live in the "real world," and yet another in which they live in the "online world." Even the simple notion of a young person playing video games has taken on a radically different quality, as the majority of such games now emphasize the online interaction with other players. Although researchers have been attempting to study how the various forms of technology may potentially affect children and adolescents, there is no denying the fact the technology is playing an ever-important role in their lives.

Recent estimates concerning the levels of usage vary, with some suggesting that adolescents are spending over eight hours each day engaged with some form of technology. Even preschoolers are not immune to the inclusion of technology, as some studies report them as using technology over four hours each day, on average. In many ways, a generational divide has developed, such that children and adolescents, who have grown up in a world where mastery of technological change is an absolute must, have a decided advantage over their parents when it comes to the ease of use of technology. Older generations often find themselves fumbling about with their technological devices, such as their smartphones, and when they are unable to comprehend how it is supposed to function, they turn to their own children for assistance. There is little doubt that contemporary children and adolescents are exposed to changing technologies, not only at home, but also in their schools. As such, they frequently demonstrate a mastery of technological devices which leaves most adults both impressed and embarrassed.

Several decades ago, researchers from a variety of disciplines began to raise alarms about children's increasing exposure to television, and how both the quantity and quality of television shows could affect them. In most instances, such concerns were made from a social problem perspective, and typically involved a range of warnings about the potentially deleterious and harmful impacts which television could have upon children. Technologies and technological change, however, do not necessarily have to be envisioned as harmful. More and more often, educators are developing new ways to integrate technology into the classroom, making it a central component of their pedagogical plans. Given that children are already drawn by the appeal of using new technologies, it is quite logical that the use of such in the classroom could significantly improve the learning environments and educational outcomes for children. On the other hand, technologies in the lives of children can, in fact, be quite problematic. Many parents live in fear of their children being lured by sex predators on the internet, of having their adolescents engaging in "sexting" with others, or of being "bullied" online. The effects of technologies on children and adolescents, then, can potentially be both good and bad, which makes our need for better understanding the nature of technology in the lives of children and adolescents even greater.

In this volume of *Sociological Studies of Children and Youth*, we attempt to pursue these issues, with investigations of a variety of technological forms, and across a broad representation of children and adolescents. In "Cyberbullying in the Era of Digital Relationships: The Unique Role of Resilience and Emotion Regulation on Adolescents' Adjustment," Giovanna Gianesini and Antonella Brighi examine the effects of peer violence in cyberspace upon adolescents' emotion regulation and socio-emotional adjustment. They find that while online victimization is quite common, resilience can often play an important role in coping with online bullying. Although social interaction in the online world appears to involve a great deal of anonymity, such interactions are, nonetheless, witnessed by others. Jessica Niblack and Jodie L. Hertzog also examine the nature of online bullying, but focus upon how others react and respond when they observe such behaviors. In "Factors that Influence Bystander Behavior in the Cyberbully Context," they note that while the majority of youth have witnessed online bullying, the reactions of "bystanders" can vary quite a bit.

The ever-changing nature of technologies can often bring about tangible changes in the interactional patterns of youth, as well as the ways in which they see themselves. In "Cell Phone Use and Youth Perceptions of

Communication in South Africa," Radhamany Sooryamoorthy examines the use of cell phones by young college students. Not only are the patterns of usage found to vary substantially by gender and race, but the perceiving meanings and forms of communication are shown to vary, as well. Using a sample of Swedish adolescents, Åsa Andersson, Margareta Bohlin, Linda Lundin, and Emma Sorbring investigate the manners by which teens use their online activities as a means of defining themselves. The internet, in the eyes of many youth, represents an opportunity to establish an identity, and the authors find that many young people are eager to do so, yet significant differences exist between boys' and girls' usage. The meaning of the internet, particularly as an element of the social lives of youth, is explored by Elmir de Almeida, Marilena Nakano, Maria Elena Villar e Villar, and Vanderlei Mariano in "Young Collegians: Between the Physical and Numerical Territories, Different Processes of Individuation." Using a sample of young college students in Brazil, their study demonstrates the relative meaning of online lives, which can vary considerably in both experience and consequence, for young females and males.

The generational divide between young people and adults, particularly in regard to the use of the internet, often leaves adults fearful of what young people are doing online. Ana Campos-Holland, Brooke Dinsmore, Gina Pol, and Kevin Zevallos investigate how children and adolescents try to steer their way through the online world in "Keep Calm: Youth Navigating Adult Authority across Networked Publics." They find that while young people are aware of adult monitoring and supervision, many children and adolescents are adept at working around such controls. As Harry T. Dyer points out in his study, "All the Web's a Stage: The Effects of Design and Modality on Youth Performances of Identity," although youth are increasingly making use of social networking sites, their usage patterns are constrained by the form and structure of the sites, themselves. Identity portrayals and social interactions depend heavily upon the design and modality of the websites, thus prompting young people to adapt their behaviors as a function of the qualities of the sites. Despite such boundaries of the online world, however, youth are able to cultivate a peer environment of their own design. In "n00bs, Trolls, and Idols: Boundary-Making among Digital Youth," Matthew H. Rafalow examines how young people create social boundaries which actually serve to encourage civility among youthful online users.

Understandably, researchers, practitioners, and parents alike are concerned about the effects of online activity upon the development and well-being of children and adolescents. In "The Impact of Information and

Communication Technology (ICT) Usage on Psychological Well-Being among Urban Youth," LaToya O'Neal Coleman, Timothy M. Hale, Shelia R. Cotten, and Philip Gibson examine the impact of online activity upon the psychological well-being of urban adolescents. Their findings suggest that both the quantity and quality of online activity needs to be considered, in regards to the well-being of youth. Beyond psychological well-being, behavioral issues must also be addressed, in relation to the use of technology by children and adolescents. In "A Longitudinal Examination of the Relationship between Technology Use and Substance Use during Adolescence," Christine McCauley Ohannessian examines how the substance use patterns of adolescents may be affected by the larger combination of cell phone, computer, and video game usage. While significant effects are shown, she also demonstrates that the relationship between technology use and substance use is very much a reciprocal one.

In some instances, a seemingly benign technology, such as the cellular phone, can be used in a manner by youth which causes great alarm within society. In "Sexting, Digital Dissent and Narratives of Innocence – Controlling the Child's Body," Brian Simpson examines how young people are sending sexualized images via their phones. Focusing upon the Australian context, he demonstrates that what many adults perceive as an entirely inappropriate activity may also present a dilemma for how the legal system deals with the same. Jennifer E. Simpson, in "Technology and the fostered child: A new social work response." offers a similar perspective. She notes that traditional perceptions of children's and adolescents' online activities have primarily come from a problem-orientation, and have largely overlooked how youth can actually exercise a substantial amount of both self-control and self-responsibility. The misperceptions of adults concerning children's and adolescents' use of technology are also made evident in the research by Sarah Tulane, J. Mitchell Vaterlaus, and Troy E. Beckert. In ""That Is SO Not True": Adolescent Perspectives of Adult Misconceptions of Teen Text Messaging," they use qualitative analyses of a sample of high students, who clearly illustrate that adults often have little understanding of what adolescents are actually doing, and their fears about such behaviors as "sexting" are typically not based upon realistic assessments of teen's actual cellular phone use.

Although emerging technologies often generate a considerable amount of alarm and fear among parents and researchers, even long-existing technologies have been shown to have substantial effects upon the lives of children and adolescents. In "Watching Television and Reading Achievement: A Study of Third Grade Language Minority Students," Gregory J. Mills examines how television viewership can affect the development of reading skills among language-minority elementary students. In his study, Mills notes that, contrary to many existing concerns, television viewing appears to provide a positive impact upon the reading achievements of language-minority children, as it may serve to facilitate their development of reading skills. In regard to other behaviors, however, the use of technology may not necessarily be beneficial. Such is shown by Yuping Mao and Lu Shi in their study, "Can Media Consumption Predict Immigrant Adolescents' Acculturation-Related Risky Health Behavior? An Analysis of Latino Sample in CHIS Survey." Using a sample of Latino adolescents in the United States, they find that higher usage rates of television, video games, and computers are associated with a tendency toward greater risk-taking behaviors by Latino teens. In "United Future Leaders: A Case of the Use of Technology in Youth Programming and Hidden Curriculum," Erin Kostina-Ritchey, Holly E. Follmer-Reece, Sara L. Dodd, Kayla Sherman, and Gloria Gonzales examine the "hidden curriculum" within technology platforms, and how these might be used by practitioners to provide more positive influences in the lives of youth.

Overall, the authors in this volume provide a very broad and enlightening examination of the role and impact of technology in the lives of children and adolescents. Their studies clearly offer a much greater understanding of how changing technologies are bringing about change in the very fabric of childhood and adolescence. Their collective research offers considerable insight into the potential effects of technological change, and also underscores the need for such research to continue. Indeed, given the pace of technological change, it is vital that researchers maintain a similar pace of investigation. Many thanks are due to the authors for their efforts herein, to the many reviewers who helped along the way, and to the editorial staff at Emerald Publishing.

Sampson Lee Blair
Patricia Neff Claster
Samuel M. Claster
Editors

CYBERBULLYING IN THE ERA OF DIGITAL RELATIONSHIPS: THE UNIQUE ROLE OF RESILIENCE AND EMOTION REGULATION ON ADOLESCENTS' ADJUSTMENT

Giovanna Gianesini and Antonella Brighi

ABSTRACT

Purpose – *In this study, we aimed at examining the unique and interactive effects of peer violence in cyberspace on adolescents' emotion regulation and socioemotional adjustment, as well as the mediational role of resilience in the link between adolescent's pathogenic relational experiences and behavioral outcomes. Specifically, we intended to explore emotion differentiation and regulation in reaction to bullying perpetration and victimization and in terms of positive (proud, confident, good) and negative (ashamed, excited, guilty), Passive (sad, embarrassed, humiliated) and Reactive (angry, scared) emotions and how it impacted and predicted positive and negative outcomes.*

Methodology/approach – *A stratified convenient sample of 494 Italian students aged 13–19 years (*M = *15.27, SD = 1.23) was selected to*

Technology and Youth: Growing Up in a Digital World
Sociological Studies of Children and Youth, Volume 19, 1–46

ISSN: 1537-4661/doi:10.1108/S1537-466120150000019001

represent all different school types in Italy and the students were administered a self-report questionnaire on school bullying involvement. General Linear Models, ANOVA, and T-*tests were employed to explore gender differences, the relationships between variables, and their contribution to the predictive model. A two-step Cluster analysis was used to profile adolescents based on patterns of resilience, health outcomes, and cyberbullying involvement.*

Findings – *Results showed significant gender differences, with females using internet and Facebook more than males and being more resilient, positive, and prosocial, but also responding to victimization with higher levels of alienation, anger, humiliation, and psychosomatic and emotional symptoms. Males perpetrated peer violence more than females, were less likely to be victimized, and were generally less emotionally impacted by it. Victimization rates (63.7%,* n = *296) were higher than perpetration rates (51.7%,* n = *233) and bully-victimization was prevalent (47.1%). Victims prevalently experienced passive emotions (sadness, humiliation, embarrassment) while perpetrators experienced negative ones (guilt and shame). Cluster analysis evidenced different pathways and trajectories of resilience and cyberbullying involvement: Resilient victims (RV), Healthy uninvolved (HU), Healthy Bullies (HB), Alienated Bully-Victims (ABV), and Resilient Bully-Victims (RBV). RV, HU, and HB resulted all well-adjusted, despite the different involvement in cyberbullying, and also RBV and despite the double involvement in cyberbullying, ABV were the only maladjusted and at-risk group in our sample characterized by very low Positivity, very low Resilience, and extremely high Alienation.*

Research implications – *This study proposes a comprehensive, developmental, ecological, relational, and self-regulatory resilience approach to cyberbullying, which represents an innovative and advanced contribution to the literature with significant implication for research and practice. Fully understanding and measuring the emotional impact of cyber peer violence and resilience following cyberbullying victimization and perpetration can help in developing targeted interventions for both victims and bullies. This study highlighted the need for a self-regulatory model of resilience for modulating emotions, arousal, and behaviors across contexts, relationships, and difficulties. It also evidenced that moderate levels of resilience and positivity are sufficient to buffer youth from involvement in cyberbullying and to predict healthy adjustment and less pathological outcomes.*

Originality/value – *By profiling adolescents based on resilience levels, health outcomes, and cyberbullying involvement, we evidenced five distinct trajectories of risk evaluation for cyberbullying beyond participating roles. Our results confirmed the fundamental importance of assessing resilience and emotion regulatory resources together with peer violence involvement in identifying and targeting adolescents at risk.*

Keywords: Cyberbullying; resilience; emotion regulation; adolescents' well-being; Youth; digital relationships

INTRODUCTION

Young people are increasingly interacting and living in an online environment characterized by eyes–hands–fingers communication that has dramatically changed peer relationships, expectations, and behaviors (Gianesini, 2013b; Cusinato & L'Abate, 2010). In a world immersed in digital, the net generation communicates across all domains by sharing a "third space" (neither home nor school/work) central to their lives. The digital environment is ideal for communicating with existing friends or starting new mix-mode relationships (Walther & Parks, 2002) that originated online but were further developed offline (Brighi, Guarini, Melotti, Galli, & Genta, 2012). The nature of these digital exchanges, being characterized by a particular language usage (i.e., short utterances, abbreviations, emoticons, and symbols), simulates face-to-face informal conversations that are particularly well suited for peer-to-peer social networking. In online forums and weblogs, in fact, language is a key means through which youth express and explore their identity (Wilding, 2006). Instant messaging may foster intimacy, self-disclosure, and feelings because it cultivates connectedness (Hu, Wood, Smith, & Westbrook, 2004). Online chatting, which includes similar features of the face-to-face interactions in terms of verbosity, assertiveness, profanity, politeness, and representations of emotions, facilitates interactive engagement (Herring, 1999). Demographic factors, like age, gender, and socioeconomic status together with digital expertise account for variations in the breadth and depth of digital and online communication in youth and inequalities in the quality of access to and use of the ICT technologies (Brighi, Guarini, & Genta, 2009; Gianesini, 2013c). Digital communication knows no geographical boundaries and time constrains, and thus it represents both an extended social opportunity and a challenge (Godwin-Jones, 2005) where flexibility and anonymity are possible and adolescents feel more comfortable expressing their identity beyond social prescriptions. However, the extensive use of the Internet for communication (Kraut et al., 1998), digital playing, or Internet surfing may also have negative effects on social involvement and psychological well-being. It has been found to be associated with poor relations with parents (Punamäki, Wallenius, Hölttö, Nygård, & Rimpelä, 2009) and it may represent a forum for aggressive dynamics and dysfunctional behaviors like internet abuse and cyberbullying (Brighi et al., 2012). Higher levels of exposure to antisocial media content significantly contributed to higher initial rates of cyberbullying behavior in a sample of 1.005 adolescents aged 11–17 in the Netherlands (den Hamer & Konijn, 2015). Youth digital interactions and behaviors may convey conflict, interpersonal victimization, and cyberbullying that adolescents somehow consider common and expected, although distressing, and highly gendered (Mishna, McInroy, & Lacombe-Duncan, 2015). In a Canadian stratified random sample of 684 students, grade 4–0th, Mishna et al. (2015) found that girls were more often cyberbullied through social media, with demeaning comments about their appearance and private or compromising photos within their friendship group ("drama"), while boys were more likely to be cyberbullied through online games and homophonic comments ("trash talk"). Although young people mostly communicate online with those they already know offline and the content of such communication tends to be about everyday issues (Reich, Subrahmanyam, & Espinoza, 2012), online experiences have amplified peer-to-peer sexual pressure in youth relationships, particularly for girls ("sexting") (Sen, 2015). Research has reliably demonstrated the importance of examining gender differences in aggressive behaviors finding that males are more physically aggressive (such as punching and kicking) and more likely than females to bully their peers. Yet, research has shown mixed and contrasting evidences regarding gender differences in cyberbullying across different age groups (Tokunaga, 2010). Mark and Ratliffe (2011) and Del Rey, Elipe, and Ortega-Ruiz (2012) found age differences to be significant only with regard to cyber victimization. In their recent meta-analysis Barlett and Coyne (2014) found that males engaged more in cyberbullying than females, but they reach comparable levels of cyberbullying frequency in early adolescence (at 11 years approximately) while prior to that girls are more likely to cyberbully peers than males, whereas males are more likely to be cyberbullied in late adolescence and into college-aged years. Their results also showed that the country of origin moderates gender effects, with gender differences found for North America, Europe, and Asia in the male direction, but not for Australia (Barlett & Coyne, 2014). Cultural differences also emerged in the work of Balakrishnan (2015) who found among Malaysian young adults (aged 17–30) that females outnumbered males as cyberbullies and cyber victims and that internet use frequency significantly predicted cyber victimization and cyberbullying. There seem to be a continuity between processes of inclusiveness, social acceptance, social exclusion or rejection, and aggressive behavior expressed by the individual offline and online, as dysfunctional relational dynamics can take place in real life as well as in a virtual environment (Ortega-Ruiz, Del Rey, & Casas, 2012). However, some of its peculiar features suggest that cyberbullying may have a distinct causal pathway (Brighi et al., 2012; Brighi et al., 2012a, 2012b).

Exposure to and involvement in cyberbullying are evident risks that contemporary youth face through adolescence. Cyberbullying frequency, participating role, and characteristics have been widely studied and found to be related differently to several emotional, psychological, and physical problems, poor academic performance, and increased suicidal ideation (Hindjua & Patchin, 2010). Moreover, the perceptions, expectations, and emotional reactions of those involved differ according to the role and the context in which cyberbullying has been experienced (Gualdo, Hunterr, Durkin, Arnaiz, & Maquilon, 2014). However, few studies have considered how adolescents manage these risks and develop resilience (Papatraianou, Levine, & West, 2014) simultaneously accounting for emotional regulation and behavioral self-control (You, Kim, & No, 2015). Raskauskas, Rubiano, Offen, and Wayland (2015) found lower peer victimization and depression, higher academic performance and emotional resilience in middle school students with above average social self-efficacy. Youth tend to seek situations that help them value themselves positively and to avoid those which make them feel bad about who they are. Their perception and acceptance of their changing self play a critical role in directing their personal and even professional growth trajectory (Andreou, 2004). Experience with cyberbullying, both as a victim and as an offender, is associated with significantly lower levels of self-esteem, even after controlling for demographic differences (Brighi et al., 2012a, 2012b). Self-esteem, in fact, is an internal representation of social acceptance and rejection and a psychological gauge monitoring the degree to which a person is included versus excluded by others (Leary & Downs, 1995). Previous European studies have assessed the emotional impact of cyber victimization classifying victims accordingly to the specific type of aggression suffered (verbal, physical, and relational) (Rigby, 2008) and its severity, mostly based on a frequency measure (Ortega, Elipe, & Monks, 2012; Ortega, Elipe, Mora-Merchan,

Calmaestra, & Vega, 2009). In those studies, the emotions identified were grouped by cluster analysis in order to establish the emotional profile of victims or classify the pattern of victimization into two groups: victims not emotionally affected and those suffering from a wide range of negative emotions (i.e., *defenseless, embarrassed, afraid, upset, angry, stressed, worried, depressed, and lonely*; Ortega et al., 2012). In the present study, the emotional impact of cyberbullying as self-reported by both perpetrators and victims was coded as "affected" (present) or "indifferent" (absent) based initially on the answer to the item "*I felt nothing*." Then, to further explore emotion differentiation and regulation we recoded the six possible emotional responses for perpetrators into the categories Positive (*proud, confident, good*) and Negative (*ashamed, excited, guilty*) and the five possible emotional responses for victims into the categories Passive (*sad, embarrassed, humiliated*) and Reactive (*angry, scared*). We used a relational and emotion regulatory resilience prospective (Gianesini, 2009, 2015), rather than a traditional deficit-based approach, to explore what factors and mechanisms promote emotional and behavioral adjustment (Bowes, Maughan, Caspi, Moffitt, & Arseneault, 2010; Jenkins, 2008) to cyberbullying and the key role of emotion regulation. Specifically, we tested the hypothesis that emotion regulation serves as a resilience factor in buffering youth from the negative influences and outcomes of peer violence in early to late adolescence. First, we compared the two different measures for Resilience (RS-14 and DS) for validity, expecting a strong correlation. Specifically, we hypothesized (H1) high levels of resilience (RS-14) to be positively related to a positive attitude (DS subscale Positivity) and negatively to helplessness (DS subscale Helpless/Alienation). Second, we intended to verify whether and how resilience can moderate the negative impact of cyberbullying, and thus predict healthy adjustment and a positive outcome in adolescents. We therefore expected (H2) high levels of resilience (RS-14) and a positive attitude (DS) to predict higher scores on the SDQ Prosocial Behavior subscale (a measure of healthy social adjustment) and lower scores on the SDQ-Emotional Symptoms subscale (a measure of psychosomatic complains) for both victims and perpetrators (or both) of cyber peer violence. Third, we were interested in understanding the diverse impact of cyberbullying on adolescents, whether due to factors characterizing cyberbullying events or to differences in the resilience of the victims and perpetrators (or both) (Fenaughty & Harre', 2013; Van Royen, Poels, Daelemans, & Vandebosch, 2015). Hence, we hypothesized (H3) high resilience levels to predict lower emotional reactivity as the presence of emotion regulatory skills.

This study proposes a comprehensive, developmental, ecological, relational, and self-regulatory resilience approach to cyberbullying, which represents an innovative and advanced contribution to the literature with significant implication for research and practice. Learning to regulate and modulate emotions, to navigate relationships, and to acquire appropriate strategies for dealing with difficult emotional and social situations, is an essential developmental task especially during adolescence. Fully understanding and measuring the emotional impact of cyber peer violence and resilience following cyberbullying victimization and perpetration can help in developing targeted interventions for both victims and bullies.

SOCIAL EXCLUSION, BULLYING, AND CYBERBULLYING

Bullying is a relationship problem (Pepler et al., 2006) comprising behaviors that harm others directly or indirectly through damage or threat to friendship or group inclusion (James et al., 2011; Perren, Gutzwiller-Helfenfinger, Malti, & Hymel, 2011) and involving peers at different levels and in different roles. These include manipulation, controlling, belittling, demeaning, rejecting, and exclusionary behaviors in social contexts, as well as damage or threat to one's reputation. Social exclusion is a subtle form of relational aggression or bullying, which entails a lack of connectedness and participation from a peer group, in which adolescents are exposed to harm through the manipulation of their social relationships and status (DeWall, Deckman, Pond & Bonser, 2011). This can take many forms, ranging from deliberate exclusion from a peer group to rumors spreading, names calling, and intentional embarrassment. Research suggests that approximately one in six children report experiences of social exclusion and that adolescents are particularly sensitive to peer rejection as belonging and connectedness to peers is particularly relevant in adolescence (Killen, 2007). Typically, adolescence is a time of increased independence from parents and family and identity development in relation to peer groups. This increased dependence on the peer group, fluid in- and out-group's rules and peer group differences make peer relationships per se complicated and causing depression and anxiety (Leets & Wolf, 2005). Moreover, studies on the neurological profile of children suggest that their brain areas for emotion (such as the Anterior Cingulate Cortex) become more activated in response to peer rejection with age, peaking in adolescence. In contrast, adolescents show significantly less activation in the brain regions, which govern emotional regulation such as the Ventrolateral Prefrontal Cortex in response to peer rejection in comparison with younger children (Bolling et al., 2011). This unique neurological profile for adolescents suggests that social exclusion at this age may be particularly distressing and that they may have significant difficulty in managing such distress. Exclusion is multifaceted, based on prejudicial judgments, stereotyping norms and status quo, social desirability, social and interpersonal dominance, interpersonal empathy, depression, self-worth, self-esteem, and self-awareness (Hoover, 2005). In the end, engaging in or experiencing relational aggression may be an unconscious reaction to dynamics in relationships (James et al., 2011), stemming from family background and related to jealousy, malicious gossip, friendship exclusion, inappropriate remarks, feelings of humiliation, fear of rejection, loneliness, and anger (Nishina, Juvonen, & Witkow, 2005).

Engagement in aggressive behaviors using digital technologies like emails, text messaging via mobile phones and social media, video clips, instant messaging and photos on personal websites mostly by making derogatory or ostracizing hostile comments about others with the intention to harm is defined as "cyberbullying" (Smith et al., 2008; Wang, Iannotti, & Luk, 2012). It represents an interpersonal problem grounded in a social context facilitated by technological tools (Cowie, 2013). Figures showed that rates of cyberbullying are lower than for traditional face-to-face bullying, although many studies indicate a significant overlap between the two (Kowalski, 2008; Kowalski, Limber & Agatston, 2008; Perren, Dooley, Shaw, & Cross, 2010; Sourander et al., 2010; Ybarra & Mitchell, 2004). Whereas traditional bullying seems to carry over into cyberbullying, cyberbullying does not appear to turn into bullying (Li, 2008; Riebel, Jaeger, & Fischer, 2009; Schneider, O'Donnell, Stueve, & Coulter, 2012; Smith et al., 2008). Ortega et al. (2012) found that victimization predicts later abuse and cyber-abuse, but that abuse correlates inversely to victimization, cyber victimization, and cyber-abuse, suggesting that a victim is more likely to turn into an aggressor/bully than vice versa. However, cyberbullying may create more damage than physical aggression among youth (Crick & Grotpeter, 1996) because it targets the context of a wider peer group network. This manipulation of social relationships requires verbal, cognitive, and social skills (Crick, Casas, & Ku, 1999) and the ability of the perpetrator to identify the social and emotional weaknesses of the targeted victims (Arsenio & Lemerise, 2001; Thornberg & Jungert, 2013). When communicating online, individuals feel that they are part of an anonymous mass and therefore tend not to take responsibility for their actions or do not perceive their actions as particularly harmful because "virtual" is not "real." In addition, the lack of face-to-face contact also allow the perpetrator not to be directly confronted with the harm caused (Hymel, Schonert-Reichl, Bonanno, Vallaincourt, & Rocke Henderson, 2010).

Research shows that in the context of peer violence, social status (Mitsopoulou & Giovazolias, 2015; Wachs, 2012) and dominance play an important role, can greatly vary by age (Pellegrini & Long, 2002) and represent a means by which individuals protect themselves from outsiders and preserve the purity of the group (Rigby, 2002). Bullying is a deliberate type of aggression (Pellegrini, 2004; Rigby, 2002) having psychological and physiological elements (Farrington, 1993; Smith & Sharp, 1994) used as a strategy to establish and maintain social dominance (Hawley, 1999; Volk, Camilleri, & Dane, 2012) by compensation (aggression) or generalization (victimization) of power differential (Nation, Vieno, Perkins, & Santinello, 2008; Pellegrini & Long, 2002). It facilitates hierarchical behavior within groups, in-group and out-group processes, by providing individuals with a sense of membership, belonging, and identity, and enhancing self-esteem. It also facilitates greater access to resources valued by a group, the achievement of one's needs (Gianesini, 2013a), and the minimization of group aggression (Pellegrini, 2004) and safety risks to the in-group (Kurzban & Leary, 2001; Brand, Felner, Shim, Seitsinger, & Dumas, 2003). Both explicit factors (e.g., overt submission behaviors) and implicit factors (e.g., greater competence, status, or influence) are involved in bullying; however, the role of power dependence, perceived interpersonal power, and empowerment is still unclear (Emerson, 1981; Gianesini, 2000; Hooper, L'Abate, Sweeney, Gianesini, & Jankowsli, 2013; Lawler & Yoon, 1996; Molm, 1997, 2000; Molm, Peterson, & Takahashi, 1999; Nation et al., 2008). A clear relationship on the contrary exists between moral disengagement (Menesini et al., 2003; Bussey, Fitzpatrick, & Raman, 2015), moral standards, levels of emotionality, empathy, and bullying participating roles (Almeida, Correia, & Marinho, 2010; Hymel, Rocke-Henderson, & Bonanno, 2005; Hymel et al., 2010; Jolliffe & Farrington, 2006). In terms of emotional responses to cyberbullying, anger may act as a form of bonding between the victim and the perpetrator and create emotional connectedness and ambivalent or negative feelings (i.e., powerlessness) while feeling powerless for victims may result in rage and hopelessness (Arcidiacono, Procentese, & Di Napoli, 2007).

EMOTIONAL RESPONSES TO CYBERBULLYING

Cyberbullying emerges most commonly from relationship problems such as break-ups, envy, intolerance, ganging up, and its victims experience powerfully negative effects because their reactive behavior is generally inappropriate, absent, or ineffective (Hoff & Mitchell, 2009). It can take multiple forms (i.e., threats, exclusion, and name-calling) in different contexts (Van Royen et al., 2015). Research has consistently identified the devastating consequences of cyberbullying: it is associated with a range of negative emotional and behavioral outcomes for adolescents with a great individual variability. Victims experience lack of acceptance in the peer groups, which results in loneliness and social isolation, social withdrawal, low self-esteem, and depression. Bullies too are at risk as they are more likely to engage in a range of maladaptive and antisocial behaviors (Cowie, 2011). As found in studies of face-to-face bullying, cyber victims report feeling unsafe and isolated, both at school and at home, and experience psychosomatic problems, such as headaches, recurring abdominal pain, and sleeplessness. Like traditional bullies, they too are engaged in a range of other antisocial behaviors, conduct disorders, and alcohol and drug abuse (Sourander et al., 2010).

Individuals differ considerably in their emotion experience, which could be highly differentiated, clearly distinguishing among a variety of negative and positive discrete emotions or relatively undifferentiated, treating a range of like-valence terms as interchangeable (Barrett, Gross, Christensen, & Benvenuto, 2001). Emotion differentiation is defined by associations, rather than an absolute co-occurrence of level, in self-reports of emotional experience by distinctive representations of felt experience (i.e., cyberbullying) across negative states such as anger, sadness, and nervousness, or between different positive states, such as happiness, relaxation, and enthusiasm (Kashdan, Ferssizidis, Collins, & Muraven, 2010). In contrast, other individuals may represent their emotional experiences in an undifferentiated fashion along a single pleasant–unpleasant dimension. Awareness of discrete emotions leads to clearer inferences about antecedents, behavioral response tendencies, and steps that can be taken to maintain or alter a situation as desired (Ekman, 1992). According to the self-regulation theory, individuals with highly differentiated emotion experience are at an advantage in regulating their emotions (Keltner & Bonanno, 1997; Bonanno, Gooring, & Coifman, 2008). This differentiation–regulation relationship is strongest in the context of intense negative emotions, where the need for emotion regulation is greatest. Intense and unresolved negative emotions have an acute effect on body functioning and trigger an increase in the risk of health problems, while the use of adaptive emotion regulation strategies, like cognitive appraisal and emotion suppression, facilitate interpersonal interactions and is associated with a healthier pattern of physical and psychological functioning (John & Gross, 2004). Although individuals report regulating both positive and negative emotions (Parrott, 1993), the greatest call for emotion regulation typically comes when there are high levels of negative emotions such as anger, sadness, and fear. In contrast, positive emotions appear to motivate an individual to broaden and build, that is, to explore new intellectual and social pursuits and store resources for future negative events that require regulatory attempts (Fredrickson, 1998). The ability to identify, understand, discriminate, process, and express emotions plays an important role in resilience (Bonanno, 2009; Gianesini, 2015). In resilient individuals emotions are competently used as coping strategies, allowing a broader flexibility to adjust to the shifting demands of different situations (Bonanno, Pat-Horenczyk, & Noll, 2011; Tugade & Fredrickson, 2007). Positive emotions and positive emotional granularity, that is the tendency to discriminate between positive emotions (e.g., joy, interest, contentment) rather than representing feelings in terms of more global states (i.e., pleasantness), are protective factors serving important short-term health-promoting functions as well as long-term advantages for coping in the future. Such a fine-tuned understanding of emotions, especially during times of stress is beneficial to direct coping and is associated with healthier outcomes (Tugade & Fredrickson, 2002).

EMOTION REGULATION AND RESILIENCE

The experience of emotions always occurs in an environmental context, not always safe and predictable, where individuals process information from multiple sources, to develop adaptive and functional responses (Davis, Zautra, & Smith, 2004; Ekman, 1992, 1993). Psychological resilience is the resourceful adaptation to changing circumstances and environmental contingencies (Lifton, 1993; Gillespie, Chaboyer, & Wallis, 2007). It is an analysis of the goodness of fit between situational demands and behavioral possibility (Masten, Best, & Garmezy, 1990) that implies a flexible invocation of the available repertoire of problem-solving, emotional responses, and emotional regulation strategies within relationships (Block & Block, 1980; Gottman, 1997; Ryff, Singer, Dienberg Love, & Essex, 1998; Werner, 1989). However, the ability to flexibly choose to express or to suppress positive and negative emotions (Bonanno, Wortman, & Nesse, 2004;

Clark & Finkel, 2005) depends upon the context (Bonanno, Papa, O'Neil, Westphal, & Coifman, 2005), can be observed very early in development, changes over time (Westphal, Seivert, & Bonanno, 2010), and can be enhanced or reduced by developmental experiences (Bonanno & Mancini, 2012). While cyberbullying inevitably varies across relationships (Nation et al., 2008; Olweus, 1991), empowerment and resilience are more stable characteristic developed when individuals are in control of their environment, life, and resources (Perkins & Zimmerman, 1995). This study offers a useful and unique relational, emotion regulatory, and developmental perspective on cyberbullying and victimization. It also shows that resilience is a process that varies across gender lines and changes throughout particular lifespan stages (i.e., adolescence), explains both functional and dysfunctional outcomes, is defined by positive and negative emotions, and implies the ability to flexibly regulate emotional expression.

METHOD

Sample and Procedure

A cross-sectional sample of 494 Italian middle- and high-school students aged 13–19 years ($M = 15.27$, SD = 1.23) from different public high schools in the city of Bologna participated in this study during the academic year 2013–2014. School represented a stratified convenient sample, selected to represent all different school types in Italy (Technical, Vocational, and Lyceum) with the approval of the local school council. Participants were 50.8% males and 49.2% females, distributed prevalently in the age groups 14 (38.4%, $n = 186$) and 16 (35.2%, $n = 174$) and belonging to intact families having a biological mother (96.9%) and a biological father (79.7%), together with siblings (71.5%). An anonymous self-report questionnaire was filled in either online or in paper-and-pencil in the presence of a trained and licensed psychologist during regular class hours. Participation was voluntary, without reward or incentive. Both students and their parents signed an informed consent.

Measures

The questionnaire comprised different scales, all previously validated in Italian language and with good reliability, previously used in a survey by Genta, Brighi and Guarin (2013) on bullying and cyberbullying.

Resilience Scales

Psychological resilience was assessed using two scales, the Resilience Scale (RS-14, Wagnild & Young, 1993) and the Dispositional Resilience Scale (Prati, 2010). The RS-14 fits the one-factor model acceptably (Nishi, Uehara, Kondo, & Matsuoka, 2010; Wagnild, 2009) and in our study, as previously conducted by Gianesini (2013, 2015) the one-dimensional RS-14 scale was scored on a five-point Likert Scale (from "absolutely disagree" to "absolutely agree") and not on the original seven-point Likert Scale, yielding a reliability of .82. The Dispositional Resilience Scale-II (Sinclair & Oliver, 2003) was used in its 17-item version scored on a five-point Likert Scale (from "absolutely disagree" to "absolutely agree"), with three factor solutions *positive attitude* (positivity), *rigidity/inflexibility*, and *helplessness/alienation* (six items) adapted by Prati (2010). The internal consistency was good only for Alienation (six items, $\alpha =$ .82) and Positivity (8 items, $\alpha =$ .74). The Inflexibility (3 items, $\alpha =$ .68) subscale was discharged due to the limited number of items and the low reliability. Similar reliability coefficients were found by Prati (2010) on a sample of adults, aged 18–75, respectively .82, .63, and .79 and were higher than those found by Sagone and De Caroli (2014) on a sample of college students, aged 20–26, which ranged from .61 to .65.

The Strengths and Difficulties Scale (SDQ)

The SDQ (Goodman, 2001), a 25-item behavioral screening questionnaire, rates the occurrence of various psychopathology symptoms within the last six months on a three-point Likert scale ("Not true," "Somewhat true," and "Certainly true") and was used in this study as a measure of the outcomes of peer victimization, in terms of psychological adjustment versus maladjustment. The original structure of the scale includes four dimensions (Emotional Symptoms, Conduct Problems, Hyperactivity-Inattention, and Peer Problems) and a single Prosocial Behavior as a protective factor. In particular, Prosocial Behavior includes being considerate of other people's feelings, ready to share with others, helpful if someone is hurt, upset, or feeling ill, kind to younger children, and often volunteering to help others. The second, Emotional Symptoms, comprises complaints of headaches, stomach aches or sickness, worries, feeling unhappy, depressed or tearful, nervous or clingy in new situations, easily scared and prone to losing temper. Discrepancies exist in the literature regarding the structural validity of this scale (Palmieri & Smith, 2007) as its subscales measure more heterogeneous content than intended (van Widenfelt, Goedhart, Treffers, & Goodman,

2003) and a cross-loading among factors may exist. In line with previous studies on the Italian population (Essau et al., 2012), we found moderately higher internal consistency coefficients for the subscale Prosocial Behavior (five items, $\alpha = .69$) and the Emotional Symptoms (5 items, $\alpha = .75$; the study from Essau et al. reported $\alpha = .68$ for Prosocial Behaviors and $\alpha = .60$ for Emotional symptoms).

Cyberbullying and Cyber victimization

Cyber peer violence and victimization were assessed using the short nine-item version of the European Cyberbullying Intervention Project Questionnaire (ECIPQ; Brighi et al., 2012) proposed by Bright et al. (2012) on a five-point Likert Scale (from "never" to "more than once a week"), in active (for bullies) and passive (for victims) forms. The reduced version of the two subscales yielded a high reliability of .82 for aggression and .81 for victimization respectively. The experience of victimization investigated included different behaviors: direct aggression (*Someone told me something nasty or threatened me via the internet or texting*), verbal attack during gaming (*Someone verbally attacked or offended me during an online game*), social exclusion (*I have been excluded or ignored on a social network or in a chat room*), posting or editing of embarrassing personal pictures or videos (*Someone posted online embarrassing pictures or videos or modified the one I posted*), identity theft (*Someone illegally logged into my email or social network account and stole personal information*), account hacking (*Someone pretended to be me and created a fake account or illegally accessed mine*) as well as indirect aggression such as spreading rumors (*Someone spread rumors about me online*), releasing personal information (*Someone released personal information on me online*), relational aggression (*Someone has told other offensive or unpleasant things about myself using internet, email or texting*), and gossiping (*Someone has gossiped about me on the internet*). The same items were used in active form, to tackle cyberbullying behaviors perpetrated by subjects.

Emotional Response to Peer Victimization and Perpetration

The emotional impact of cyberbullying on victims and perpetrators was assessed by asking students to answer the question "*How did you feel*?" by choosing "*nothing*" or/and specific emotions from a given list that included for perpetrator *proud*, *confident*, *good*, *ashamed*, *excited*, and *guilty* and for victims *sad*, *embarrassed*, *humiliated*, *angry*, and *scared*. All responses were coded as 0–1 (YES–NO) and were NOT mutually exclusive. Hence, a subject could answer YES to the first emotion ("nothing") but also answer YES to one or more single emotions (e.g., "*proud*"), although this is somehow contradictory. However, the first emotion being indicative of an overall emotional impact, was recoded separately from the single emotions into "indifferent" (YES, I felt nothing) and "affected" (NO, I felt something). In addition, to further explore emotion differentiation and regulation we recoded the six possible emotions for perpetrators into two categories: Positive (*proud, confident, good*) and Negative (*ashamed, excited, guilty*). Similarly, we recorded the five possible emotions for victims into the categories Passive (*sad, embarrassed, humiliated*) and Reactive (*angry, scared*). We created within-person emotion differentiation variables adding the individual's scores on the negative and positive emotions for perpetrators and within-person emotion differentiation variables adding scores on the passive and reactive emotions for victims. Higher scores would be indicative of greater differentiation, a measure of emotional reactivity and regulation. (Barrett et al., 2001; Campbell, Slee, Spears, Butler, & Kift, 2013). Moreover, we used the two measures of resilience (RS-14 and DS) as independent variables to test their unique and combined effects on the dependent variables Prosocial behavior and Emotional symptoms (SDQ scale) for both victims and perpetrators of cyberbullying, as indices respectively of social adjustment and of psychosomatic complains. Further, we profiled adolescents by cluster analysis based on patterns of resilience, health outcomes, and cyberbullying involvement into: Resilient victims (RV), Healthy uninvolved (HU), Healthy Bullies (HB), Alienated Bully-Victims (ABV), and Resilient Bully-Victims (RBV). This study proposes a comprehensive, developmental, ecological, relational, and self-regulatory resilience approach to cyberbullying, which represents an innovative and advanced contribution to the literature with significant implication for research and practice.

ANALYSIS

Using SPSS version 22 the scales internal consistency and reliability, all Pearson's correlations, and the significant interactions between all variable of interest were verified. A series of *t*-tests were conducted to determine gender and age differences in cyberbullying behavior, its emotional impact on victims and perpetrators, resilience score, and outcomes (see Table 1).

Table 1. Significant Gender Differences (*T*-Test for Independent Samples).

	Male ($n = 247$) *M* (SD)	Female ($n = 239$) *M* (SD)	*p* value
Internet use	3.63 (1.43)	4.03 (1.54)	.003
FB use	3.06 (1.57)	3.49 (1.52)	.004
Victimization			
Previous years	1.13 (.33)	1.27 (.44)	.000
Perceived impact emotions:	.49 (1.08)	1.03 (1.08)	.007
None	.53 (.50)	.16 (.36)	.000
Anger	.22 (.42)	.51 (.50)	.002
Humiliation	.11 (.31)	.35 (.48)	.004
Perpetration			
Previous years	1.27 (1.08)	1.87 (1.4)	.000
Perceived impact	.49 (1.08)	1.03 (1.08)	.007
Bullying	11.46 (4.11)	10.65 (2.48)	
Resilience			
RS-14	53.45 (6.67)	50.39 (7.62)	.000
DS-alienation	29.41 (4.41)	27.34 (4.73)	.000
DS-positivity	12.31 (4.26)	16.26 (5.23)	.000
Outcomes			
SQD-prosocial behavior	11.80 (2.13)	12.70 (1.79)	.000
SDQ-emotional symptoms	7.33 (1.82)	9.36 (2.30)	.000

General Linear Models and ANOVA were employed to explore the relationships between variables and their contribution to the model.

RESULTS

Description of the Sample

Participants were 50.8% males and 49.2% females, aged 13–19 ($M = 15.27$, $SD = 1.23$), with the majority of students in the age group 14 (38.4%, $n = 186$) and 16 (35.2%, $n = 174$). Ethnically, 83.2% of the students were Western European speaking exclusively or prevalently the Italian language (88.3% cumulative) at home with a minority of Asians, North Africans, and Easter Europeans speaking another language. Fathers had a middle or high school diploma (66.1% cumulative), performed white-collar jobs (24.9%), were professional (19.3%), self-employed (19.2), or farmers

(16%). The unemployment rate for fathers was low at 4.4%. A higher percentage of mothers, on the other hand, held a college degree (32%) and were employed in white-collar jobs (31.7 %,) or as farmers (17.4%) or professionals (16.2%). However, their unemployment rate was also higher than their male partners (18.9%). Students came from intact families composed of a biological mother (96.9%) and a biological father (79.7%), together with siblings (71.5%). Rarely were additional family members (2.9%) other than grandparents (8.8%) or a new partner for either the mother (5.1%) or the father (1.0 %) present. In terms of school performance, 15% of the students had failed a grade in school, 51.4% currently attended high school, while 34.7% had changed school after failing a grade. Family demographic characteristics of the sample were representative of the Italian population.

Digital Relationships

Internet use averaged around 1–2 (24.4%) and 2–3 (24.9%) hours a day with 8.9% surfing the web for more than 5 hours a day. Students in our sample had a Facebook (FB) account (91.2%) on which they mostly spent 30 min–1 hour (28.5%) or 10–30 minutes (23.85%) a day. Only 6.5% of the students sampled spent more than 3 hours a day on FB. More than 400 FB friends were reported by the majority of students (52.7%), 12.9% between 300 and 400, and just a cumulative 8.1% had a number of friends from 10 or less to 100. A *t*-test for independent samples evidenced significant gender difference with girls spending more time on the Internet ($t = -2.97$, $p = .003$) and Facebook ($t = -2.91$, $p = .004$) as well as significant age differences with regard to the number of FB friends, higher in the age group 15–16 compared to younger (13–14) and older (17–19) students.

Cyber Victimization

Its frequency and intensity were measured on a five-point Likert scale from none (1) to more than once a week (5). The mean values for the nine items ranged from a minimum of 1.15 (SD = .49) for *identity theft* to a maximum of 1.52 (SD = 1.12) for *spreading rumors*. The most widespread and frequent forms of cyber victimization were direct verbal attack or offence during an online game (6.9% more than once a week), indirect offences (27.1% 1 or 2 times), direct offences or threats (23.1%), spreading rumors (16.2%), account hacking (14%), and illicit posting (13.4%). Victims believed they knew who perpetrated the acts (61.5%), supposedly an individual male (40%), female (27.1%), or a gender-mixed group (8.2%). Victims supposed they did not directly know their bullies (28.1%) or they were acquaintances (19.1%), friends (14.6%), and best friends (4.5%), ex-partners (6.7%), or classmates (13.5%). The duration of victimization ranged from 1 to 2 weeks (65%), to 1 month (19%) to many years (6.8%). In 20.4% of the cases, victimization was also experienced in previous years. We recoded the victimization variable into four categories according to both intensity and number of aggressive behaviors, from none to very high risk. The students in our sample were distributed into the categories no risk (32.4%), moderate risk (23.15%), low risk (15.9%), high risk (21.8%), and very high risk (6.7%). In terms of the participating roles, uninvolved students account for 23.3% ($n = 103$), pure victims were 20.6% ($n = 89$) compared with a much higher 47.1% ($n = 204$) for bully-victims.

Cyberbullying

The experience of cyberbullying perpetration investigated in this study included the same items as peer victimization in the active form (i.e., *I told someone something nasty or threatened him/her via the internet or texting*), measured on a five-point Likert scale from none (1) to more than once a week (5). The mean values for the nine items ranged from a minimum of 1.09 (SD = .40) for *spreading rumors* to a maximum of 1.49 (SD = 1.13) for *verbal attack or offence during an online gaming* which resulted in being the most widespread and frequent forms of cyber bulling (7.5% more than once a week and 9.9% one or two times). Gossiping or offending via internet, email, or text messaging was the second most frequent (23.5% one or two times) form of peer violence perpetrated. Social exclusion from a social network or a chat room (19.4% one or two times), verbal offence or threat on internet or through text messaging or email (18.6% one or two times), and account hacking (8.3% one or two times) were slightly less frequent. Only a small percentage (3.7%) reported having perpetrated cyberbullying in previous years also. The duration of the cyber violence perpetration was assessed on a five-point Likert scale from 1 (1–2 weeks) to 5 (many years) and yielded a mean value of 1.48 (SD = 1.10), with 1–2 weeks being the most frequent duration (78.9%), followed by 1 month (9.9%) and many years (5.6%). The target of the violence was an individual male (39.7%) or a female (37.9%), most frequently an acquaintance (23.7%), someone unknown to the bully (23.3%), a friend (16.7%), or a classmate (13.8%). Gender differences were found for peer violence with boys perpetrating more than girls ($t=2.55$, $p=.011$). We recoded the cyberbullying perpetration variable into four categories according to both frequency and number of behaviors, from none to very high. The students in our sample were distributed into the categories no risk (44.6%, $n=201$), moderate risk (18.2%), low risk (16.6 %,), high risk (14.2%), and very high risk (6.4%).

Participant Roles

Looking at the participating roles, uninvolved students account for 23.3% ($n=103$), pure victims were 20.6% ($n=89$) compared with a much higher 47.1% ($n=204$) for bully-victims, while pure perpetrators were only 8.5% ($n=37$). No significant age differences emerged for victimization and perpetration. Gender differences were found for peer violence with boys perpetrating more than girls ($t=2.55$, $p=.011$). While no significant gender differences were found for the duration of the victimization, girls were more likely to have been victimized in previous years than boys ($t=-3.94$, $p=.000$)

General Impact of Cyberbullying and Cyber victimization

The impact of the cyberbullying was measured on a five-point Likert Scale ranging from no impact (0) to enormous (4) for victims (*What impact did the cyberbullying events you experienced had on your life*?) and perpetrators (*When you have cyberbullied others, what impact do you think such behaviors have had on the life of that person?*). It yielded a mean value of .81 (SD = .00, $n=127$) for victims and .57 (SD =00, $n=87$) for bullies. Cross-tabulation for victims by emotional reactions shows them reporting mostly no impact (55.1%) with only 3.4% reporting the highest impact on their lives. Similarly, bullies perceived their behavior to have no impact (66.7%, $n=70$) or a low impact (22%) on the life of their victims. No significant gender differences were found for the length of the victimization, however, girls were more likely to have been previously victimized (before the year assessed) than boys ($t=-3.94$, $p=.000$) and the Victimization general impact was stronger for girls than boys ($t=-2.73$, $p=.007$). Significant gender differences were also found for perpetrators: girls perceived the impact of their actions on their victims more ($t=-2.73$, $p=.000$) than boys but were also more likely to have perpetrated cyberbullying in previous years. No significant age differences emerged for victimization and perpetration.

Emotional Response to Peer Victimization and Perpetration
When students were asked to disclose their emotional reaction to cyber victimization on a 0–1 (YES–NO) scale 30.4% of them reported no reaction (*I felt nothing*). Among the most frequent emotional responses *anger* ranked first (39.7%) followed by *humiliation* (25.9%), *sadness* (17.2%), *embarrassment* (13.5%), and *fear* (9.5%). An additional 16.4% reported feelings of *helplessness*, *abuse*, and *insecurity*. When we asked bullies about their feeling, 45.6% of them reported no reaction (*I felt nothing*), 26.6% were feeling guilty, 15.2% felt good, followed by ashamed (8.9%) and confident (8.9%). A *t*-test for independent samples showed significant gender difference in the emotional response only for peer victimization (and not for perpetration) with girls experiencing more *anger* ($t = -3.15$, $p = .002$) and *humiliation* ($t = -2.97$, $p = .004$) than boys who more than girls reported no feelings ($t = 4.62$, $p = .004$). No age difference was found.

The emotional impact of cyberbullying disclosed by students, either perpetrators or victims, was coded in dichotomous terms as "affected" (present) or "indifferent" (absent) based on the first answer to the item "*I felt nothing*." In addition, to further explore emotion differentiation and regulation we recoded the six possible emotional responses for perpetrators into the categories Positive (*proud*, *confident*, *good*) and Negative (*ashamed*, *excited*, *guilty*) and the five possible emotional responses for victims into the categories Passive (*sad*, *embarrassed*, *humiliated*) and Reactive (*angry*, *scared*). Responses were NOT mutually exclusive, thus more than one emotion could be experienced, in both categories. Among perpetrators, results indicated the prevalence of a nonemotional response ("indifferent"), higher for those students reporting positive emotions (78.5%) than those in the negative emotion category (67.1%). However, exclusively negative emotions prevailed (54.1%) compared to only positive (29.7) or mixed (16.2%). Even for victims, a nonemotional response was prevalent, with a higher percentage (60.3%) among those in the category aggressive (58.6%) compared to passive (41.4%) and reactive (39.7%) emotions. Exclusively passive emotions prevailed (37%) compared to only reactive (34.2%) or mixed (28.8%). To fully understand the implications of emotional indifference (a nonemotional response) to peer violence we compared the resilience, positivity, and alienation levels of those who were emotionally reactive (i.e., "affected") versus those who were not (i.e., "indifferent") and found significant differences between perpetrators and victims (Fig. 1).

For perpetrators, overall Resilience (RS-14) and Positivity levels (DS) were positively correlated. A high level of resilience corresponded to high levels of Positivity, regardless of the emotional impact reported

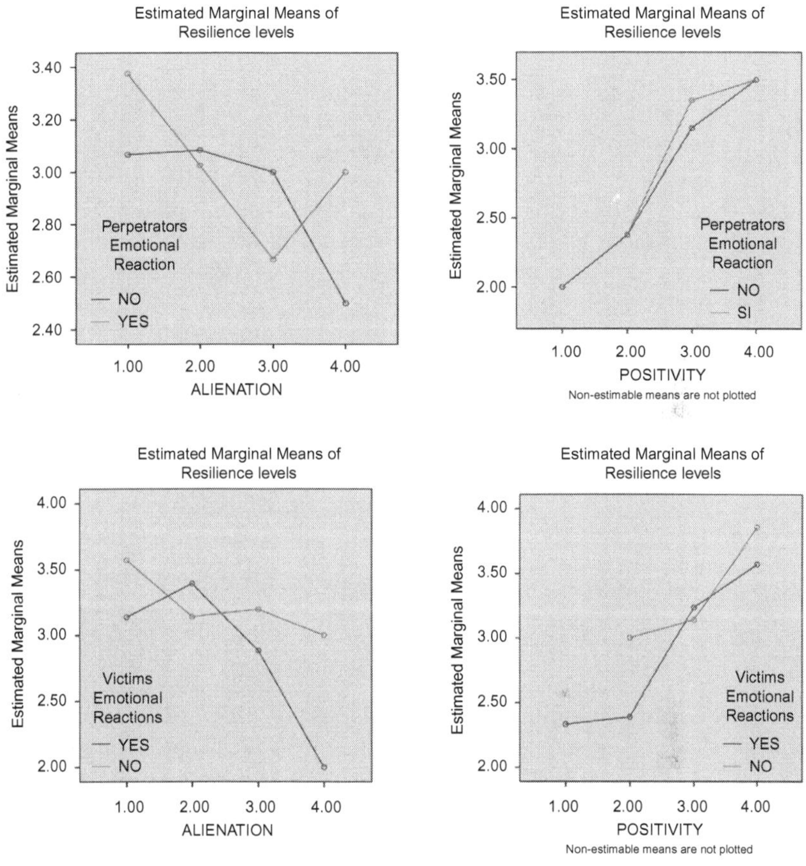

Fig. 1. Resilience, Alienation and Positivity levels in Perpetrators and Victims by Emotional Reaction (GLM).

(i.e., affected vs. indifferent). The same was not true for Victims. Although even in this case Positivity grew in the same direction as resilience, students who were "indifferent," that is reported having NO emotional reaction (green line) to peer victimization, showed higher initial and final levels of resilience and positivity compared to those who were emotionally "affected" by it (blue line). On the other hand, Alienation scores were negatively correlated to resilience levels, with low levels of the former corresponding to high levels of the latter. This pattern was similar for both perpetrators and victims but only when emotionally "affected" (yes for emotional reaction; blue line). The results of the interception between resilience and alienation scores for students who were not emotionally reactive (green line) are quite different. They showed a higher initial level of resilience at low levels of alienation, which remains high despite the rise of alienation. The gap between emotionally reactive (i.e., "affected") and non-reactive (i.e., "indifferent") victims is wider in the condition of high Alienation, with low scores for those who were emotionally responsive versus high scores for those who did not react emotionally to peer victimization. For perpetrators once again the pattern is similar but only in the condition of emotional involvement (blue line). For students who are NOT emotionally responsive (i.e., "indifferent") to peer victimization the resilience level drops steadily as alienation grows to rise again at the highest level of alienation.

Resilience

The resilience levels measure on the RS-14 scale (RS-14, Wagnild & Young, 1993) in our sample ranged from a low 16 to a high 69 ($M = 51.9$, SD = 7.30, range 14–70). Our mean value was lower than previously found, for example by Losoi et al. (2013) on the Finnish adult population (54.5 or 76.3 in the original seven-point Likert Scale). Individual scores were recoded into four categories, 14–28 as low, 28–42 as average, 43–56 as high, and 57–70 as very high which resulted in a sample distribution in the high (67%, $n = 309$) and very high (24%, $n = 113$) categories. The mean scores were low for Alienation ($M = 14.23$, SD = 5.14, range 6–30) and high for Positivity ($M = 28.37$, SD = 4.67, range 8–40). Recoded into four categories (low, moderate, high, very high) they lay in the lower end of the distribution as low (41.4%) or average scores (36.2%) and only 2.7% scored very high. The opposite was true for the Positivity subscale recoded scores, which were distributed into the high (63.4%), average (18.3%), and very high (17%) categories with only 1.3% in the low cluster. Significant gender differences were found: boys showed higher levels of resilience (RS-14, $p = .000$) and also higher levels of dispositional resilience alienation ($t = 4.83$, $p = .000$), while girls scored higher on dispositional resilience positivity ($t = -9.02$, $p = .000$; see Table 1). GLM evidenced age differences in the resilience level of the students sampled with a drop between the age 13 and 14, and a rise after 17 years of age, with the lower levels registered in the age range 14–17 (Fig. 2).

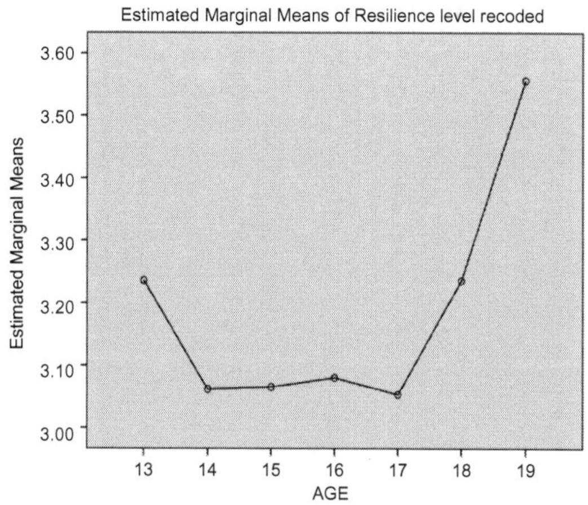

Fig. 2. Resilience Level by Age (GLM).

To negotiate the developmental transition between youth and adulthood successfully, adolescents must acquire skills necessary for independence including increased social behavior (Spear, 2000). Resilience represents an asset and a resource enabling some adolescents to overcome the negative effects of risk exposure (Fergus & Zimmerman, 2004). Interestingly, in this study similar patterns of resilience were found for both victimization and perpetration (Fig. 3), which could be explained by the high percentage of bully-victims (47.1%, $n=204$) in our sample, with the highest level of peer violence involvement corresponding with the lowest resilience level. However, the rates of cyber victimization and bullying suddenly drop as resilience rises, indicating a significant cutoff at level 2 (moderate).

The Role of Resilience on Peer Violence Involvement (Victimization and Perpetration)

Although the resilience scores in our study resulted in higher values, with a percentage of 67% at level 3 (high resilience), 24% at level 4 (very high), and only 7.6% at level 2, a moderate level of resilience was the significant cutoff for both victimization and perpetration. In other words, resilience worked as a protective factor against peer violence involvement, active and passive, starting at moderate to high scores.

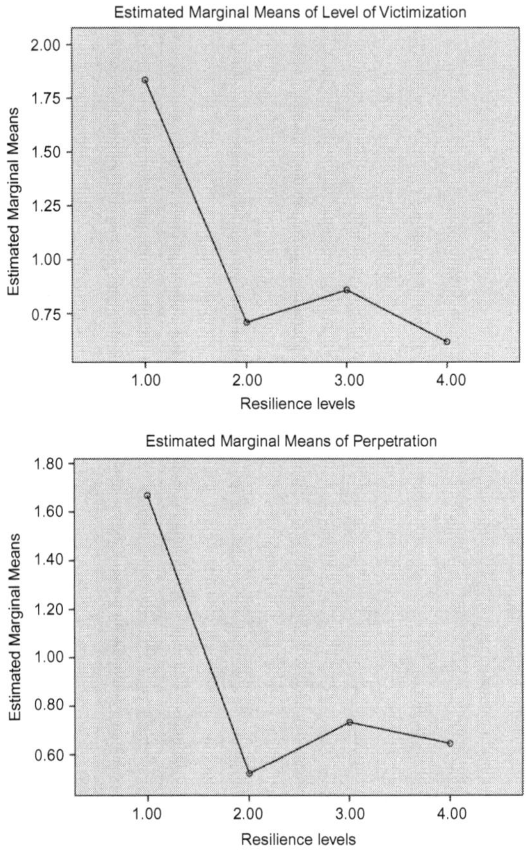

Fig. 3. Resilience Level in Victimization versus Perpetration (GLM). *Note*: 1 = Low (14–28), 2 = Moderate (28–42), 3 = High (43–56), 4 = Very High (57–70).

The Role of Resilience and Emotion Regulation on Adolescents' Adjustment

The scores for both SDQ subscales (Prosocial Behavior and Emotional Symptoms) were differently recoded into three risk categories (no risk, low risk, and high risk) according to the Department of Health indications. For Prosocial Behavior scores the range 6–10 is considered as moderate, 5 is slightly low, and 0–4 is low with increased risk; for the Emotional Symptoms scores between 0–5 are moderate, 6 is slightly raised, and 5–10 is high with increased risk. The mean scores for the SDQ subscale

Prosocial behavior ($M = 12.25$, SD = 2.02, range 5–15) and Emotional Symptoms ($M = 8.33$, SD = 2.30, range 5–15) were low and high respectively. Students in our samples showed many emotional symptoms exposing them to high (56.8%, $n = 260$), very high (19%, $n = 87$), and potential (24.2%, $n = 66$) risk of clinically significant problems. In addition, their scores for Prosocial Behavior were low exposing them once again to a high risk (80.8%, $n = 367$) of clinically significant problems. Significant gender differences were found with girls scoring higher on both SDQ subscales (prosocial behaviors and emotional symptoms; see Table 1).

Pearson's correlations were significant for the three resilience variables (overall RS-14, DS-Alienation, and DS-Positivity subscales) with the two SDQ subscales (Emotional Symptoms and Prosocial Behavior) but Alienation and Prosocial Behavior were in the expected direction. Victimization was positively correlated to Alienation ($r = .522$), and Emotional Symptoms ($r = .197$) and negatively correlated with Resilience ($r = -.375$), Positivity ($r = .-306$), and Prosocial Behavior ($r = -.167$) indicating a high risk for victims deriving from a lack of emotion regulatory and social skills and low resilience levels. Aggression was negatively correlated with Prosocial Behavior ($r = -.237$) and Resilience ($r = -.101$; see Table 2). In other words, students who showed a positive adjustment (Prosocial Behavior) seemed to be those who had higher levels of Resilience and Positivity and were low in both forms of peer violence, perpetration, and victimization.

On the other hand, students who did not adjust and were at risk due to Emotional Symptoms had higher scores in both alienation and victimization and low scores in Resilience and Positivity. General Linear Modeling evidenced a similar but opposite pattern of resilience levels for the adjustment outcomes measure by the SQD scale, positive (Prosocial Behavior) and negative (Emotional Symptoms; Fig. 4).

A Two-Step Cluster Analysis identified five distinct patterns of adolescents' resilience and cyberbullying involvement (Table 3). RV were high in

Table 2. Significant Pearson's Correlation Coefficients for Positive and Negative Outcomes.

	Resilience	Alienation	Positivity	Victimization	Aggression
Emotional symptoms	$-.275^{**}$	$.522^{**}$	-306^{**}	$.197^{**}$	
Prosocial behavior	$.244^{**}$		$.194^{**}$	$-.167^{**}$	$-.237^{**}$

**correlation is significant at the 0.01 level (2-tailed).

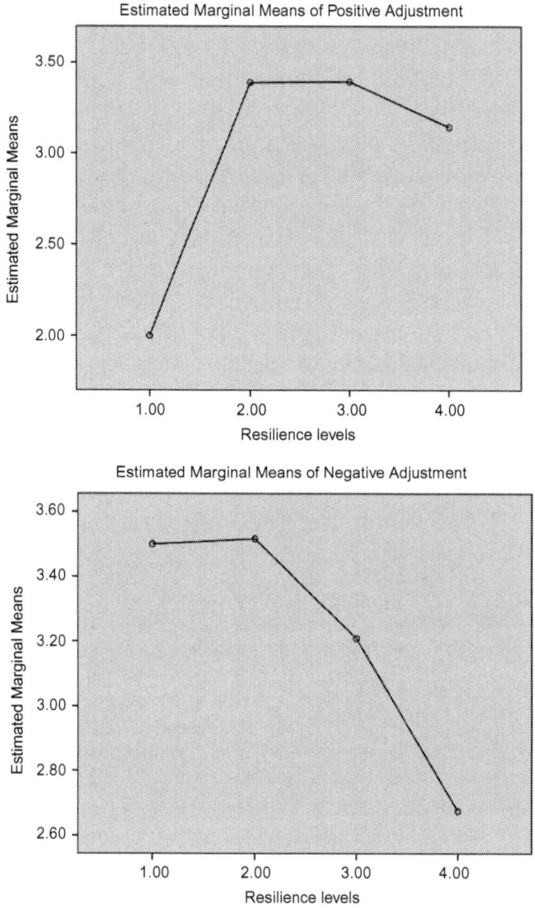

Fig. 4. Resilience Level in Negative vs. Positive Adjustment (GLM). *Note*: 1 = Low (14–28), 2 = Moderate (28–42), 3 = High (43–56), 4 = Very High (57–70).

resilience, average in Positivity, and low in Alienation. HU were average in resilience, average in Positivity, and extremely low in Alienation. HB were average in Resilience and Positivity, low in Alienation; ABV were Low in Positivity, Very low in Resilience, but extremely high in Alienation. RBV were high in Positivity, extremely high in Resilience, and average in Alienation.

Table 3. Two-Step Clustering of Resilience by Cyberbullying Roles.

Cluster	SDQ Alienation		SDQ Positivity		Resilience Level		Not Involved		Pure Bullies		Pure Victims		Bully-Victims	
	Mean	SD	Mean	SD	Mean	SD	*n*	%	*n*	%	*n*	%	*N*	%
1	14.20	4.32	29.16	4.04	53.37	7.31	0	0.0	0	0.0	77	98.7	0	0.0
2	12.45	5.06	28.97	4.55	52.87	7.11	93	100	0	0.0	0	0.0	0	0.0
3	13.34	4.86	28.14	4.83	51.62	6.46	0	0.0	35	97.2	0	0.0	0	0.0
4	20.77	4.49	22.63	3.87	42.57	7.87	0	0.0	1	2.8	1	1.3	47	26
5	13.61	4.31	30.27	3.40	54.52	4.89	0	0.0	0	0.0	0	0.0	134	74

RV, HU, and HB were all well-adjusted (Average Positivity and low to very low Alienation), no matter the role in cyberbullying, and were characterized by average to high levels of Resilience. Surprisingly, RBV were also well-adjusted, showing the highest level of Positivity and an average level of Alienation, but were also high in resilience (as RV), despite the double involvement in cyberbullying,. ABV were the only maladjusted and at-risk group in our sample characterized by very low Positivity, very low Resilience, and extremely high Alienation. Previous research (Sagone & De Caroli, 2014) has found high levels of positive attitude to be correlated positively with strategies of reinterpretation and problem solving and all dimensions of psychological well-being, and negatively with avoidance coping, while high levels of alienation were related positively to avoidance and negatively with psychological well-being. Our results confirmed the fundamental importance of assessing resilience resources together with peer violence in identifying and targeting adolescents at risk.

DISCUSSION

No matter the definition, characteristics, and subtypes, negative virtual interactions are real to the young people engaged with them (Smith, Polenik, Nakasita, & Jones, 2012). Their effects on the victims have been extensively researched. They includes conduct disorder, oppositional defiant disorder, attention deficit hyperactivity disorder (ADHD), passive-aggressive personality disorder (Coolidge, DenBoer, & Segal, 2004), depression, low self-esteem and high loneliness (Brighi et al., 2012), self-harm, suicide, and suicidal ideation (Barker, Arseneault, Brendgen, Fontaine, & Maughan, 2008; Coolidge et al., 2004; Klomek et al., 2009; Salmon, James, & Smith, 1998; Seals & Young, 2003). Children who perpetrate cyberbullying have also increased conduct problems and hyperactivity symptoms (Gini, 2008; Viding, Simmonds, Petrides, & Frederickson, 2009), are more aggressive in both reactive (as a response to provocation) and proactive (deliberate and provocative) terms (Camodeca & Goossens, 2005; Crick & Dodge, 1996; Salmivalli & Nieminen, 2002). Individual differences exist in peer violence behaviors, which can be attributed to deficits regarding morality (Hymel, Schonert-Reichl, Bonanno, Vaillancourt, & Henderson, 2010; Thornberg & Jungert, 2014), lack of empathy in social interactions (Brighi, Gianesini, & Melotti, 2015; Crick & Dodge, 1999), or acting outside the norms of appropriate conduct (Finkelhor, 2008; Finkelhor, Turner, & Hamby, 2012). In order to understand the complexity of peer group relationships and networks and become aware of their risks as well as their benefits, both in the context of the digital and the real world, young people need to be equipped with the critical tools of awareness, strategies for successful coping and emotional regulation (Cowie, 2011). Resilience and emotional well-being are of fundamental importance in inhibiting the prevalence of peer violence in cyberspace. This subtle and insidious form of aggression mostly affects relationships, which may lead to low self-esteem, school avoidance, or psychological damages (Owens, Shute, & Slee, 2000; Rigby, 2005). Our findings confirmed that victims and perpetrators vary in emotion regulation skills, which greatly influence resilience levels especially under negative circumstances thus buffering youth from the negative influences and outcomes of peer violence in adolescence. Significant gender differences were found in terms of resilience with boys showing higher levels of resilience (RS-14) and also higher levels of dispositional resilience alienation (DS), while girls scored higher on dispositional resilience positivity. These findings confirmed Prati's (2010) results but in a different direction. In his sample of 1,192 adults aged 18–75, men scored higher than women on the positivity dimension and women scored higher than men on alienation. Losoi et al. (2013) found no significant difference in resilience between genders, which concurs with previous findings on the Swedish population (Lundman, Strandberg, Eisemann, Gustafson, & Brulin, 2007). However, Abiola and Udofia (2011) have also reported a significant gender difference in Nigerian adults and Nygren et al. (2005) on the elderly population. Interesting age differences also emerged in this study pertaining to resilience levels, with a drop between the age 13 and 14, and a rise after 17 years of age, with the lowest levels registered in the age range 14–17. Other studies have found that increases in individual resilience factors such as self-esteem are age-dependent among children and adolescents (Bolognini, Plancherel, Bettwshart, & Halfon, 1996; Frost & McKelvie, 2004) and that in early to mid-adolescence resilient students aged 12–13 years displayed fewer behavioral and emotional difficulties, an easier temperamental style, and more social competence (Rigby & Slee, 2002). Although the resilience scores in our study resulted toward higher values, with a percentage of 67% at level three (high resilience), a moderate level of resilience was the significant cutoff for both victimization and perpetration. Our cutoff was below the score of 45 (64 on the original seven-point Likert scale) found by Tian and Hong (2013) on a sample of Chinese adult hospital cancer patients. The peer group plays a critical role in the psychological and physiological health of adolescents (Chessor, 2008). Adolescent peers are dynamic, fluid, and characterized by the changes of roles of individual member, popularity, acceptance, and position within the group. Status within the group represents a need to belong and give a sense of importance within the group. Peer violence in cyberspace aims at undermining or ridiculing a person (Simmons, 2004), and is quite a deliberate and hurtful process damaging relationships within the group (Rigby, 2005) which allows conflicts to continue. Acquiring appropriate coping strategies for dealing with difficult emotional and social situations, such as cyberbullying, implies competencies: emotion self-regulation, assertiveness, supportive friends, looking out for vulnerable classmates, knowing one's rights, knowing where to find help, how to mobilize other peers and adults, and a strong belief in the self and one's right (Smith, Talamelli, Cowie, Naylor, & Chauhan, 2004). Moreover, the type of emotions expressed plays a critical role in the responses of peer (Cowie & Berdondini, 2002). Internalizing and externalizing emotions were examined as risk factors for peer victimization by Hanish et al. (2004). Anger, specifically, predicted being victimized differently for boys and girls and across time while support for internalizing variables was weak (Thompson & Calkins, 1996). Our results highlight the need for accounting in peer violence research for positive versus negative emotions and passive versus aggressive responses. In victims, anger can turn inward and become self-destructive or foster destructive rage and retaliation, which turns adolescents into bully-victims. However, anger may also cover other emotions as hurt, disappointment, jealousy, fear, shame, frustration. The emotional reaction of the victim or "target," no matter which emotion is displayed, serves the perpetrator and allows him/her to avoid responsibility. Similarly, for perpetrators the dynamics of shame and guilt could be internalized, become self-destructive, and negatively affect self-worth. Feelings and emotions bond and tie victims and perpetrators in an ongoing circle of destructive pattern of mistreatment and reciprocal feelings of inadequacy (Namie & Namie, 2003). Successfully navigating emotions is complex and implies both producing and regulating them (Riebel et al., 2009) choosing the one to express depending on the context and relationships (Newberry, Gallant, & Riley, 2013).

In this study we explored the unique and interactive effects of peer violence in cyberspace on adolescents' emotion regulation and socioemotional adjustment, as well as the mediational role of resilience in the link between adolescent's pathogenic relational experiences and behavioral outcomes. We expected high levels of resilience and a positive attitude to predict healthy adjustment and fewer psychosomatic complains for both victims and perpetrators and to be related to a lower emotional reactivity as a sign of emotion regulatory skills. All our hypotheses were confirmed. High levels of resilience were positively related to Positivity and negatively to Helpless/Alienation (H1); high levels of resilience and a positive attitude predicted healthy adjustment and fewer psychosomatic complains (H2) for both victims and perpetrators of cyber peer violence; high resilience levels were related to a lower emotional reactivity as a sign of emotion regulatory skills (H3). Moreover, our findings evidenced significant gender differences: despite being more resilient and prosocial than boys, girls nevertheless reacted more negatively to victimization with higher levels of Alienation and Emotional Symptoms and more anger and humiliation than males, and were at higher risk for clinical problems. Boys perpetrated peer violence more than girls but were less likely to be victimized, were generally less emotionally responsive to it, and responded with less anger and humiliation. Involvement in cyber peer violence as either a victim or perpetrator was associated with low resilience levels, and resilience mitigated the negative emotional effect of victimization favoring positive adjustment (Prosocial Behavior) Differences in emotion regulation processes that emerged between victims and perpetrators and affected resilience levels. A moderate resilience level was sufficient to buffer adolescents from peer violence involvement. Early adolescent years appear to be an important transition point in the progression to peer violence. Between the ages of 13–14 clear differences in resilience levels emerged, which coincided with the transition from middle school to secondary school and may have resulted in a change of peer groups for some. Students older than 17 have passed puberty and are becoming increasingly capable of complex cognitive processes and emotion regulatory processes, which significantly influenced both the individuals' views of themselves and their interpersonal relationships. Students in our sample were passive toward victimization showing emotions like sadness, embarrassment, and humiliation, and experienced negative moral emotions like shame and guilt when perpetrating peer violence. Adolescents' emotion processes and emotion regulation may underlie some of the individual differences that have been found in empathy (Gini, Albiero, Benelli & Altoè, 2007), social information processing, and in reactive and proactive aggressive and bullying patterns (Arsenio & Lemerise, 2002; Little, Jones, Henrich, & Hawley, 2003). Failure in emotional regulation was found to be related to lower resilience levels and raised the risk for psychosocial and emotional dysfunctions in response to peer violence in cyberspace (Barrett et al., 2001). Peer violence in cyberspace is a complex social phenomenon (Sawyer, Mishna, Pepler, & Wiener, 2011) that requires interventions at different levels to encourage positive alternatives to negative power struggles. This paper has provided evidence that certain kinds of relational resources are critical in developing resilience at every stage of development and in different contexts. Overall, the ability to build relationships and regulate positive and negative emotions is fundamental and a fuel for resilience (Gianesini, 2015). The full potential for resilience can only be realized by competently fulfilling relational and emotion regulatory functions (Kinnear, 2002; Masten & Coatsworth, 1998).

CONCLUSION

In the digital and real word, all interpersonal relationships and interactions influence and are influenced by other's behaviors and interactions (Scholte & Aken, 2006) and in this interpersonal context, victimization can occur and involves not only peers but also adults (James et al., 2011; Leadbeater & Hoglund, 2006). Social websites (MySpace, Facebook, etc.), email, chat rooms, mobile phone texting and cameras, picture messages, IM (instant messages), and/or blogs offer powerful ways to communicate, interact with peers and learn how to cooperate, to take different perspectives, and to satisfy growing needs for intimacy (Jones, Mitchell, & Finkelhor, 2010). However, the Internet also frees individuals from traditionally constraining pressures of society, conscience, morality, and ethics thus providing a greater advantage for the bully who can tease, harass, harm, and threat a victim at home anonymously (Jones, Manstead, & Livingstone, 2011) with the possibility of reaching a larger number of victims (Reece, 2012). Technology is the chosen means for exerting power and control over others and reach their targets at any time, day or night (Notar, Padgett, & Roden, 2013). Its double-edged nature, between risks and opportunities, manifests itself clearly in adolescence in the context of peer relationships (Livingstone & Helsper, 2010). Social interaction with peers provides a forum for learning and refining socioemotional skills needed for enduring relationships (Notar et al., 2013) and also for the interaction between the bully and the victim in a reciprocal and dynamic process of dominant relational behavior (Dukes, Stein, & Zane, 2010; Lines, 2008). In cyberbullying, the anonymity features of the medium, the presence of an unlimited audience, and the individual and peer group characteristics are all factors that influence the severity of its impact (Dredge, Gleeson, & de la Piedad, 2014; Mesch, 2009) which is difficult for adults to control (Smith, Salmivalli, & Cowie, 2012). Research indicates that the most at-risk are adolescents who are both targets and perpetrators, the cyberbully-victims who represented the majority (47.1%, $n=204$) in our sample. However, a few studies have considered how adolescents manage these risks and develop resilience (Papatraianou et al., 2014) accounting for their emotional regulation skills and behavioral self-control (You et al., 2015). A deeper evaluation of the complex interaction of personal, relational, and situational factors within which cyberbullying occurs, and their influence on the individual cognitive, affective, and arousal states (Bushmann & Anderson, 2002) is fundamental for understanding the phenomenon of cyberbullying. Its impact ultimately depends on the specific acts, the statuses of the actors, and the type of relationship in which the aggression occurred (Finkelhor, Turner, Hamby, & Ormrod, 2011) and is greatly influenced by gender, age, social intelligence, and parent–child communication (Wallenius, Punamaki, & Rimpela, 2006) and moderated by resilience and emotion regulatory processes (Gianesini, 2015). Building relational and emotion regulatory resilience (Gianesini, 2015) in children and young people is vital in the battle against cyberbullying. An emphasis on the relational and emotional strengths of young people provides resilience against adversity, challenging situations, and the stressors of cyberbullying and helps to prevent behavioral and emotional difficulties later in life (Chadwick, 2014). Smith et al. (2004) found that "escaped victims" (no longer victimized after a period of two years) did not differ from nonvictims in terms of personal characteristics or the type of aggression experienced, but were more resilient in developing effective coping strategies. Learning to regulate emotions and navigate relationships is an essential developmental task for adolescents to cope with the complex array of situations, some very stressful, which they will inevitably encounter in their lives. As children grow older, they become increasingly reliant on relationships with peers, and thus it is necessary to acquire appropriate strategies for dealing with difficult emotional and social situations, such as peer violence in cyberspace (Cowie, 2011). The complex process that led to adaptive outcomes versus dysfunctional behaviors in adolescents is of strategic importance as well as the simultaneous assessment of age-relevant stressors and the often neglected positive social-emotional assets, resources, and competencies including resilience. These factors increase the probability of positive outcomes in the face of underestimated, everyday adversities and empower both individuals and families, promoting optimism and enhancing collaboration (Merrel, Felver-Gant, & Tom, 2011). Investigating and defining resilience as a relational and emotion regulatory process has important implications for both research and clinical practice (Gianesini, 2015). It may result in specific intervention strategies aimed at assisting individuals and families in managing both the turbulences and transitions of their life stages as well as unexpected violent and traumatic events. This study further confirms that cyberbullying needs to be addressed in broader social terms looking at relational, social, and emotional resources and weaknesses (Pepler et al., 2006; Salmivalli, 1999). It also proposes that resilience enhancement programs for youth are not to be based on building a singular or a set of skills but on a more comprehensive, developmental, ecological, relational, and self-regulatory model of resilience for modulating emotions, arousal, and behaviors across contexts, relationships, and difficulties.

LIMITATIONS AND SUGGESTIONS FOR FUTURE RESEARCH

In this study, we considered all cyber behaviors as equivalent. Further analysis is needed to correctly weigh each single behavior within each category (i.e., Direct vs. Indirect; Communication vs. Entertainment; public vs. private) and discriminate between the different emotions they elicited, as different coping strategies may be used for different types of behaviors perpetrated and/or suffered. We used a variable-based approach because of the cross-sectional nature of our data and the reliance on self-report measures. A person-based approach, whereby subgroups of individuals who possess similar constellations of characteristics at one age, but who have differing outcomes at a later age, can shed light on differing development pathways and facilitates the identification of key ages or stages of development in which significant changes in pathways occur (see Haapasalo, Tremblay, Boulerice, & Vitaro, 2000). More sophisticated data analysis procedures (SEM, LGMM) are needed to confirm our findings.

Although youth digital aggressive interactions fit the scholarly definition of cyberbullying, adolescents may not perceive it as such. Attitudes and expectations toward cyberbullying, concerns, and vulnerabilities of victims and perpetrators need to be differentiated and interpreted within cultures, as evident discrepancies have emerged between objective criteria and individual outcomes that have salient implications for assessment and intervention (Mishna et al., 2015). Moreover, the identification of cyberbullying and of the potential risks involved in digital social media use is complex and highly personalized and contextual in nature (Sen,

2015). Because of the increasingly diverse population due to globalization, there is a need for investigating cyberbullying from a multicultural context (Lim & Hoot, 2015). Stereotypical perspectives of cyberbullying often sustain scholastic, academic, and workplace aggressive engagements as adults often seem to struggle to recognize and respond to it (Migliaccio, 2015).

ACKNOWLEDGMENT

The authors thank Francesca Carpani for her help with collecting the data.

REFERENCES

Abiola, T., & Udofia, O. (2011). Psychometric assessment of the Wagnild and Young's resilience scale in Kano, Nigeria. *BMC Research Notes*, *4*(1), 509.

Andreou, E. (2004). Bully/victim problems and their association with Machiavellianism and self-efficacy in Greek primary school children. *British Journal of Educational Psychology*, *74*, 297–309.

Arcidiacono, C., Procentese, F., & Di Napoli, I. (2007). Youth, community belonging, planning and power. *Journal of Community and Applied Social Psychology*, *17*, 280–295. doi:10.1002/casp.935

Arsenio, W. F., & Lemerise, E. A. (2001). Varieties of childhood bullying: Values, emotion processes, and social competence. *Social Development*, *10*(1), 59–73.

Arsenio, W. F., & Lemerise, E. A. (2002). Varieties of childhood bullying: Values, emotion processes, and social competence. *Social Development*, *10*(1), 59–73.

Balakrishnan, V. (2015). Cyberbullying among young adults in Malaysia: The roles of gender, age and internet frequency. *Computers in Human Behavior*, *40*, 149–157.

Barker, E. D., Arseneault, L., Brendgen, M., Fontaine, N., & Maughan, B. (2008). Joint development of bullying and victimization in adolescence: Relations to delinquency and self-harm. *Journal of the American Academy of Child & Adolescent Psychiatry*, *47*(9), 1030–1038.

Barlett, C., & Coyne, S. M. (2014). A meta-analysis of sex differences in cyber-bullying behavior: The moderating role of age. *Aggressive Behavior*, *40*(5), 474–488.

Barrett, L. F., Gross, J., Christensen, T. C., & Benvenuto, M. (2001). Knowing what you're feeling and knowing what to do about it: Mapping the relation between emotion differentiation and emotion regulation. *Cognition and Emotion*, *15*(6), 713–724.

Block, J. H., & Block, J. (1980). The role of ego-control and ego-resiliency in the organization of behavior. In W. A. Collins (Ed.), *Development of cognition, affect, and social relations: The Minnesota symposia on child psychology* (Vol. 13). Hillsdale, NJ: Erlbaum.

Bolling, D., Pitskel, N., Deen, B., Crowley, M., Mayes, L., & Pelphrey, K. (2011). Development of neural systems for processing social exclusion from childhood to adolescence. *Developmental Science*, *14*(6), 1431–1444.

Bolognini, M., Plancherel, B., Bettwshart, W., & Halfon, O. (1996). Self-esteem and mental health in early adolescence: Development and gender difference. *Journal of Adolescence*, *19*, 233–245.

Bonanno, G. A., & Mancini, A. D. (2012). Beyond resilience and PTSD: Mapping the heterogeneity of responses to potential trauma. *Psychological Trauma: Theory, Research, Practice, and Policy*, *4*(1), 74.

Bonanno, A. G. (2009). *The other side of sadness*. New York, NY: Basic Books.

Bonanno, G. A., Goorin, L., & Coifman, K. G. (2008). Sadness and grief. In M. Lewin, J. Haviland-Jones, & L. Feldman Barrett (Eds.), *The handbook of emotions* (3rd ed.). New York, NY: Guilford.

Bonanno, G. A., Papa, A., O'Neil, K., Westphal, M., & Coifman, K. (2005). The importance of being flexible; the ability to enhance and suppress emotional expression predicts long-term adjustment. *Psychological Science*, *15*, 482–487.

Bonanno, G. A., Pat-Horenczyk, R., & Noll, J. (2011). Coping flexibility and trauma: The Perceived Ability to Cope With Trauma (PACT) scale. *Psychological Trauma: Theory, Research, Practice, and Policy*, *3*(2), 117.

Bonanno, G. A., Wortman, C. B., & Nesse, R. M. (2004). Prospective patterns of resilience and maladjustment during widowhood. *Psychology and Aging*, *19*, 260–271.

Bowes, L., Maughan, B., Caspi, A., Moffitt, T. E., & Arseneault, L. (2010). Families promote emotional and behavioural resilience to bullying: Evidence of an environmental effect. *Journal of Child Psychology and Psychiatry*, *51*(7), 809–817.

Brand, S., Felner, R., Shim, M., Seitsinger, A., & Dumas, T. (2003). Middle school improvement and reform: Development and validation of a school-level assessment of climate, cultural pluralism, and school safety. *Journal of Educational Psychology*, *95*(3), 570–588. doi:10.1037/0022-0663.95.3.570

Brighi, A., Ortega, R., Scheitauer, H., Smith, P. K., Tsormpatzoudis, C., Barkoukis, V., Del Rey, R. (2012a). *European Cyberbullying Intervention Project Questionnaire (ECIPQ)*. University Of Bologna. Unpublished Manuscript.

Brighi, A., Ortega, R., Scheitauer, H., Smith, P. K., Tsormpatzoudis, C., Barkoukis, V., Del Rey, R., et al. (2012b). *European bullying and cyberbullying: Overlapping and predictive value of the co-occurrence*. Intervention Project Questionnaire (EBIPQ). University of Bologna. Unpublished Manuscript.

Brighi, A., Gianesini, G., & Melotti, G. (2015). Bullying and moral reasoning: The influence of life values and culture on Italian teenagers' self-assertion. In P. Triggs (Ed.), *Handbook on bullying: Prevalence, psychological impacts and intervention strategies*. New York, NY: Nova Science Publisher.

Brighi, A., Guarini, A., & Genta, M. L. (2009). Bullying in the digital age. *Bullying and Cyberbullying in Adolescence* (pp. 14–39). Roma: Carocci.

Brighi, A., Guarini, A., Melotti, G., Galli, S., & Genta, M. L. (2012). Predictors of victimization across direct bullying, indirect bullying and cyberbullying. *Emotional and Behavioral Difficulties*, *17*, 375–388.

Bushman, B. J., & Anderson, C. A. (2002). Violent video games and hostile expectations: A test of the general aggression model. *Personality and Social Psychology Bulletin*, *28*, 1679–1686.

Bussey, K., Fitzpatrick, S., & Raman, A. (2015). The role of moral disengagement and self-efficacy in cyberbullying. *Journal of School Violence*, *14*(1), 30–46.

Camodeca, M., & Goossens, F. A. (2005). Aggression, social cognitions, anger and sadness in bullies and victims. *Journal of Child Psychology and Psychiatry, 46*(2), 186–197.

Campbell, M. A., Slee, P. T., Spears, B., Butler, D., & Kift, S. (2013). Do cyberbullies suffer too? Cyberbullies' perceptions of the harm they cause to others and to their own mental health. *School Psychology International, 34*(6), 613–629.

Chadwick, S. (2014). Introduction. In *Impacts of cyberbullying, building social and emotional resilience in schools* (pp. 1–10). New York, NY: Springer International Publishing.

Chessor, D. (2008). Developing student wellbeing and resilience using a group process. *Educational and Child Psychology, 25*(2), 82–90.

Clark, M. S., & Finkel, E. J. (2005). Willingness to express emotion: The impact of relationship type, communal orientation, and their interaction. *Personal Relationships, 12*(2), 169–180.

Coolidge, F. L., DenBoer, J. W., & Segal, D. L. (2004). Personality and neuropsychological correlates of bullying behavior. *Personality and Individual Differences, 36*(7), 1559–1569.

Cowie, H., & Berdondini, L. (2002). The expression of emotion in response to bullying. *Emotional and Behavioral Difficulties, 7*(4), 207–214.

Cowie, H. (2011). Coping with the emotional impact of bullying and cyberbullying: How research can inform practice. *The International Journal of Emotional Education, 3*(2), 50–56.

Cowie, H. (2013). Cyberbullying and its impact on young people's emotional health and wellbeing. *The Psychiatric Bulletin, 37*, 167–170.

Crick, N. R., & Dodge, K. A. (1996). Social information-processing mechanisms in reactive and proactive aggression. *Child Development, 67*(3), 993–1002.

Crick, N. R., & Dodge, K. A. (1999). Superiority is in the eye of the beholder: A comment on Sutton, Smith, and Swettenham. *Social Development, 8*, 128–131.

Crick, N. R., & Grotpeter, J. K. (1996). Children's treatment by peers: Victims of relational and overt aggression. *Development and Psychopathology, 8*(02), 367–380.

Crick, N. R., Casas, J. F., & Ku, H. C. (1999). Relational and physical forms of peer victimization in preschool. *Developmental Psychology, 35*(2), 376.

Cusinato, M., & L'Abate, L. (Eds.). (2010). *Advances in relational competence theory: With special attention to Alexithymia*. New York, NY: Nova Science Publishers.

Davis, M. E., Zautra, A. J., & Smith, B. W. (2004). Dynamic model of affect relationships within interpersonal relations. *Journal of Personality, 72*, 1105–1132.

Del Rey, R., Elipe, P., & Ortega-Ruiz, R. (2012). Bullying and cyberbullying: Overlapping and predictive value of the co-occurrence. *Psicothema, 24*(4), 608–613.

den Hamer, A. H., & Konijn, E. A. (2015). Adolescents' media exposure may increase their cyberbullying behavior: A longitudinal study. *Journal of Adolescents Health, 56*(2), 203–208.

DeWall, C., Deckman, T., Pond, R., & Bonser, I. (2011). Belongingness as a core personality trait: How social exclusion influences social functioning and personality expression. *Journal of Personality, 79*(6), 979–1012.

Dredge, R., Gleeson, J. F. M., & de la Piedad, G. X. (2014). Risk factors associated with impact severity of cyberbullying victimization: A qualitative study of adolescent online social networking. *Cyberpsychology, Behavior, and Social Networking, 17*(5), 287–291.

Dukes, R. L., Stein, J. A., & Zane, J. I. (2010). Gender differences in the relative impact of physical and relational bullying on adolescent injury and weapon carrying. *Journal of School Psychology*, *48*, 511–532.

Ekman, P. (1992). Are there basic emotions? *Psychological Review*, *99*, 550–553.

Ekman, P. (1993). Facial expression and emotion. *American Psychologist*, *48*(4), 384.

Emerson, R. M. (1981). Social exchange theory. In M. Rosenberg & R. H. Turner (Eds.), *Social psychology sociological perspectives* (pp. 30–65). New York, NY: Basic Books.

Essau, C. A., Olaya, B., Anastassiou-Hadjicharalambous, X., Pauli, G., Gilvarry, C., Bray, D., … Ollendick, T. H. (2012). Psychometric properties of the strength and difficulties questionnaire from five European countries. *International Journal of Methods in Psychiatric Research*, *21*(3), 232–245.

Farrington, D. (1993). Understanding and preventing bullying. In M. Tonry (Ed.), *Crime and justice: A review of research* (Vol. 17). Chicago, IL: University of Chicago Press.

Fenaughty, J., & Harre', N. (2013). Factors associated with distressing electronic harassment and cyberbullying. *Computers and Human Behavior*, *29*, 803–811.

Fergus, S., & Zimmerman, M. A. (2004). Adolescent resilience: A framework for understanding healthy development in the face of risk. *Annual Review of Public Health*, *26*, 399–419. doi:10.1146/annurev.publhealth.26.021304.144357

Finkelhor, D. (2008). *Childhood victimization: Violence, crime, and abuse in the lives of young people*. New York, NY: Oxford University Press.

Finkelhor, D., Turner, H. A., & Hamby, S. (2012). Let's prevent peer victimization, not just bullying. *Child Abuse and Neglect*, *36*(4), 271–274.

Finkelhor, D., Turner, H., Hamby, S. L., & Ormrod, R. (2011). *Polyvictimization: Children's exposure to multiple types of violence, crime, and abuse*. National Survey of Children's Exposure to Violence.

Fredrickson, B. (1998). What good are positive emotions? *Review of General Psychology*, *2*, 300–319.

Frost, J., & McKelvie, S. (2004). Self-esteem and body satisfaction in male and female elementary school, high school, and university students. *Sex Roles*, *51*(1–2), 45–54.

Genta, M. L., Brighi, A., & Guarini, A. (2013). *Cyberbullismo: Ricerche e strategie di intervento*. Milan: F. Angeli.

Gianesini, G. (2000). *A multivariate approach to the dynamic of power in intimate relationship. An application of Molm's theory of coercive power in exchange to domestic violence*. Master thesis, University of Central Florida.

Gianesini, G. (2009). Ten years of family resilience. New theoretical and empirical perspectives. *Interdisciplinary Journal of Family*, *2*, 37–53.

Gianesini, G. (2013). Negotiating family challenges by transforming traditional gender roles in new identities: How patterns of resilience vary by gender, employment status and age of first child. In L. Blair (Ed.), *Visions of the 21st century family: Transforming structures and identities* (Vol. 7). Contemporary Perspective in Family Research. Bingley, UK: Emerald Group Publishing Limited.

Gianesini, G. (2013c). E-counseling in the era of Digital Families. Presented at the International Conference: Counselling and Career Counselling. Padua, May 19, 2013.

Gianesini, G. (2015). Model RERR. Resilience as a relational Construct. Theoretical and Empirical Evidences. Edizioni Accademiche Italiane, VDM-Verlag.

Gillespie, B. M., Chaboyer, W., & Wallis, M. (2007). Development of a theoretically derived model of resilience through concept analysis. *Contemporary Nurse*, *25*, 124–135.

Gini, G. (2008). Associations between bullying behaviour, psychosomatic complaints, emotional and behavioural problems. *Journal of Paediatrics and Child Health, 44*(9), 492–497.

Gini, G., Albiero, P., Benelli, B., & Altoè, G. (2007). Does empathy predict adolescents' bullying and defending behavior? *Aggressive Behavior, 33*(5), 467–476. doi:10.1002/ab.20204

Godwin-Jones. (2005). Emerging technologies messaging, gaming, peer-to-peer sharing: Language learning strategies and tools for the millennial generation. *Language Learning and Technology, 9*(1), 17–22.

Goodman, R. (2001). Psychometric properties of the strengths and difficulties questionnaire (SDQ). *Journal of the American Academy of Child and Adolescent Psychiatry, 40*, 1337–1345.

Gottman, J. (1997). *Raising an emotionally intelligent child.* New York, NY: Fireside/Simon and Schuster, Inc.

Gualdo, M. G., Hunterr, S. C., Durkin, K., Arnaiz, P., & Maquilon, J. J. (2014). The emotional impact of cyberbullying: Differences in perceptions and experiences as a function of role. *Computers and Education, 82*, 228–235.

Haapasalo, J., Tremblay, R. E., Boulerice, B., & Vitaro, F. (2000). Relative advantages of person-and variable-based approaches for predicting problem behaviors from kindergarten assessments. *Journal of Quantitative Criminology, 16*(2), 145–168.

Hanish, L. D., Eisenberg, N., Fabes, R. A., Spinrad, T. L., Ryan, P., & Schmidt, S. (2004). The expression and regulation of negative emotions: Risk factors for young children's peer victimization. *Development and Psychopathology, 16*(2), 335–353.

Hawley, P. H. (1999). The ontogenesis of social dominance: A strategy based evolutionary perspective. *Development Review, 19*, 97–132.

Herring, S. (1999). Interactional coherence in CMC. *Journal of Computer-Mediated Communication, 4.* doi:10.1111/j.1083-6101.1999.tb00106.x

Hindjua, S., & Patchin, J. W. (2010). Bullying, cyberbullying, and suicide. *Archives of Suicide Research, 14*(3), 206–221.

Hoff, D. L., & Mitchell, S. N. (2009). Cyberbullying: Causes, effects, and remedies. *Journal of Educational Administration, 47*(5), 652–665.

Hooper, L., L'Abate, L., Sweeney, L. G., Gianesini, G., & Jankowsli, P. (2013). *Models of psychopathology: Generational process and relational roles.* New York, NY: Springer-Science.

Hu, Y., Wood, J. F., Smith, V., & Westbrook, N. (2004). Friendships through IM: Examining the relationship between instant messaging and intimacy. *Journal of Computer-Mediated Communication, 10.* doi:10.1111/j.1083-6101.2004.tb00231.x

Hymel, S., Rocke-Henderson, N., & Bonanno, R. A. (2005). Moral disengagement: A framework for understanding bullying among adolescents. *Journal of Social Sciences, 8*(1), 1–11.

Hymel, S., Schonert-Reichl, K. A., Bonanno, R. A., Vallaincourt, T., & Rocke Henderson, N. (2010). Bullying and morality: Understanding how good kids can behave badly. In S. R. Jimerson, S. M. Swearer, & D. L. Espelage (Eds.), *The international handbook of school bullying* (pp. 101–108). Mahwah, NJ: Lawrence Erlbaum Associates, Inc.

James, D., Flynn, A., Lawlor, M., Courtney, P., Murphy, N., & Henry, B. (2011). A friend in deed? Can adolescent girls be taught to understand relational bullying? *Child Abuse Review, 20*, 439–454. doi:10.1002/car.1120

Jenkins, J. M. (2008). Psychosocial adversity and resilience. In M. Rutter, D. Bishop, D. Pine, S. Scott., J. Stevenson, E. A. Taylor, & A. Thapar (Eds.), *Rutter's handbook of child and adolescent psychiatry* (pp. 377–391). Oxford: Blackwell.

John, O. P., & Gross, J. J. (2004). Healthy and unhealthy emotion regulation: Personality processes, individual differences, and life span development. *Journal of Personality*, *72*(6), 1301–1334.

Jolliffe, D., & Farrington, D. P. (2006). Development and validation of the basic empathy scale. *Journal of Adolescence*, *29*(4), 589–611. doi:10.1016/j.adolescence.2005.08.010

Jones, S. E., Manstead, A. S. R., & Livingstone, A. G. (2011). Ganging up or sticking together? Group processes and children's responses to text-message bullying. *British Journal of Psychology*, *102*(1), 71–96.

Jones, L. M., Mitchell, K. J., & Finkelhor, D. (2010). Online harassment in context: Trends from three youth internet safety surveys (2000, 2005, 2010). *Psychology of Violence*, *3*(1), Special issue: Technology and Violence, 53–69.

Kashdan, T. B., Ferssizidis, P., Collins, R. L., & Muraven, M. (2010). Emotion differentiation as resilience against excessive alcohol use an ecological momentary assessment in underage social drinkers. *Psychological Science*, *21*(9), 1341–1347.

Keltner, D., & Bonanno, G. A. (1997). A study of laughter and dissociation: The distinct correlates of laughter and smiling during bereavement. Journal of Personality and Social Psycho*logy*, *73*, 687–702.

Killen, M. (2007). Children's social and moral reasoning about exclusion. *Current Directions in Psychological Science*, *6*(1), 32–36.

Kinnear, P. (2002). *New families for changing times*. The Australia Institute, Discussion Paper No. 47.

Klomek, A. B., Sourander, A., Niemelä, S., Kumpulainen, K., Piha, J., Tamminen, T., … Gould, M. S. (2009). Childhood bullying behaviors as a risk for suicide attempts and completed suicides: A population-based birth cohort study. *Journal of the American Academy of Child & Adolescent Psychiatry*, *48*(3), 254–261.

Kowalski, R. M. (2008). Cyber bullying: Recognizing and treating victim and aggressor. *Psychiatric Times*, *25*(11), 45–47.

Kowalski, R. M., Limber, S. E., & Agatston, P. W. (2008). *Cyber bullying: Bullying in the digital age*. Malden, MA: Blackwell Publishers.

Kraut, R., Patterson, M., Lundmark, V., Kiesler, S., Mukophadhyay, T., & Scherlis, W. (1998). Internet paradox: A social technology that reduces social involvement and psychological well-being? *American Psychologist*, *53*(9), 1017–1031.

Kurzban, R., & Leary, M. R. (2001). Evolutionary origins of stigmatization: The functions of social exclusion. *Psychological Bulletin*, *127*(2), 187–208.

Lawler, E. J., & Yoon, J. (1996). Commitment in exchange relations: Test of a theory of relational cohesion. *American Sociological Review*, *61*, 89–108.

Leadbeater, B. J., & Hoglund, W. L. (2006). Changing the social contexts of peer victimization. *Journal of the Canadian Academy of Child and Adolescent Psychiatry*, *15*(1), 21–26.

Leary, M. R., & Downs, D. L. (1995). Interpersonal functions of the self-esteem motive: The self-esteem system as a sociometer. In M. H. Kernis (Ed.), *Efficacy, agency, and self-esteem* (pp. 123–144). New York, NY: Plenum.

Leets, L., & Wolf, S. (2005). Adolescent rules for social exclusion: When is it fair to exclude someone else? *Journal of Moral Education*, *34*(3), 343–362.

Li, Q. (2008). A cross-cultural comparison of adolescents' experience related to cyberbullying. *Educational Research*, *50*(3), 223–234.

Lifton, R. J. (1993). *The protean self: Human resilience in an age of transformation*. New York, NY: Basic Books.

Lim, S. J. J., & Hoot, J. L. (2015). Bullying in an increasingly diverse school population: A socio-ecological model analysis. *School Psychology International*. 014303431557 1158.

Lines, D. (2008). *The bullies: Understanding bullies and bullying* (p. 224). London: Jessica Kingsley.

Little, T. D., Jones, S. M., Henrich, C. C., & Hawley, P. H. (2003). Disentangling the "whys" from the "whats" of aggressive behaviour. *International Journal of Behavioral Development*, *27*, 122–133.

Livingstone, S., & Helsper, E. (2010). Balancing opportunities and risks in teenagers' use of the internet: The role of online skills and internet self-efficacy. *New Media and Society*, *12*, 309–329.

Losoi, H., Turunena, S., Wäljasa, M., Helminencd, M., Öhmana, J., & Julkunenb, J. (2013). Psychometric properties of the Finnish version of the resilience scale and its short version. *Psychology, Community and Health*, *2*(1), 1–10. doi:10.5964/pch.v2i1.40

Lundman, B., Strandberg, G., Eisemann, M., Gustafson, Y., & Brulin, C. (2007). Psychometric properties of the Swedish version of the resilience scale. *Scandinavian Journal of Caring Sciences*, *21*, 229–237.

Mark, L., & Ratliffe, K. T. (2011). Cyber worlds: New playgrounds for bullying. *Computers in the Schools*, *28*(2), 92–116. doi:10.1080/07380569.2011.575753

Masten, A. S., & Coatsworth, J. D. (1998). The development of competence in favorable and unfavorable environments. Lessons from research on successful children. *American Psychologist*, *53*(2), 205–220.

Masten, A. S., Best, K. M., & Garmezy, N. (1990). Resilience and development: Contributions from the study of children who overcome adversity. *Development and Psychopathology*, *2*, 425–444.

Menesini, E., Sánchez, V., Fonzi, A., Ortega, R., Costabile, A., & Lo Feudo, G. (2003). Moral emotions and bullying: A cross-national comparison of differences between bullies, victims and outsiders. *Aggressive Behavior*, *29*(6), 515–530.

Merrell, K. W., Felver-Gant, J. C., & Tom, K. M. (2011). Development and validation of a parent report measure for assessing social-emotional competencies of children and adolescents. *Journal of Child and Family Studies*, *20*(4), 529–540.

Mesch, G. S. (2009). Parental mediation, online activities, and cyberbullying. *Cyber Psychology & Behavior*, *12*(4), 387–393.

Migliaccio, T. (2015). Teacher engagement with bullying: Managing an identity within a school. *Sociological Spectrum*, *35*(1), 84–108.

Mishna, F., McInroy, L., & Lacombe-Duncan, A. (2015). *Bridging the Gap: A Mixed-Method Investigation of Cyberbullying*. Abstract from the SSWR (Society for Social Work and Research) Conference, New Orleans. Retrieved from www.//sswr.confex.con/sswr/2015/webprogram/Paper22728.html. Accessed on January 14–18, 2015.

Mitsopoulou, E., & Giovazolias, T. (2015). Personality traits, empathy and bullying behavior: A meta-analytic approach. *Aggression and Violent Behavior*, *21*, 61–72.

Molm, L. D. (1997). *Coercive power in social exchange*. Cambridge: Cambridge University Press.

Molm, L. D. (2000). Theories of social exchange and exchange networks. In G. Ritzer & B. Smart (Eds.), *Handbook of social theory* (pp. 260–272). Thousand Oaks, CA: Sage.

Molm, L. D., Peterson, G., & Takahashi, N. (1999). Power in negotiated and reciprocal exchange. *American Sociological Review*, *64*, 876–890.

Namie, G., & Namie, R. (2003). *Anger and shame: Emotions of bullying. The bully at work: What you can do to stop the hurt and reclaim your dignity on the job* (pp. 197–207). Naperville, IL: Sourcebooks, Inc.

Nation, M., Vieno, A., Perkins, D. D., & Santinello, M. (2008). Bullying in school and adolescent sense of empowerment: An analysis of relationships with parents, friends, and teachers. *Journal of Community and Applied Social Psychology*, *18*, 211–232.

Newberry, M., & Gallant, A. (Eds.). (2013). *Emotion and school: Understanding how the hidden curriculum influences relationships, leadership, teaching, and learning* (Vol. 18). Bingley, UK: Emerald Group Publishing.

Nishi, D., Uehara, R., Kondo, M., & Matsuoka, Y. (2010). Reliability and validity of the Japanese version of the Resilience Scale and its short version. *BMC Research Notes*, *3*(1). Article 310.

Nishina, A., Juvonen, J., & Witkow, M. R. (2005). Sticks and stones may break my bones, but names will make me feel sick: The psychosocial, somatic, and scholastic consequences of peer harassment. *Journal of Clinical Child and Adolescent Psychology*, *34*(1), 37–48.

Notar, C. E., Padgett, S., & Roden, J. (2013). Cyberbullying: A review of the literature. *Universal Journal of Educational Research*, *1*(1), 1–9.

Nygren, B., Aléx, L., Jonsén, E., Gustafson, Y., Norberg, A., & Lundman, B. (2005). Resilience, sense of coherence, purpose in life and self-transcendence in relation to perceived physical and mental health among the oldest old. *Aging and Mental Health*, *9*, 354–362. doi:10.1080/1360500114415

Olweus, D. (1991). Bully/victim problems among schoolchildren: Basic facts and effects of a school based intervention program. *The Development and Treatment of Childhood Aggression*, *17*, 411–448.

Ortega, R., Elipe, P., & Monks, C. P. (2012). The emotional responses of victims of cyberbullying: Worry and indifference. *The British Psychological Society*, *11*(9), 139–153.

Ortega, R., Elipe, P., Mora-Merchan, J. A., Calmaestra, J., & Vega, E. (2009). The emotional impact of victims of traditional bullying and cyberbullying. A study of Spanish adolescents. Zeitsschrift fur Psychologie/Journal of Psychology, *217*(4), 197–2004.

Ortega, R., Elipe, P., Mora-Merchan, J. A., Genta, M.-L., Brighi, A., Guarini, A., ... Tippett, A. (2012). The emotional impact of bullying and cyberbullying on victims: A European cross-national study. *Aggressive Behavior*, *38*, 341–356.

Ortega-Ruiz, R., Del Rey, R., & Casas, J. A. (2012). Knowing, building and living together on internet and social networks: The ConRed cyberbullying prevention program. *International Journal of Conflict and Violence*, *6*(2), 302–312.

Owens, L., Shute, R., & Slee, P. (2000). Guess what I just heard: Indirect aggression among teenage girls in Australia. *Aggressive Behavior*, *26*, 67–83.

Palmieri, P. A., & Smith, G. C. (2007). Examining the structural validity of the Strengths and Difficulties Questionnaire (SDQ) in a U.S. sample of custodial grandmothers. *Psychological Assessment*, *19*(2), 189–198.

Papatraianou, L. H., Levine, D., & West, D. (2014). Resilience in the face of cyberbullying: An ecological perspective on young people's experiences of online adversity. *Pastoral Care in Education*, *32*(4), 264–283.

Parrott, W. G. (1993). Beyond hedonism: Motives for inhibiting good moods and for maintaining bad moods. In D. M. Wegner & J. W. Pennebaker (Eds.), *Handbook of mental control* (pp. 278–308). Englewood Cliffs, NJ: Prentice Hall.

Pellegrini, A. D., & Long, J. A. (2002). A longitudinal study of bullying, dominance, and victimization during the transition from primary to secondary school. *The British Journal of Developmental Psychology, 20*, 259–2280. doi:10.1348/026151002166442

Pellegrini, A. D. (2004). Bullying during the middle school years. In C. E. Sanders & G. D. Phye (Eds.), *Bullying: Implications for the classroom*. San Diego, CA: CalifElsevier.

Pepler, D. J., Craig, W. M., Connolly, J. A., Yuile, A., McMaster, L., & Jiang, D. (2006). A developmental perspective on bullying. *Aggressive Behavior, 32*, 376–384. doi:10.1002/ab.20136

Perkins, D. D., & Zimmerman, M. A. (1995). Empowerment theory: Research and applications. *American Journal of Community Psychology, 23*, 569–579.

Perren, S., Dooley, J., Shaw, T., & Cross, D. (2010). Bullying in school and cyberspace: Associations with depressive symptoms in Swiss and Australian adolescents. *Child and Adolescent Psychiatry and Mental Health, 4*(28), 1–10.

Perren, S., Gutzwiller-Helfenfinger, E., Malti, T., & Hymel, S. (2011). Moral reasoning and emotion attributions of adolescent bullies, victims, and bully-victims. *British Journal of Developmental Psychology*. Early online publication. doi:10.1111/j.2044-835X.2011.02059.x

Prati, G. (2010). Psychometric properties of the dispositional resilience scale. *Journal of Psychology, 4*(3), 252–261.

Punamäki, R. L., Wallenius, M., Hölttö, H., Nygård, C. H., & Rimpelä, A. (2009). The associations between Information and Communication Technology (ICT) and peer and parent relations in early adolescence. *International Journal of Behavioral Development, 33*(6), 556–564.

Raskauskas, J., Rubiano, S., Offen, I., & Wayland, A. K. (2015). Do social self-efficacy and self-esteem moderate the relationship between peer victimization and academic performance? *Social Psychology of Education*, 1–18.

Reece, T. (2012). Cyberbullying 411. *Current Health Teens, 38*(5), 7–9.

Reich, S. M., Subrahmanyam, K., & Espinoza, G. (2012). Friending, IMing, and hanging out face-to-face: Overlap in adolescents' online and offline social networks. *Developmental Psychology, 48*(2), 356.

Riebel, J. R. S. J., Jaeger, R. S., & Fischer, U. C. (2009). Cyberbullying in Germany–an exploration of prevalence, overlapping with real life bullying and coping strategies. *Psychology Science Quarterly, 51*(3), 298–314.

Rigby, K., & Slee, P. T. (2002). Bullying among Australian schoolchildren: Reported behavior and attitudes toward victims. *The Journal of Social Psychology, 131*, 615–627.

Rigby, K. (2002). *New perspective on bullying*. London: Jessica Kingsley Publishers.

Rigby, K. (2005). Bullying in schools and the mental health of children. *Australian Journal of Guidance and Counselling, 15*(2), 195–208.

Rigby, K. (2008). *Children and Bullying. How parents and educators can reduce bullying in schools*. Malden, MA: Blackwell/Wiley.

Ryff, C. D., Singer, B., Dienberg Love, G., & Essex, M. J. (1998). Resilience in adulthood and later life. In J. Lomaranz (Ed.), *Handbook of aging and mental health: An integrative approach* (pp. 69–96). New York, NY: Plenum Press.

Sagone, E., & De Caroli, M. E. (2014). A correlational study on dispositional resilience, psychological well-being, and coping strategies in university students. *American Journal of Educational Research*, *2*(7), 463–471.

Salmivalli, C., & Nieminen, E. (2002). Proactive and reactive aggression among school bullies, victims, and bully-victims. *Aggressive Behavior*, *28*(1), 30–44.

Salmivalli, C. (1999). Participant role approach to school bullying: Implications for intervention. *Journal of Adolescence*, *22*(4), 453–459.

Salmon, G., James, A., & Smith, D. M. (1998). Bullying in schools: Self reported anxiety, depression, and self esteem in secondary school children. *British Medical Journal*, *317*(7163), 924–925.

Sawyer, J.-L., Mishna, F., Pepler, D., & Wiener, J. (2011). The missing voice: Parents' perspectives of bullying. *Children and Youth Services Review*, *33*, 1795–1803.

Schneider, S. K., O'Donnell, L., Stueve, A., & Coulter, R. W. (2012). Cyberbullying, school bullying, and psychological distress: A regional census of high school students. *American Journal of Public Health*, *102*(1), 171–177.

Scholte, R., & Aken, M. (2006). Peer relations in adolescence. In S. Jackosn & L. Goossens (Eds.), *Handbook of adolescent development* (pp. 175–199). New York, NY: Psychology Press.

Seals, D., & Young, J. (2003). Bullying and victimization: Prevalence and relationship to gender, grade level, ethnicity, self-esteem, and depression. *Adolescence*, *38*(152), 735.

Sen, R. (2015). Not all that is solid melts into air? Care-experienced young people, friendship and relationships in the 'digital age'. *British Journal of Social Work*, 1–17. doi:10.1093/bjsw/bcu152

Simmons, R. (2004). Odd girl speaks out: Girls write about bullies, cliques, popularity and jealo*usy*. San Diego, CA: Harcourt Books.

Sinclair, R. R., & Oliver, C. M. (2003). *Development and validation of a short measure of hardiness*. Defense Technical Information Center Report. Portland, OR: Portland State University.

Smith, P. K., & Sharp, S. (1994). *School Bullying: Insights and Perspectives*. London: Routledge.

Smith, B. C., Dalen, J., Wiggins, K., Tooley, E., Christopher, P., & Bernard, J. (2008). The brief resilience scale: Assessing the ability to bounce back. *International Journal of Behavioral Medicine*, *15*, 194–200.

Smith, H., Polenik, K., Nakasita, S., & Jones, A. P. (2012). Profiling social, emotional and behavioral difficulties of children involved in direct and indirect bullying behaviors. *Emotional and Behavioral Difficulties*, *17*(3–4), 243–257.

Smith, P. K., Salmivalli, C., & Cowie, H. (2012). Effectiveness of school-based programs to reduce bullying: A commentary. *Journal of Experimental Criminology*, *8*(4), 433–441.

Smith, P. K., Talamelli, L., Cowie, H., Naylor, P., & Chauhan, P. (2004). Profiles of non-victims, continuing victims and new victims in school bullying. *British Journal of Educational Psychology*, *24*, 565–581.

Sourander, A., Klomek, A. B., Ikonen, M., Lindroos, J., Luntamo, T., Koskelainen, M. (2010). Psychosocial risk factors associated with cyberbullying among adolescents. *Archives of General Psychiatry*, *67*(7), 720–728.

Spear, L. P. (2000). The adolescent brain and age-related behavioral manifestations. *Neuroscience and Biobehavioral Reviews*, *24*(4), 417–463.

Tee, K., Brush, A. J. B., & Inkpen, K. M. (2009). Exploring communication and sharing between extended families. *International Journal of Human-Computer Studies*, *67*(2), 128–138.

Thompson, R. A., & Calkins, S. D. (1996). The double-edged sword: Emotional regulation for children at risk. *Development and Psychopathology*, *8*(1), 163–182.

Thornberg, R., & Jungert, T. (2013). Bystander behavior in bullying situations: Basic moral sensitivity, moral disengagement and defender self-efficacy. *Journal of Adolescence*, *36*(3), 475–483.

Thornberg, R., & Jungert, T. (2014). School bullying and the mechanisms of moral disengagement. *Aggressive Behavior*, *40*, 99–108. doi:10.1002/ab.21509

Tian, J., & Hong, J. S. (2013). Validation of the Chinese version of the resilience scale and its cutoff score for detecting low resilience in Chinese cancer patients. *Support Care Cancer*, *21*(5), 1497–1502. doi:10.1007/s00520-012-1699-x

Tokunaga, R. S. (2010). Following you home from school: A critical review and synthesis of research on cyberbullying victimization. *Computers in Human Behavior*, *26*(3), 277–287.

Tugade, M. M., & Fredrickson, B. L. (2002). Positive emotions and emotional intelligence. In L. F. Barrett & P. Salovey (Eds.), *The wisdom of feelings: Psychological processes in emotional intelligence* (pp. 319–340). New York, NY: Guilford Press

Tugade, M. M., & Fredrickson, B. L. (2007). Regulation of positive emotions: Emotion regulation strategies that promote resilience. *Journal of Happiness Studies*, *8*(3), 311–333.

Van Royen, K., Poels, K., Daelemans, W., & Vandebosch, H. (2015). Automatic monitoring of cyberbullying on social networking sites: From technological feasibility to desirability. *Telematics and Informatics*, *32*(1), 89–97.

van Widenfelt, B. M., Goedhart, A. W., Treffers, P. D. A., & Goodman, R. (2003). Dutch version of the Strengths and Difficulties Questionnaire (SDQ). *European Child and Adolescent Psychiatry*, *12*, 281–289.

Viding, E., Simmonds, E., Petrides, K. V., & Frederickson, N. (2009). The contribution of callous-unemotional traits and conduct problems to bullying in early adolescence. *Journal of Child Psychology and Psychiatry*, *50*(4), 471–481.

Volk, Camilleri, J. A., Dane, A. V., et al. (2012). Is adolescent bullying an evolutionary adaptation? *Aggressive Behavior*, *38*(3), 222–238.

Wachs, S. (2012). Moral disengagement and emotional and social difficulties in bullying and cyberbullying: Differences by participant role. *Emotional and Behavioural Difficulties*, *17*(3–4), 347–360.

Wagnild, G. M., & Young, H. M. (1993). Development and psychometric evaluation of the resilience scale. *Journal of Nursing Measurement*, *1*, 165–178.

Wagnild, G. M. (2009). *The resilience scale user's guide for the US English version of the resilience scale and the 14-item resilience scale (RS-14)*. Worden, MT: The Resilience Center.

Wallenius, M., Punamaki, R. L., & Rimpela, A. (2006). Digital game playing and direct and indirect aggression in early adolescence: The role of age, social intelligence and parent-child communication. *Journal of Youth Adolescence*, *36*, 325–336.

Walther, M.-T., & Parks, M. R. (2002). *Cues filtered out, cues filtered in compute-mediated communication and relationships*. In M. L. Knapp & J. A. Daly (Eds.), *Handbook of interpersonal communication* (pp. 529–563). Thousand Oaks, CA: Sage.

Wang, J., Iannotti, R. J., & Luk, J. W. (2012). Patterns of adolescent bullying behaviors: Physical, verbal, exclusion, rumor, and cyber. *Journal of School Psychology*, *50*(4), 521–534.

Werner, E. E. (1989). High risk children in young adulthood: A longitudinal study from birth to 32 years. *American Journal of Orthopsychiatry*, *59*, 72–81.

Westphal, M., Seivert, N. H., & Bonanno, G. A. (2010). Expressive flexibility. *Emotion*, *10*(1), 92.

Wilding, R. (2006). 'Virtual' intimacies? Families communicating across transnational contexts. *Global Networks*, *6*(2), 125–142.

Ybarra, M. L., & Mitchell, K. J. (2004). Online aggressor/targets, aggressors, and targets: A comparison of associated youth characteristics. *Journal of Child Psychology and Psychiatry*, *45*(7), 1308–1316.

You, S., Kim, E., & No, U. (2015). Impact of violent video games on the social behaviors of adolescents: The mediating role of emotional competence. *School Psychology International*, *36*(1), 94–111.

FACTORS THAT INFLUENCE BYSTANDER BEHAVIOR IN THE CYBERBULLY CONTEXT☆

Jessica Niblack and Jodie L. Hertzog

ABSTRACT

Purpose – *Despite growing attention to the prevalence and consequences of cyberbullying within the social sciences, research on cyber-bystander reactions has been largely overlooked. Drawing from Latane and Darley's (1970) bystander engagement model, the current study sought to fill this gap by exploring how common it is for adolescents to encounter cyberbullying on social networking sites (SNS), how youth react to the cyberbullying witnessed on SNS, and most importantly to uncover factors that may be related to two potential bystander trajectories on SNS, namely traditional bystanding and prosocial bystander engagement.*

Methodology/approach – *Data was drawn from the 2011 Pew Research Center's Internet and American Life Project (Princeton*

☆A previous version of this manuscript was presented at the 2013 Midwest Sociological Society's Annual Meeting in Chicago, IL.

Technology and Youth: Growing Up in a Digital World
Sociological Studies of Children and Youth, Volume 19, 47–72

ISSN: 1537-4661/doi:10.1108/S1537-466120150000019002

Survey Research Associates International, 2011). The secondary analysis was restricted to only adolescents who ranged in age from 12 to 17. Grounded in existing research on face-to-face bystander behaviors, two Ordinary Least Squares regression models were run testing which independent variables (age, gender, frequency of SNS use, perceived peer norms, and prior cyberbully victimization) were related to traditional and to prosocial bystander behavior online.

Findings – *Approximately 88% of youth reported they've witnessed a cyberbullying exchange on an SNS. Among these witnesses, the majority reported engaging in both prosocial (62%) and traditional (74%) bystander practices. Based on the regression analyses, a key factor for bystander practice online appears to be observed peer behavior.*

Originality/value – *The findings from this research provide an initial exploration into cyber-bystander behavior, with potential implications for both future research directions and cyberbully prevention programming.*

Keywords: Cyberbullying; bystander engagement; Adolescent technology use

In the last 20 years electronic communication has evolved from receiving a page on one's beeper to electronic mail, Social Networking Sites (SNSs) (like Facebook, MySpace, and Twitter), instant messaging, chat rooms, and listservs, just to name a few. The options for communication in this modern era are seemingly endless and new tech options are being created at a rapid pace. Adolescents have latched onto these forms of communication with 97% of adolescents aged 12–18 years reporting that they regularly use the internet (Kowalski & Limber, 2007). In addition, 90% of adolescents report that they use technology mediated modes of communication to chat and instant message friends, post status updates on SNSs, and post comments to friend's statuses (Lenhart et al., 2011). Finally, Mishna, Saini, and Solomon (2009) assert that electronic forms of communication are now viewed as "critical tools" in the social lives of the majority of adolescents.

While technological advancements bring many positive aspects, the popular media has become increasingly inundated with stories of potential risks youth may face in the cyber sphere. At the forefront of these discussions are the ways in which technologies are being used to change the face of bullying. As Twyman, Saylor, Taylor, and Comeaux (2010) point out, "being anonymous allows for reduced social accountability which may encourage individuals to engage in inappropriate behavior online" (p. 195). According to the Center for Disease Control and Prevention, electronic aggression, more commonly referred to as cyberbullying, can be defined as "any type of harassment or bullying (teasing, telling lies, making fun of someone, making rude or mean comments, spreading rumors, or making threatening or aggressive comments) that occurs through e-mail, a chat room, instant messaging, a website (including blogs), or text messaging" (Hertz & David-Ferdon, 2008, p. 3). Existing research suggests cyberbullies tend to use a combination of technical outlets to bully someone. In comparison to traditional bullying where the bully is often known, Kowalski and Limber (2007) assert many victims of cyberbullying do not know the identity of their bully. In the cyber world, bullies can hide behind ambiguous screen names and even hijack others' accounts. This same anonymity, however, can also create an environment in which those witnessing cyberbullying behavior speak out.

Within social psychology, audience members who witness negative interactions such as bullying are known as bystanders (Latane & Darley, 1970; Polanin, Espelage, & Pigott, 2012). While increasing scholarly attention has been devoted to exploring the ways in which traditional bullying behaviors have entered the cyber sphere, little is known about the reactions of potential bystanders who witness bullying online. Bystander reactions are an important factor in fully understanding the dynamics of cyberbullying. Based on Latane and Darley's model of bystander behavior, there are several types of bystanders to be considered: those who act in a prosocial manner by helping or defending the cyber victim, those who join in on the harassment and further bully the victim, and those who simply ignore the victimization taking place. Take the case in Stuebenville, Ohio, for example, where a group of teens at a party shared pictures and video through text and SNSs of a 16-year-old girl being raped by two high school athletes but did nothing to intervene during the rape (Dahl, 2013). The Stuebenville case provides not only an opportunity to consider how social media became a medium of expressing initial bystander reactions but also to explore internet bystander reactions to the materials the witnessing bystanders posted. Gaining a better understanding of factors influencing bystander practices in various settings, including the cyber sphere, is necessary for developing programs that promote prosocial bystander engagement practices.

LITERATURE REVIEW

The exponential growth of electronic and computer-based communication and information sharing during the last decade has drastically altered individuals' social interactions. Websites devoted specifically to social updating (the most popular being Twitter) or SNSs with updating features (such as Facebook) have become popular among adolescents and adults alike. However, adolescents today are a unique population of SNS users because they are the first to have grown up entirely surrounded by communication technologies. Almost 50% of the teenage population in the United States uses cell phones and 97% uses the Internet, with 51%, or approximately 45 million adolescents, using it on a daily basis (Kowalski & Limber, 2007). Finally, 74% of adolescents report having access to the Internet from their homes with the average time per day adolescents spent online being 84.9 minutes (Lee & Chae, 2007). Thus, it's not surprising that for adolescents raised in an internet-enabled world, blogs and SNSs are competing with face-to-face and telephone communication as the dominant means and methods through which personal interaction takes place (Hinduja & Patchin, 2008; Pujazon-Zazik & Park, 2010).

While technology has advanced our society in more ways than can be listed, it has also altered the face of bullying. The nature of adolescent peer-to-peer aggression has evolved along with the proliferation of information and communications technologies. Due to their unique characteristics, online environments might be perceived by youth as a liberating platform on which to express themselves, creating an environment where adolescents are increasingly susceptible to negative social interactions (Erdur-Baker, 2010; Hinduja & Patchin, 2010; Mesch, 2009). As Ahn (2011) points out, however, "technology is a structuring factor. Features of technology, not the technology itself, enable and constrain how one uses that tool" (p. 1436). Take for example the popular app Snapchat which may make it increasingly easy to engage in bullying behaviors without leaving an electronic trail as photographic and text posts, known as "Snaps," are only visible for up to 10 seconds and then deleted from the application's server (NoBullying.com, 2014).

Existing research suggests that frequently using common technologies, such as sending text messages, instant messages, and e-mails to friends, may increase an adolescent's risk of victimization (Mesch, 2009; Wolak, Mitchell, & Finkelhor, 2007). Additionally, how often one uses the computer to navigate various online worlds is a potentially important predictor for determining victimization risk. Research has shown that offenders tend to use the Internet significantly more frequently and with more proficiency than individuals who have been victimized; however, the more time respondents spent on the Internet the more likely they experienced cyberbullying (Ahlfors, 2010; Hinduja & Patchin, 2008). Thus, using the Internet can be seen as a two-sided coin. The more one uses technology the more skilled they become, but "if students are not allowed to use new technologies and participate in online communities like SNSs, they will not be able to develop the necessary skills and technical literacy that will be vital in the future" (Ahn, 2011, pp. 1447–1448).

The Bully, Victim, Bystander Triangle

Research on traditional bullying suggests that in many instances bullying is not just a dyadic process between a bully and a victim but is more like a triangle, frequently involving others who become witness to the bullying behaviors either directly or indirectly (Barhight, Hubbard, & Hyde, 2013; Coloroso, 2011; Evers, Prochaska, Van Marter, Johnson, & Prochaska, 2007; Gini, Albiero, Benelli, & Altoe, 2008; Gini, Pozzoli, Borghi, & Franzoni, 2008; Law, Shapka, Hymel, Olson, & Waterhouse, 2012; Lodge & Frydenberg, 2005; McNamee & Mercurio, 2008; Salmivalli, 2010; Salmivalli & Voeten, 2004). The three main players within the bullying triangle consist of the bully, the victim/target, and the bystander.

The Bully

Bullying behavior often involves an imbalance of power between the bully, the victim, and potentially the bystander based on physical strength, popularity, or peer group status attainment, and academics (Coloroso, 2011; Faris & Felmlee, 2011, 2014; Jerome & Segal, 2003; McNamee & Mercurio, 2008; Oh & Hazler, 2009; Pozzoli, Gini, & Vieno, 2012; Smith, Mahdavi, Carvalho, & Tippett, 2006). Bullying has historically affected children and teenagers in specific contexts: while at school, while traveling to or from school, or in public places such as playgrounds and bus stops or other venues such as malls, restaurants, or at neighborhood hangouts (Mesch, 2009; Patchin & Hinduja, 2006). Technology, however, has allowed cyberbullying to have a pervasive presence in the life of a victim beyond these limited contexts. Whereas in past decades, one could duck out on a bully, take a different route to school, or hide out in their bedroom all weekend, technology has changed the face of bullying. Patchin and Hinduja (2006) state that "because of the advent and continued growth of technological advances, the transmutation of bullying has moved from physical to the virtual. Physical separation of the bully and victim is no longer a limitation in the frequency, scope, and depth of harm experienced and doled out. As instances for bullying are no longer restricted to real-world settings, the problem has matured" (p. 155).

The Victim/Target

With the way SNSs are set up, it is easy for an adolescent to end up as a victim or target of cyberbullying. With one post, one status update, or one tweet, an adolescent's reputation can be challenged and witnessed by numerous voyeurs in a matter of seconds. According to Li (2006), 62% of adolescents have been victimized by a cyberbully at least once and about 38% were victimized more than three times. For over a third of victims, cyber harassment "involves information being posted or sent to someone else about the adolescent" (Ybarra, Mitchell, Wolak, & Finkelhor, 2006). Studies further show that as many as "60% of adolescents have been ignored while online, 50% have reported being disrespected, about 30% have been called names, 21% have been threatened, 20% were picked on" (Patchin & Hinduja, 2006, p. 158), and "29.3% have had rumors spread about them by others" (Hinduja & Patchin, 2010, p. 214). Through various mediums, cyberbullying can occur at any time of day or night (Mesch, 2009; Patchin & Hinduja, 2006; Pujazon-Zazik & Park, 2010); and digital storage archives provide the potential for episodes of cyberbullying to be forever available online meaning that at any given time someone can reread, relook at, or rewatch the negative information posted about a target with the potential consequence of a victim reliving the victimization over and over again (Law, Shapka, Domene, & Gagné, 2012).

The Bystander

When assessing the bullying triangle it is found that bystanders tend to make up the vast majority of a school's population (McNamee & Mercurio, 2008). Studies have shown that bystanders are present about 85% of the time during instances of face-to-face bullying (Barhight et al., 2013; Lodge & Frydenberg, 2005; McNamee & Mercurio, 2008). The bystander in the triangle of bullying plays many different roles. There are the reinforcers, the traditional bystanders, and the engaged prosocial bystanders. The reinforcer engages in behaviors that support the bully. So, basically, reinforcers side with the bully and may participate by laughing, joining in themselves, or encouraging the bully to continue the victimization (Oh & Hazler, 2009; Salmivalli & Voeten, 2004; Thornberg et al.,

2012). The traditional bystander is the second role a bystander can play in the bullying triangle. Though aware of the bullying incident, these bystanders tend to remain uninvolved, do not appear to take sides, and do not take steps to defend the victim (Oh & Hazler, 2009; Stueve et al., 2006; Thornberg et al., 2012). The final role a bystander can play in the triangle is the defender or prosocial bystander. Engaged bystanders help the victim by siding with the victim, comforting or encouraging the victim to take a stand, or by displaying conflict resolution skills by trying to diffuse the bullying situation (Oh & Hazler, 2009). Such defending actions have been shown to develop when adolescents have a greater sense of empathy. The goal of the defender is to promote prosocial change (Cowie, 2000; Dunn, 2009; Espelage, Green, & Polanin, 2012; Salmivalli & Voeten, 2004; Thornberg et al., 2012).

Bystander Engagement

Based on existing bystander engagement models, there are several precursors to becoming an engaged prosocial bystander (Banyard, Moynihan, & Crossman, 2009; Banyard, Plante, & Moynihan, 2004; Latane & Darley, 1970; Stueve et al., 2006). To begin, the bystander must first realize something is happening. At this point the bystander needs to determine if what is happening is cause for concern. Next the bystander must decide if it is their personal responsibility or duty to act and if so, determine what type of intervention method is needed. Finally, the bystander needs to act.

The process of cognitive assessment is further influenced by social factors that may affect bystander practices. For example, adolescent bystanders may not want to get involved for fear of bodily harm, humiliation in front of peers, or concerns about potential legal problems (Darley & Latane, 1968). In fact, the presence of others in a crisis situation has been found to be a key factor. Numerous studies have found that the more people who are present in a crisis situation the less likely the victim will receive help and it boils down to the belief that someone else will certainly do something (Banyard et al., 2004; Choi & Cho, 2013; Critelli & Keith, 2003; Darley & Latane, 1968; Hudson & Bruckman, 2004; Obermann, 2011; Stueve et al., 2006). This line of practice may also carry over into the virtual world, specifically in chat rooms. Markey (2000) found that victims did not receive help as quickly or at all when there were several people involved in the chat thread. Likewise, Hudson and Bruckman (2004) suggest that the lack of social cues in the cyber realm can affect social awareness and may even lead to inhibitive behaviors in either positive or negative ways.

To date, little research, especially in North America, has explored bystander reactions to bullying in the cyber sphere. Research on face-to-face interactions, however, suggests there are several measures that may promote prosocial bystander engagement. Bystanders will feel a greater sense of urgency to act prosocially if they are asked for help by name, if the bystander feels obligated for some reason, if they are empathetic, if they have commonalities with the victim, and if they belong to a group that promotes intervention (Banyard, 2008; Banyard et al., 2004; Barlińska, Szuster, & Winiewski, 2013; Bastiaensens et al., 2014; Choi & Cho, 2013; Markey, 2000; Stueve et al., 2006). In one of the few published articles investigating cyberbully bystander reactions, DeSmet et al.'s (2014) focus group with adolescents in Belgium found that the determinants of defending online were similar to those in face-to-face situations and interestingly that defenders preferred to intervene in cyberbully situations when offline. Similarly, in an experimental study with Flemish adolescents, Bastiaensens et al. (2014) found that "when acquaintances defended the victim, [*cyber*] bystanders had significantly higher behavioural intentions to help the victim in private than in public" (p. 7). Macháčková, Dedkova, Sevcikova, and Cerna (2013) also found that existence of a relationship with the victim or the cyberbully affected whether and in what way bystanders engaged.

Contextual Factors Mediating Bystander Reactions

In addition to social influences, existing research suggests contextual factors may further impact how a bystander reacts, particularly age, gender, and level of parental monitoring experienced.

Age. Age influences bystander reaction in two ways. First, as youth get older, they increase their online use, which increases their potential for encountering cyberbullying as either a victim or a bystander (Kowalski & Limber, 2007; Mesch, 2009; Vandebosch & Van Cleemput, 2009; Ybarra & Mitchell, 2004); with the transition to high school appearing to be a particularly salient factor for increased exposure (Ahlfors, 2010; Vandebosch & Van Cleemput, 2009). Secondly, as an adolescent gets older the decision to defend has been found to decrease, with students in higher grades admitting they would prefer to not get involved against the bully or for the victim (Salmivalli & Voeten, 2004). In fact, as an adolescent gets older both sympathy for and liking of the victim have been found to decrease

(Gini, Pozzoli et al., 2008), which is antithetical to prosocial bystander engagement.

Gender. Like age, gender has also been found to influence exposure, participation, and reactions to cyberbullying. Girls are often members of online communities earlier, use instant messaging and e-mail at higher rates than boys, and thus have increased potential for exposure to cyberbullying (Espinoza & Juvonen, 2011; Jackson, Cassidy, & Brown, 2009; Lee & Chae, 2007; Pujazon-Zazik & Park, 2010). Furthermore, because girls tend to use their words as weapons instead of their fists (Li, 2006), the cyber world presents a wide range of opportunities for engaging in and witnessing such negative interactions. Boys, on the other hand, tend to be more aggressive toward their male peers overall including friends, so it may be more difficult to distinguish acts of cyberbullying from what may be perceived as "typical boy antics" (Wilson-Simmons, Dash, Tehranifar, O'donnell, & Stueve, 2006). While many types of goading seem acceptable among males in society, males acting in an aggressive manner toward females are not viewed as acceptable (Laner, Benin, & Ventrone, 2001). Thus, the gender of the victim, the perpetrator, and the bystander themselves may be factors mediating bystander assessment of a given situation.

Additionally, gender expression may influence prosocial bystander engagement more directly. Behaviors such as empathy and self-efficacy, which are often stronger among girls, have been positively related to prosocial bystander engagement (Gini, Albiero et al., 2008; Lwin, Li, & Ang, 2012). Victim liking, another possible determinant of bystander engagement, is also found to be higher among girls. Moreover, females have been found to exhibit higher levels of protective intentions than males overall (Gini, Pozzoli et al., 2008; Lwin et al., 2012). Thus, while girls may be primed toward cyberbullying (Ahmed, 2008), girls have also been shown to step into helper or defender roles more than boys (Choi & Cho, 2013). In face-to-face bully interactions, males are more likely to be the aggressors or ignore the situation while females are more likely to defend the victim in a prosocial manner (Ahmed, 2008; Banyard, 2008; Choi & Cho, 2013; Erdur-Baker, 2010; Lenhart et al., 2011; Li, 2006; Obermann, 2011). In sum, being female has been related to a range of positive bystander outcomes such as showing support for the victim, defending behavior, initiating constructive conflict resolution, and/or initiating third-party interventions (Ahmed, 2008; Banyard, 2008; Choi & Cho, 2013; Cowie, 2000; Gini, Pozzoli et al., 2008; Lodge & Frydenberg, 2005; Obermann, 2011; Oh & Hazler, 2009; Pozzoli et al., 2012; Salmivalli & Voeten, 2004).

Parental Monitoring. One approach to guarding negative online experiences is parental monitoring. Research suggests there are several techniques parents can use to control or monitor their adolescent's online activities. These techniques generally consist of four elements: parental supervision in cyberspace, communicating with the adolescent, tracking what and where the adolescent is going online, and disclosure of what the adolescent is doing (Liau, Khoo, & Ang, 2008; Mesch, 2009). Specific measures consist of discussions about internet use, discussions of online content, computer placement, restriction of certain websites, creating rules together about internet use, checking what websites the adolescents have been on, and what types of information adolescents have about themselves online, as well as the less effective method of "looking over the shoulder" monitoring (Lenhart et al., 2011; Liau et al., 2008; Mesch, 2009).

One problematic finding related to parental monitoring is the type of advice given. While schools urge students to engage in prosocial bystander engagement by providing information about cyberbullying events, many parents are instructing their adolescents to stay out of the problem unless the incident directly involves a family member (Wilson-Simmons et al., 2006). In addition, there are mixed findings regarding the outcomes of parental monitoring as a protective factor. If the monitoring is not frequent enough, chances of victimization increase (Ybarra & Mitchell, 2004). Furthermore, there appears to be a disconnect between what parents report they are doing and what youth report their parents are doing in terms of monitoring. Approximately 34–54% of parents state they are in fact using some sort of parental monitoring technique; however only 19–39% of adolescents agree that their parents practice monitoring (Lenhart et al., 2011). Intentions versus action may explain some of the discrepancies in what parents say they are doing and what adolescents are aware their parents are doing. Many adolescents feel that adults are out of touch with technology and the prevalence of cyberbullying. Parents may feel that technology is simply intimidating or frightening (Ahlfors, 2010; Mishna et al., 2009). Parental monitoring, nonetheless, can greatly influence how adolescents engage one another on the internet (Mesch, 2009).

Current Study

There is ever growing empirical knowledge around the issue of bullying online, particularly in regards to victims and perpetrators. Yet, there has been limited investigation of the third member of the bully triad, namely the bystander (see Barlińska et al., 2013; Bastiaensens et al., 2014; DeSmet et al., 2014 as exceptions). The goal of the current exploratory study, then, is to increase the current knowledge base regarding online bystander behavior by exploring the following research questions:

1. How common is cyberbullying on SNSs frequented by youth?
2. How do youth react to cyberbullying instances on popular SNSs?
3. What factors influence traditional bystanding versus prosocial bystander engagement online?
4. Do bystander outcomes vary across different types of SNS (i.e., Facebook, MySpace, Twitter)?

METHODOLOGY

The data used for this study was supplied by the Pew Research Center's Internet and American Life Project and funded by the Pew Charitable Trusts (Princeton Survey Research Associates International, 2011). Adolescents were eligible to participate in this random digit dial survey selected from telephone exchanges if they were a resident of the United States between the ages of 12 and 17, had parental consent to participate, and lived in a household with a landline. The sample design uses a random generation of the last two digits of phone numbers chosen on the basis of their area code, telephone exchange, and bank number. The sample for the 2011 Internet and American Life Project was conducted between the dates of April 19th to July 1st 2011 and consisted of 800 parent–teen pairs, with an oversampling of African-American and Latino families. A weight for the data was created based on the most recent data from the Census Bureau's 2011 American Community Survey. The sample is weighted from the Census Bureau's population parameters that consist of gender by age, gender by education, age by education, region race, and Hispanic origin. These parameters are then compared with the sample characteristics to construct sample weights. The final weight was created to balance all weighting parameters (Princeton Survey Research Associates International, 2011). With the weight in effect, the original sample size of nearly 800 respondents was nearly cut in half yielding a sample of 412 for prosocial bystander engagement and 413 for traditional bystander behavior (for more details see Niblack, 2013).

Dependent Variables

Prosocial Bystander Engagement

Prosocial bystander engagement consists of an adolescent defending the cyber victim. An index was created combining responses to two items: (1) whether the adolescent has ever stood up for the victim and (2) whether the adolescent ever acted on the victim's behalf to stop the cyberbullying. Before indexing, the variables were recoded so that zero represented the response of never and three represented frequently. After indexing, the responses were relabeled so that zero represented never and six represented frequently engaged in both modes of prosocial bystander engagement.

Traditional Bystanding

A single item asking how frequently an adolescent responds by ignoring a cyberbullying incident was used to represent traditional bystander behavior. The variable was recoded so that zero represented never and three represented frequently.

Independent Variables

Age

Age was operationalized using a chronological variable ranging from 12 to 17.

Gender

Gender was a dichotomous variable recoded so that zero represented males and one represented females.

Technology

Four variables were used to assess technology use. The first, *frequency of social networking site use*, assessed the general frequency in which adolescents logged onto a SNS. This variable was recoded so one represented less often while six represented several times a day. The second set of variables assessed which of the most popular SNS reported in the Pew data, *Facebook, MySpace, or Twitter User*, the adolescent used. All three of the SNS variables were recoded so that zero represented the fact that the adolescent did not use the SNS and one represented the fact that they did use that SNS.

Parental Monitoring

The amount of parental monitoring techniques adolescents believed their parents were using was assessed by creating an index of four different items: whether the parent used parental controls, checked or visited websites, checked SNSs profiles, and/or used cell phone restrictions. The four techniques were combined and recoded in such a way that zero represented the fact that no parental monitoring techniques were used and four represented the fact that all techniques were used.

Norms

Based on existing research related to determinants of prosocial bystander engagement, contextual norms were assessed using four variables that provide insight into what may be considered acceptable or unacceptable behavior on SNSs. The first variable, *Perceptions of Cruel On-line Environment*, assessed how often an adolescent witnessed what they believe to be cruel behavior online. This variable was recoded so that zero represented adolescents never witnessing other adolescents acting cruel and three represented adolescents witnessing cruel behavior frequently.

Three variables were used to assess how often adolescents observed the three prevalent bystander reactions: prosocial engagement, reinforcing behavior, and traditional bystanding. The first index, *Witnessing Pro-Social Behavior*, was created from two items: (1) How frequently witnessed others telling the bully to stop and (2) How often witnessed others defending a victim. Before indexing each prosocial item was first recoded so that zero represented never witnessing and three represented frequently witnessing. After indexing, the responses options were then relabeled so that zero represented never and six represented frequently witnessing both prosocial engagement modes.

Reinforcing behavior was assessed using a single item that measured how often adolescent's witnessed others joining in with the cyberbullying. Responses to the *Witness Joining In* variable were recoded so that zero represented never witnessing and three represented frequently witnessing.

Observations of traditional bystanding behavior were also assessed using a single item that measured how often an adolescent witnesses others ignoring cyberbullying incidents. Responses to the *Witness Ignoring* variable were recoded so that zero represented never and three represented frequently witnessing cyberbullying being ignored.

Prior Cyberbully Victimization

Two variables were used to assess experience with cyberbully victimization. The first variable, *Social Networking Site Victim in Past 12 Months*, consisted of the response to the question, "In the past 12 months when you have been on a SNS, has anyone been mean or cruel to you?" The response items were recoded so that zero represented no and one represented yes. The second measure, an index of *Overall Cyber Victimization*, assessed whether an adolescent has been "bullied" in the past 12 months through any of the following electronic mediums: cell phone, text, or online. Each of the three items were individually recoded so that zero represented no experience and one represented affirmatively experienced. The items were then indexed and responses were relabeled so that zero represented experiencing cyberbullying through none of the mediums and three represented experiencing cyberbullying via all three mediums.

RESULTS

Univariate Analysis

Tables 1 and 2 depict the univariate results for all variables. As can be seen in the tables, there were nearly equal numbers of male and female participants. Participants ranged in age from 12 to 17 with an average age of 14.5. Forty percent of participants reported checking their SNS several times a day, with the most popular SNS being Facebook. Fifteen percent of participants reported experiencing cyberbullying on an SNS in the last year. Finally, participants reported observing prosocial or traditional bystanding behaviors more frequently than reinforcing behaviors online.

Table 1. Univariate Analysis Descriptives.

Variables	*N*	Median	Mean	SD+/−	Range
Age	800	14.5	14.5	1.72	12–17
Parental monitoring techniques	459	2	1.67	1.30	0–4
Perceptions of cruel online environment	610	2.5	1.41	.84	0–3
Witness prosocial behavior	535	3	3.20	1.72	0–6
Witness joining in	538	1.5	1.28	1.12	0–3
Witness ignoring	539	1.5	2.34	.86	0–3
Prosocial bystander engagement	531	3	3.31	1.89	0–6
Traditional bystander behavior	535	1.5	2.00	.94	0–3

Table 2. Univariate Analysis Frequencies.

Variables	*N*	Percent
Gender		
Male	411	51.4
Female	389	48.6
Total	*800*	*100%*
Social networking site use		
Less often	33	5.4%
Every few weeks	34	5.5%
One to two times a day	73	11.8%
3–5 days	80	13.0%
Once a day	148	24.1%
Several times a day	246	40.1%
Total	*614*	*100%*
SNS victimization in last 12 months		
Yes	93	15.1%
No	522	84.9%
Total	*614*	*100%*
Overall cyberbully victimization		
None	691	86.6%
One	41	5.2%
Two	45	5.6%
All	21	2.6%
Total	*798*	*100%*
Facebook user		
Yes	571	71.4%
No	229	28.6%
Total	*800*	*100%*
MySpace user		
Yes	150	18.7%
No	650	81.3%
Total	*800*	*100%*
Twitter user		
Yes	76	9.5%
No	724	90.5%
Total	*800*	*100%*

Multivariate Analysis

Two regression models were run to explore the factors that may predict each bystander outcome (i.e., prosocial bystander engagement vs. traditional bystander behavior). Based on existing cyberbullying research, each model included age, gender, frequency of SNS use, Facebook, MySpace, or Twitter users, parental monitoring techniques, perceptions of cruel online environment, witnessing prosocial bystander behavior online, witness joining in with bullying, witnessing ignoring the bullying, victimization on a SNS in the last 12 months, and experiencing overall cyberbully victimization in the past 12 months.

Tests for Assumptions
For model 1 the dependent variable, does the bystander act in a prosocial manner, was not normally distributed. There was a total of 413 participants. Tests of residuals suggest that this might be an issue. However, none of the independent variables were correlated over .70. Tests for outliers were conducted. The maximum value found in the Mahalonobis distance test was 50, but the maximum for the Cook's distance test was less than 1. Outliers were less than 1% of the total sample and therefore no outliers were removed.

Model 2, does the bystander engage in traditional bystander engagement, was also found not to be normally distributed. The total sample for this model is 414. Tests of residuals suggest that this might be an issue. However, none of the independent variables were correlated over .70. Tests for outliers were conducted. The maximum found in the Mahalonobis distance test was 49, but the maximum for the Cook's distance test was less than 1. Outliers were less than 1% of the total sample and therefore no outliers were removed.

OLS: Model 1, Prosocial Engagement
Table 3 presents the results for the multiple regression analysis that was conducted for model 1 to evaluate how well the independent variables predict prosocial bystander engagement. Findings indicate that the 16 variables in the model only accounted for approximately 27% of variation in prosocial behavior when encountering cyberbullying on an SNS. Three variables in the model were statistically significant predictors of prosocial engagement: parental monitoring, witnessing prosocial behavior from others, and overall prior cyberbullying victimization. For each increase in perceived monitoring techniques used by a parent, there is a slight (.19) increase in an adolescent's likelihood of engaging in prosocial behavior online ($p \leq .01$). Those who witness others practicing prosocial bystander engagement online are more likely to report practicing prosocial bystander engagement ($p < .001$). And finally, those who had suffered overall cyberbully victimization are more likely to report engaging in prosocial bystander behaviors ($p < .05$).

Table 3. Model 1: Ordinary Least Squares Regression for Prosocial Bystander Engagement.

Variables	B	SE B	*B*
Age	.09	.05	.08^
Gender	.32	.16	.09^
Frequency of SNS use	−.04	.06	−.03
Facebook	−.48	.40	−.05
MySpace	.19	.18	.05
Twitter	.21	.22	.04
Parental monitoring techniques	.19	.06	.13**
Perceptions of cruel online environment	.13	.12	.05
Witness prosocial behavior	.49	.05	.45***
Witness joining In	.10	.08	.06
Witness ignoring	−.03	.10	−.01
SNS victimization in past 12 months	−.16	.24	−.03
Overall cyberbully victimization	.29	.12	.13*
R^2	*.27*		
F	*12.58****		
N	*412*		

**p* < .05; ***p* < .01; ****p* < .001; ^*p*≤.10.

Table 4. Model 2: Ordinary Least Squares Regression for Traditional Bystander Behavior.

Variables	B	SE B	*B*
Age	−.01	.03	−.01
Gender	−.15	.09	−.08
Frequency of SNS use	.03	.04	.04
Facebook user	.38	.23	.08^
MySpace user	.06	.11	.03
Twitter user	.28	.13	.11*
Parental monitoring techniques	−.06	.04	−.09^
Perceptions of cruel online environment	.07	.07	.06
Witness prosocial behavior	−.02	.03	−.03
Witness joining in	.04	.04	.05
Witness ignoring	.29	.06	.25***
SNS victimization in past 12 months	.08	.14	.03
Overall cyberbully victimization	−.14	.07	−.12*
R^2	*.10*		
F	*4.59****		
N	*413*		

**p* < .05; ***p* < .01; ****p* < .001; ^*p*≤.10.

OLS: Model 2, Traditional Bystander Behavior

Table 4 presents the results for the multiple regression analysis that was conducted for model 2 to evaluate how well the independent variables predict traditional bystander behavior. The adjusted R^2 for model 2 is .10, so only 10% of the variation in traditional bystanding is explained by the variables included. Like in model 1, three variables were found to be statistically significant predictors of traditional bystander behavior: Twitter use, witnessing others ignoring cyberbullying, and experiences of overall bullying victimization. Twitter users were found to be slightly more likely to practice traditional bystander behavior compared to those who do not use Twitter ($p < .05$). Adolescents who witnessed others practice traditional bystanding by ignoring cyberbullying incidents are more likely to practice traditional bystanding themselves ($p < .001$). On the other hand, those who had suffered overall cyberbully victimization were shown to have a .14 decrease in likelihood of engaging in traditional bystander behavior ($p < .05$).

DISCUSSION

Drawing from Darley and Latane's Bystander Intervention model, the focus of this exploratory study was to examine different trajectories of bystanding on SNSs. Considering face-to-face studies indicate that bystanders are present in the majority of bullying incidents, it is equally important to gain an understanding of the frequency of bystander presence during cyberbullying incidents and to begin to identify factors that may influence different modes of bystander engagement online. The three main modes of engagement for bystanders during face-to-face interactions include simply watching or ignoring the incident (i.e., traditional bystanding), joining in or supporting the bullying in some way (i.e., reinforcing), and lastly standing up by defending the victim or challenging the bully (i.e., prosocial bystander engagement). While not without limitations, this study is among the first using a North American sample to explore bystander engagement in the cyber sphere and is unique in trying to uncover factors that may influence different modes of bystander engagement in relation to cyberbullying among adolescents.

The first question explored in this research was how common cyberbullying is on SNSs frequented by youth. Based on the findings of the current study, it appears that 13% of adolescents have been bullied in the past 12 months through at least one type of electronic medium: cell phone, text, or internet. Additionally, 15% of adolescents stated that someone had been mean or cruel to them in the last 12 months on a SNS specifically. These findings are consistent with previous research done by Lenhart et al. (2011) who also found that about 15% of adolescents had experienced cyberbullying in their sample. Unfortunately, the current research was unable to explore how common cyberbullying was on specific social networking sties, an avenue that should be investigated through future research.

The second research question explored in this study was the ways in which youth react to cyberbullying on popular SNSs. The findings indicate that bystanders are indeed present online, with 41% of adolescents reporting that they witnessed others acting cruel on SNSs. The findings further show that adolescents witnessing cyberbullying incidents are acting prosocially, with 61% stating that they sometimes to frequently told the bully to stop and 62% of adolescents stating they sometimes to frequently defended the victim. Compared to previous research on traditional bullying by Lenhart et al. (2011), these numbers suggest the possibility that there are higher instances of prosocial bystander engagement in the cyber realm. It is worth noting, however, that the current research also showed that the majority of adolescents (74%) reported acting as traditional bystanders reporting they sometimes to frequently ignored the cyberbullying, a rate consistent with prior research (Lenhart et al., 2011). As DeSmet et al. (2014) suggest, contrary to offline bullying research, the roles of bystanders online may be more fluid and context specific. Further research assessing the specific characteristics of a cyberbullying incident that elicits one form of engagement or another is warranted. Our analyses were limited in this regard as there were no variables in the Pew data that asked about contextual factors that may elicit different bystander responses such as the relationship of the participant to the victim, perceived severity of the incident (i.e., a textual post versus a video or photo post), and whether there were repeated incidents, all of which are determinants that DeSmet et al.'s qualitative sample, and both Macháčková et al.'s (2013) and Bastiaensens et al.'s (2014) experimental research found important when assessing bystander intentions online among European adolescents.

What is potentially promising in the current study, nonetheless, is the finding that only 9% of adolescents stated that they sometimes to frequently joined in with the cyberbully. This result is contrary to Barlińska et al.'s (2013) finding that negative bystander behavior, or reinforcing, may occur more often in cyberspace than offline. The difference in findings may be related to differences in methodologies with Barlińska et al. using hypothetical situations (though based on actual Facebook incident reports) in their experimental method to assess bystander reactions and the Pew data relying solely on self-reports. Unfortunately, the small sample size for those who actually acknowledged reinforcing cyberbullying in our study limited our ability to explore specific factors that may predict this form of bystander engagement in the current study. Future research should include more specific measures of a variety of types of reinforcing behaviors to gain a more accurate picture of negative bystanding online. We suspect, as Faris and Felmlee (2011, 2014) assert regarding face-to-face bully behavior, it is likely that bystanders may engage in a combination of prosocial, traditional bystanding, and reinforcing practices in order to attain and maintain group status positioning. This conjecture is further supported by Bastiaensens et al.'s (2014) conclusion that adolescents prefer private (e.g., sending the victim a supportive e-mail) versus public (e.g., commenting on an SNS post directly) helping interventions as it may afford them more audience control and less social risk. Despite the potential modeling influence that our findings suggest direct interventions can have on other bystander's behavior, it would seem cyberbully prevention programs should help adolescents develop a variety of prosocial helping skills including private helping behaviors in addition to public (DeSmet et al., 2014).

The third research question in this study explored factors that may predict which adolescents engage in traditional bystanding and which participate in prosocial bystander engagement. Based on model 1, the factors that were shown to influence prosocial bystander engagement were witnessing others act in a prosocial manner, having parents who monitor their cyber activities, and having personally experienced overall cyberbully victimization. Based on model 2, it was found that witnessing others ignoring cyberbullying, being a Twitter user, and having personally been a victim of overall cyberbullying were factors that influence traditional bystander behavior. Therefore, the current research was able to determine some initial factors and SNS contexts that may influence different modes of bystander engagement. Across models, the strongest influencers on modes of engagement appeared to be perceived group norms (i.e., modeling behaviors witnessed) and personal experiences of cyberbullying. Both of these factors may also influence how adolescents define cyberbullying, a clear prerequisite for action. As Lenhart et al. (2011) point out, interaction on SNSs between adolescents may seem like bullying to an adult, but to the adolescents it could be viewed as typical banter. Further research on perceived online behavioral norms and which behaviors adolescents would identify as cyberbullying is warranted. Regardless, the association between own behavior and peer behavior in cyber-bystanding trajectories is consistent with research on peer influence in face-to-face interactions

(Faris & Felmlee, 2011) reinforcing the idea that addressing peer culture is critical to bullying prevention initiatives both on- and offline. While meta-analysis of the effects of prevention programming on bystander intentions shows promising results (Polanin et al., 2012), future research analyzing whether bystanders are being addressed in cyberbullying prevention and media literacy campaigns is warranted. One such study conducted by Ahlfors (2010) found that most cyberbully prevention websites focus on providing tips to victims and parents of victims, though a promising finding is nearly a third provided ideas for outreach using peer interaction programs.

Review of our regression results further suggests several factors approaching significance worth noting due to their potential influence on the mode of bystander engagement an adolescent may practice. First, age was found to be approaching significance for practicing prosocial bystander engagement, with adolescents increasing their likelihood of enacting prosocial behaviors with older age. The direction of this finding contradicts previous research conducted by Salmivalli and Voeten (2004), who found that as adolescents get older they prefer to ignore traditional bullying in offline incidents. Because age is tied to technology use, further research is needed to more fully assess the possible relationship of age with prosocial engagement in the cyberbully context. A second factor approaching significance for prosocial bystander engagement was gender, with the findings suggesting that females may be more likely to participate in prosocial bystander engagement online. Again, this finding is consistent with previous research. For example, Ahmed (2008) found that "girls were more likely to participate in bullying prevention" overall (p. 209). It is suggested that the role of gender in bystander engagement in the context of cyberbullying should be explored further. While the Pew data used in the current study drew from a national sample, the findings of the current study suggest that larger national data sets are necessary in order to more fully understand the determinants of bystander engagement online.

The final research question explored in the current study was whether bystander engagement outcomes vary from Facebook to MySpace to Twitter. The current research was only partially able to answer this question due to our inability to measure how much cyberbullying was witnessed or experienced on each SNS specifically. The regression results suggest, however, that it may not be uncommon for Twitter and potentially Facebook users, as opposed to nonusers, to engage in traditional bystanding behavior online. Future research should explore how specific acts of cyberbullying may vary among different sites as well as explore potential bystander responses on standard (Facebook, Twitter) and emerging social networking technologies such as Instagram, Tumblr, and Snapchat. It is likely that rates of cyberbullying and reactions to such cyberbullying vary across different social networking mediums that continue to pop up every day. Take for example the app Yik Yak that is garnering growing media attention as a potential tool for cyberbullying among both college and high school populations (Valencia, 2014). In this vein, a clear challenge for researchers is keeping up with rapid technological change, especially when using secondary data. For example, Madden (2013) found a significant drop in adolescent use of MySpace between 2011, when the data for the current study was collected by Pew, and 2012; while adolescent use of Twitter actually doubled during this timeframe. It is likely that the salience of a particular SNS will vary by age group and even geographic region as the diffusion of innovation often varies along these lines. For this reason, research exploring cyber-bystanding behavior may need to draw from a variety of methodological approaches including more contextualized experiments of bystander intentions as European scholars are now doing (Barlińska et al., 2013; Bastiaensens et al., 2014; Macháčková et al., 2013) to cyber-ethnographic observations of bystander behavior in a specific technological medium.

In conclusion, this study sought to identify potential elements that may lead to a better understanding of the various influences behind bystander behavior in reaction to cyberbullying episodes on social network sites. The findings from each model aid initial understandings of the nature and actions of bystander responses in regards to cyberbullying. However, considering the low amount of variance accounted for in each model, several other influencing variables remain to be discovered. Continuing this line of research is important for not only understanding what motivates adolescents when they choose a certain type of engagement, but results may also suggest avenues for developing effective prevention strategies when combating cyberbullying as a whole. Based on the current findings, it would seem that peer actions are an important determinant of bystander outcomes and thus prevention efforts should focus on discussing perceived norms as a form of media literacy while providing direct and indirect tools for youth to engage in a variety of prosocial practices for combating cyberbullying when encountered.

REFERENCES

Ahlfors, R. (2010). An internet based catalog of cyberbullying websites. Psy. D., Massach*usetts School of Professional Psychology*, *117*, 3415694.

Ahmed, E. (2008). Stop it, that's enough: Bystander intervention and its relationship to school connectedness and shame management. *Vulnerable Children and Youth Studies*, *7*(3), 203–213.

Ahn, J. (2011). The effect of social network sites on adolescents' social and academic development: Current theories and controversies. *Journal of the American Society for Information Science and Technology*, *62*(8), 1435–1445.

Banyard, V. (2008). Measurement and correlates of prosocial bystander behavior: The case of interpersonal violence. *Violence and Victims*, *23*, 83–97.

Banyard, V., Moynihan, M., & Crossman, M. (2009). Reducing sexual violence on campus: The role of student leaders as empowered bystanders. *Journal of College Student Development*, *50*(4), 446–457.

Banyard, V. L., Plante, E. G., & Moynihan, M. M. (2004). Bystander education: Bringing a broader community perspective to sexual violence prevention. *Journal of Community Psychology*, *32*(1), 61–79.

Barhight, L. R., Hubbard, J. A., & Hyde, C. T. (2013). Children's physiological and emotional reactions to witnessing bullying predict bystander intervention. *Child Development*, *84*(1), 375–390.

Barlińska, J., Szuster, A., & Winiewski, M. (2013). Cyberbullying among adolescent bystanders: Role of the communication medium, form of violence, and empathy. *Journal of Community and Applied Social Psychology*, *23*(1), 37–51.

Bastiaensens, S., Vandebosch, H., Poels, K., Van Cleemput, K., DeSmet, A., & De Bourdeaudhuij, I. (2014). "Can I afford to help?" How affordances of communication modalities can guide bystanders' helping intentions in cases of harassment on social network sites. *Behavior and Information Technology*. Retrieved from http://www.tandfonline.com/eprint/J3vbfqIaimAhBKuhkZY5/full#.VPYjHbN4py4

Choi, S., & Cho, Y. I. (2013). Influence of psychological and social factors on bystanders' roles in school bullying among Korean-American students in the United States. *School Psychology International*, *34*(1), 67–81.

Coloroso, B. (2011). Bully, bullied, bystander ... and beyond. *Education Digest*, *77*(4), 36–39.

Cowie, H. (2000). Bystanding or standing by: Gender issues in coping with bullying in English schools. *Aggressive Behavior*, *26*(1), 85–97.

Critelli, J. W., & Keith, K. W. (2003). The bystander effect and the passive confederate: On the interaction between theory and method. *Journal of Mind and Behavior*, *24*(3, 4), 255–264.

Dahl, J. (2013, March 18). Steubenville Rape Update: Bystanders, Coaches, and Parents will now be the Focus of Ohio AG Mike Dewine. *CBS News*. Retrieved from http://www.cbsnews.com/8301-504083_162-57574936-504083/steubenville-rape-update-bystanders-coaches-and-parents-will-now-be-focus-of-ohio-ag-mike-dewine/

Darley, J. M., & Latane, B. (1968). Bystander intervention in emergencies: Diffusion of responsibility. *Journal of Personality and Social Psychology*, *8*(4p1), 377.

DeSmet, A., Veldeman, C., Poels, K., Bastiaensens, S., Van Cleemput, K., Vandebosch, H., & De Bourdeaudhuij, I. (2014). Determinants of self-reported bystander behavior in cyberbullying incidents amongst adolescents. *Cyberpsychology, Behavior, and Social Networking*, *17*(4), 207–215.

Dunn, S. T. M. (2009). *Upstanders: Student experiences of intervening to stop bullying*. Doctoral dissertation, University of Alberta.

Erdur-Baker, Ö. (2010). Cyberbullying and its correlation to traditional bullying, gender and frequent and risky usage of internet-mediated communication tools. *New Media and Society*, *12*(1), 109–125.

Espelage, D., Green, H., & Polanin, J. (2012). Willingness to intervene in bullying episodes among middle school students individual and peer-group influences. *The Journal of Early Adolescence*, *32*(6), 776–801.

Espinoza, G., & Juvonen, J. (2011). The pervasiveness, connectedness, and intrusiveness of social network site use among young adolescents. *Cyberpsychology, Behavior, and Social Networking*, *14*(12), 705–709.

Evers, K. E., Prochaska, J. O., Van Marter, D. F., Johnson, J. L., & Prochaska, J. M. (2007). Transtheoretical-based bullying prevention effectiveness trials in middle schools and high schools. *Educational Research*, *49*(4), 397–414.

Faris, R., & Felmlee, D. (2011). Status struggles network centrality and gender segregation in same-and cross-gender aggression. *American Sociological Review*, *76*(1), 48–73.

Faris, R., & Felmlee, D. (2014). Casualties of social combat school networks of peer victimization and their consequences. *American Sociological Review*, *79*(2), 228–257.

Gini, G., Albiero, P., Benelli, B., & Altoe, G. (2008). Determinants of adolescents' active defending and passive bystanding behavior in bullying. *Journal of Adolescence*, *31*(1), 93–105.

Gini, G., Pozzoli, T., Borghi, F., & Franzoni, L. (2008). The role of bystanders in students' perception of bullying and sense of safety. *Journal of School Psychology*, *46*(6), 617–638.

Hertz, M. F., & David-Ferdon, C. (2008). *Electronic media and youth violence: A CDC issue brief for educators and caregivers*. Atlanta, GA: Centers for Disease Control.

Hinduja, S., & Patchin, J. W. (2008). Cyberbullying: An exploratory analysis of factors related to offending and victimization. *Deviant Behavior*, *29*(2), 129–156.

Hinduja, S., & Patchin, J. W. (2010). Bullying, cyberbullying, and suicide. *Archives of Suicide Research*, *14*(3), 206–221.

Hudson, J. M., & Bruckman, A. S. (2004). The bystander effect: A lens for understanding patterns of participation. *The Journal of the Learning Sciences*, *13*(2), 165–195.

Jackson, M., Cassidy, W., & Brown, K. (2009). "You were born ugly and you'll die ugly too": Cyber-bullying as relational aggression. *Technology and Social Media (Special Issue, Part 1)*, *15*(2). Retrieved from http://ineducation.ca/article/you-were-born-ugly-and-youl-die-ugly-too-cyber-bullying-relational-aggression. Accessed on February 21, 2012.

Jerome, L., & Segal, A. (2003). Bullying by internet (to the Editor). *Journal of the American Academy of Child and Adolescent Child Psychology*, *42*(7), 751.

Kowalski, R. M., & Limber, S. P. (2007). Electronic bullying among middle school students. *Journal of Adolescent Health*, *41*(6), S22–S30.

Laner, M. R., Benin, M. H., & Ventrone, N. A. (2001). Bystander attitudes toward victims of violence: Who's worth helping? *Deviant Behavior*, *22*(1), 23–42.

Latane, B., & Darley, J. M. (1970). *The unresponsive bystander: Why doesn't he help?* New York, NY: Meredith Corporation.

Law, D. M., Shapka, J. D., Domene, J. F., & Gagné, M. H. (2012). Are cyberbullies really bullies? An investigation of reactive and proactive online aggression. *Computers in Human Behavior*, *28*(2), 664–672.

Law, D. M., Shapka, J. D., Hymel, S., Olson, B. F., & Waterhouse, T. (2012). The changing face of bullying: An empirical comparison between traditional and internet bullying and victimization. *Computers in Human Behavior*, *28*(1), 226–232.

Lee, S. J., & Chae, Y. G. (2007). Children's internet use in a family context: Influence on family relationships and parental mediation. *CyberPsychology and Behavior*, *10*(5), 640–644.

Lenhart, A., Madden, M., Smith, A., Purcell, K., Zickuhr, K., & Rainie, L. (2011). Teens, kindness and cruelty on social network sites: How American teens navigate the new world of "digital citizenship". *Pew Internet and American Life Project*.

Li, Q. (2006). Cyberbullying in schools: A research of gender differences. *School Psychology International*, *27*(2), 157–170.

Liau, A. K., Khoo, A., & Ang, P. H. (2008). Parental awareness and monitoring of adolescent internet use. *Current Psychology*, *27*(4), 217–233.

Lodge, J., & Frydenberg, E. (2005). The role of peer bystanders in school bullying: Positive steps toward promoting peaceful schools. *Theory into Practice*, *44*(4), 329–336.

Lwin, M. O., Li, B., & Ang, R. P. (2012). Stop bugging me: An examination of adolescents' protection behavior against online harassment. *Journal of Adolescence*, *35*(1), 31–41.

Macháčková, H., Dedkova, L., Sevcikova, A., & Cerna, A. (2013). Bystanders' support of cyberbullied schoolmates. *Journal of Community and Applied Social Psychology*, *23*(1), 25–36.

Madden, M. (2013). *Teens haven't abandoned Facebook (yet)*. Retrieved from Pew Research Center http://www.pewinternet.org/2013/08/15/teens-havent-abandoned-facebook-yet/

Markey, P. M. (2000). Bystander intervention in computer-mediated communication. *Computers in Human Behavior*, *16*(2), 183–188.

McNamee, A., & Mercurio, M. (2008). School-wide intervention in the childhood bullying triangle. *Childhood Education*, *84*(6), 370–378.

Mesch, G. S. (2009). Parental mediation, online activities, and cyberbullying. *CyberPsychology and Behavior*, *12*(4), 387–393.

Mishna, F., Saini, M., & Solomon, S. (2009). Ongoing and online: Children and youth's perceptions of cyber bullying. *Children and Youth Services Review*, *31*(12), 1222–1228.

Niblack, J. E. (2013). *Factors that influence bystander behavior in the cyberbully context*, Wichita State University, Wichita Kansas. Unpublished master's thesis.

NoBullying.com (2014). *SnapChat bullying tactics*. Retrieved from http://nobullying.com/snap-chat-bullying-tactics/

Obermann, M. L. (2011). Moral disengagement among bystanders to school bullying. *Journal of School Violence*, *10*(3), 239–257.

Oh, I., & Hazler, R. J. (2009). Contributions of personal and situational factors to bystanders' reactions to school bullying. *School Psychology International*, *30*(3), 291–310.

Patchin, J. W., & Hinduja, S. (2006). Bullies move beyond the schoolyard a preliminary look at cyberbullying. *Youth Violence and Juvenile Justice*, *4*(2), 148–169.

Polanin, J. R., Espelage, D. L., & Pigott, T. D. (2012). A meta-analysis of school-based bullying prevention programs' effects on bystander intervention behavior. *School Psychology Review*, *41*(1), 47–65.

Pozzoli, T., Gini, G., & Vieno, A. (2012). The role of individual correlates and class norms in defending and passive bystanding behavior in bullying: A multilevel analysis. *Child Development*, *83*(6), 1917–1931.

Princeton Survey Research Associates International for the Pew Research Centers. (2011). *Internet and American life project teens and digital citizenship survey [Data file]*. Retrieved from http://www.pewinternet.org/Shared-Content/Data-Sets/2011/July-2011-Teens-and-Online-Behavior.aspx

Pujazon-Zazik, M., & Park, M. J. (2010). To tweet, or not to tweet: Gender differences and potential positive and negative health outcomes of adolescents' social internet use. *American Journal of Men's Health*, *4*(1), 77–85.

Salmivalli, C. (2010). Bullying and the peer group: A review. *Aggression and Violent Behavior*, *15*(2), 112–120.

Salmivalli, C., & Voeten, M. (2004). Connections between attitudes, group norms, and behaviour in bullying situations. *International Journal of Behavioral Development*, *28*(3), 246–258.

Smith, P. K., Mahdavi, J., Carvalho, M., & Tippett, N. (2006). An investigation into cyberbullying, its forms, awareness and impact, and the relationship between age and gender in cyberbullying. In *Research brief no. RBX03-06. London: DfES*. Retrieved from https://www.education.gov.uk/publications/eOrderingDownload/RBX03-06.pdf

Stueve, A., Dash, K., O'Donnell, L., Tehranifar, P., Wilson-Simmons, R., Slaby, R. G., & Link, B. G. (2006). Rethinking the bystander role in school violence prevention. *Health Promotion Practice*, *7*(1), 117–124.

Thornberg, R., Tenenbaum, L., Varjas, K., Meyers, J., Jungert, T., & Vanegas, G. (2012). Bystander motivation in bullying incidents: To intervene or not to intervene? *Western Journal of Emergency Medicine*, *8*(3), 247–252.

Twyman, K., Saylor, C., Taylor, L. A., & Comeaux, C. (2010). Comparing children and adolescents engaged in cyberbullying to matched peers. *Cyberpsychology, Behavior, and Social Networking*, *13*(2), 195–199.

Valencia, N. (2014, March 10). Yik Yak chat app stirring up trouble in high schools. *CNN*. Retrieved from http://www.cnn.com/2014/03/07/tech/yik-yak-app-high-school-problems/

Vandebosch, H., & Van Cleemput, K. (2009). Cyberbullying among youngsters: Profiles of bullies and victims. *New Media & Society*, *11*(8), 1349–1371.

Wilson-Simmons, R., Dash, K., Tehranifar, P., O'donnell, L., & Stueve, A. (2006). What can student bystanders do to prevent school violence? Perceptions of students and school staff. *Journal of School Violence*, *5*(1), 43–62.

Wolak, J., Mitchell, K. J., & Finkelhor, D. (2007). Does online harassment constitute bullying? An exploration of online harassment by known peers and online-only contacts. *Journal of Adolescent Health*, *41*(6), S51–S58.

Ybarra, M. L., & Mitchell, K. J. (2004). Online aggressor/targets, aggressors, and targets: A comparison of associated youth characteristics. *Journal of child Psychology and Psychiatry*, *45*(7), 1308–1316.

Ybarra, M. L., Mitchell, K. J., Wolak, J., & Finkelhor, D. (2006). Examining characteristics and associated distress related to internet harassment: Findings from the second youth internet safety survey. *Pediatrics*, *118*(4), e1169–e1177.

CELL PHONE USE AND YOUTH PERCEPTIONS OF COMMUNICATION IN SOUTH AFRICA

Radhamany Sooryamoorthy

ABSTRACT

Purpose – *The adoption of mobile technology by the youth has been gaining momentum as a growing area of exploration for scholars. Several aspects of this technology, including its adoption and usage patterns, have been explored both theoretically and in empirical research. The purpose of this paper is to understand the cell phone usage of the young university students, the effects of this usage of the cell phone on their personal communication needs, their perceptions of cell phone communication in the society, and gender similarities and differences in these matters.*

Methodology/approach – *This paper relies on primary data of randomly selected engineering students in a university in South Africa. Data was collected from 200 respondents using face-to-face interviews with the help of a structured interview schedule.*

Technology and Youth: Growing Up in a Digital World
Sociological Studies of Children and Youth, Volume 19, 73–103

ISSN: 1537-4661/doi:10.1108/S1537-466120150000019003

Findings – *The analysis shows the relationship between the measures of cell phone usage, personal communication, and perceptions, across different racial groups. The findings of the study have relevance and significance for the understanding of the communication patterns and their social consequences for the youth in South Africa.*

Originality/value – *Most of the research in this area has originated from the developed regions and countries in the Nordic region and Japan. This paper presents the findings on the cell phone use of students in South Africa.*

Keywords: Cell phone use; communication patterns; youth; South Africa

INTRODUCTION

Unlike many other media of information and communication technologies (ICTs) the rapid adoption of the cell phone has evoked serious interest among academics (Atkin, 1993; Cohen & Lemish, 2003; Özcan & Koçak, 2003; Rogers, 2003). Much of the available empirical research on mobile telephony is about the salient characteristics of technology users and the variables that determine both the adoption and nonadoption of this technology (Leung & Wei, 1999; Özcan & Koçak, 2003; Rice & Katz, 2003; Rogers, 2003; Thurlow & McKay, 2003; Vishwanath & Goldhaber, 2003). As a technological device for "any time" or "perpetual" communication, the cell phone has changed the ways in which people communicate and establish their networks of contacts. The cell phone is transforming not only society but also the framework in which society lives, modifying the social conception of space and time, and redefining social relations (Fortunati, 2002). It has changed the way people communicate. This can now be achieved at all times irrespective of their location.

Amongst cell phone users there is variation in use, both in degree and form. Licoppe's (2004) analysis of "connected" management of relationships speaks about the physical absence of a person which is rendered present by the multiplication of mediated communication. It shows how the use of mobile communication devices and different forms of textual exchange corresponds to a connected mode of relationship maintenance. Examining the interrelationship between cell phone use and social connectedness, Wei and Lo (2006) noted that the cell phone has evolved into a facilitator of social relationships and supplements the landline to strengthen family bonds by expanding the psychological neighborhoods of the users. The cell phone expands the user's "psychological neighborhoods"[1] and facilitates "maintenance of symbolic proximity" (Wei & Lo, 2006). The selective interpersonal relationship theory of Matsuda (2000, cited in Igarashi, Takai, & Yoshida, 2005) presupposes that the cell phone can change social networks among young people. Hutchby and Barnett (2005) also believed that the cell phone is bringing about new forms of social interaction and new structures of conversation and practices of interaction. These are evident in the potential of the cell phone to make or receive calls or send and receive messages at any time and from any location. However, the receiver exercises control on communication by choosing to accept or neglect a call at his/her discretion.

Media use by the youth, in comparison to the use by the general population, has received the increased attention of scholars. The new media, particularly in the use of the cell phone, is significant in the life of the youth but their experience differs from the previous generation (Oksman & Turtiainen, 2004). Communication for them has become a 24-hour activity and is an integral part of their daily routine. They are now able to connect to new contacts and maintain those contacts via the application of the new media. These also offer them new means of communication, from voice calls to texting. Compared to the previous generation communication is now limitless, transcending time, space, and location. Citing the Kaiser Family report, Morimoto and Friedland (2011) pointed out that the media use by children between the ages of 8 and 18 years in the United States in 2010 reached 7.38 hours a day which was 2.25 hours higher than in 2005. This included watching television, listening to music, using a computer, playing video games, reading print media, and watching movies. It excluded the time spent on using mobile media which could be another 20% of the total time spent on this activity (Morimoto & Friedland, 2011).

The youth constitute the major constituency of mobile technology and are considered to be the forerunners in its adoption (Thulin & Vilhelmson, 2007). However, the purposes of using media vary among the youth. These include: to gratify social and instrumental needs, fashion, accessibility, recreation, or entertainment (Leung & Wei, 2000; Wei, 2006; Wei & Lo, 2006 cited in Campbell & Kwak, 2010).

Robinson and Stubberud (2012) studied the preferred communication methods of university students in both a Norwegian college and an American public university. The students were asked to rank their communication methods from a list of face-to-face, telephone, paper, email, online chat, texting, and Facebook, for different situations of work/school, and social communication. There were different ranking orders for the two different situations. Higher mean values were observed for Facebook, paper communication, short messages (sms) texting, and chat in descending order, while face-to-face had the least mean value in the work/school communication situation. In the social communication situation, the highest mean values were reported for email and paper communication. Face-to-face was the least used means for social communication. Telephone had also been ranked high as a communication method for both work/school and social communication situations. Axelsson's (2010) study of Swedish teenagers and adolescents examined the way teenagers used and thought about cell phones. A comparison between young and senior adults has also been made in this study. Axelsson's (2010) research specifically investigated a set of questions: the extent of the use of the cell phone for making and receiving voice calls and for texting; situations and places where the cell phone was used; the partners of people involved in cell phone communication; cell phone functions they mostly and normally used; and the social and health concerns of the Swedish teenagers in regard to the use of the cell phone. The study found that the use of cell phones by young and senior adults and their attitudes toward usage differ. For young adults the cell phone serves to perform their social and individual needs, amongst which interpersonal communication and coordination activities were quite prominent (Axelsson, 2010).

Wei, Lo, Xu, Chen, and Zhang (2014) observed that the news reading behavior on the cell phone is on the increase among the Asian students who adopted a variety of news applications packaged for the cell phone. As for the motivation of the students, the study reported that the greater their belief that the cell phone helps in information-seeking, the more they read the news on the cell. Mihailidis (2014) examined how university students were using and perceiving the role of cell phones for their communication and information needs. Drawing on a sample from four countries, the study showed how young people learn to use the cell phone finally dictates their daily social and civic life. Castells, Fernández-Ardèvol, Qiu, and Sey (2007, p. 127) hypothesized that "there is a youth culture that finds in mobile communication an adequate form of express and reinforcement." Bond (2010) explored how the use of the cell phone by children is affecting their friendship and intimate relationships. They use cell phones to form friendships, and to reinforce their emotional ties (Bond, 2010). They now actively control several aspects of their relationships by means of the cell phone which serves to fulfill both their emotional and functional needs (Bond, 2010). Thulin and Vilhelmson (2007) probed the patterns of communication and contacts of Swedish youth and reported how they were affected by their increased use of the cell phone. Focusing on Irish teenagers, Cawley and Hynes (2010) explored cell phone usage, the role of the cell phone in mediating their social and economic relations, and its use as a central point for media and information consumption. Goggin (2013) noted that youth and youth culture in matters of cell phones have become complicated because cell phones turn into media, and the advent of the mobile internet changes the perceptions and conceptions of mobile media and communication.

The nature and distribution of cell phone use among the youth in Africa remain largely undocumented. Most of the available literature on cell phone youth culture is concentrated in Northern and Western Europe where the technology has diffused very widely and rapidly (Castells et al., 2007). Castells et al. (2007) observed that the fast rate of diffusion of cell phone communication among the youth is a result of the openness of the young people to new technology, their ability to appropriate it, and their capacity to use the technology for their own purposes. Current studies on cell phones have generally and more frequently dealt with regions such as Scandinavia, North America, and Japan (Ilahiane & Sherry, 2009).

Africa was not in the picture of academic studies that dealt with cell phone usage among the youth in particular. This is despite the increasing popularity of the cell phone in Africa in the last few years. Both the access and the use of mobile telephony have recently increased in sub-Saharan Africa (Aker & Mbiti, 2010). The adoption of the cell phone has seen significant growth in Africa, particularly in the Southern African Development Community (SADC) region (Mutula, 2002). The growth has actually been phenomenal. In 2008 the cell phone coverage in the continent was about 60% of the population as against 10% for 1999 (Aker & Mbiti, 2010). In sub-Saharan Africa the cell phone is the most widely used ICT (Harvey & Sturges, 2010) and the reduction in the costs since its introduction has further accelerated the expansion of cell phone use in Africa (Aker & Mbiti, 2010). Between 2003 and 2008 the cell phone subscription in Africa grew by 550% (Napolitano, 2010).

In South Africa, the adoption rate of the cell phone has been significantly high. South Africa had registered a higher level of adoption of 88% in 2008 and this rapid growth was accompanied by wide social distribution (Bratton, 2013). According to the International Telecommunication Union figures (ITU, 2014), there were 147 mobile phone subscriptions for every

100 inhabitants in South Africa in 2013. In 2000 there were only 18 subscriptions per 100 individuals and by 2010 it had exceeded 100 subscriptions per 100 individuals (123). These figures (for 2013) can be compared to 78.4 (Canada), 94.4 (the US), 104.61 (Australia), and 72.07 (China) (ITU, 2014).

A study conducted among adolescents in South Africa showed that 96% used a cell phone daily and 87% made a call or sent a text every day (Kreutzer, 2009, cited in Napolitano, 2010). Rivera-Sánchez and Walton's (2013) study of the use of the cell phone by the youth who were in the stage of transitioning from their high school to tertiary institutions or full time employment in a selected town in South Africa showed how the respondents used the cell phone to resist a sense of inferiority, redefined their everyday life, and expanded their understanding of living in the modern networked world. The study found that the cell phone is an extremely helpful tool to cope with the period between transition, serving them in many different ways.

According to Peters and Allouch (2005), understanding the behavior of the users of ICT is one of the important aspects in the process of diffusion of ICTs like mobile communication technology. Investigating media use by the youth can help understand the future forms of media reception and production (Drotner, 2000). A better and deeper understanding of the access, usage, and social impact of the cell phone for the youth is warranted (Cotten, Anderson, & Tufekci, 2009). Given this background of the studies among the youth in South Africa, which is fast becoming a cell phone society, this study of university students in South Africa is guided by the following questions.

Studies have examined the dimensions of the usage of the cell phone and its interrelationships with communication. Axelsson (2010) provides a very useful theoretical orientation that is relevant to this study. While considering the use and attitudes of the cell phone, she found the life stages of individuals to be important in cell phone use. Those in a particular life stage have similar needs to communicate, coordinate, and express themselves, and the cell phone serves them well for these purposes. According to her, the factors which include cost, that determine whether and how cell phones and other ICTs are used, and the use and attitudes toward cell phones are culturally or life-stage dependent. Four aspects of cell phone use are pertinent here – the patterns of use, effects of the use, perceptions, and the differences in relation to demographic variables. Based on these theoretical propositions, the following research questions have been formulated to guide this study.

1. What trends can be deduced from the usage of the cell phone by the selected sample of university students?
2. What are the prominent effects of cell phone use among students on their personal communication?
3. How does the use of the cell phone by students affect their perceptions on cell phone communication in the society in which they are living?
4. Are there notable variations in the perceptions of cell phone communication between young men and women in the selected sample?

METHODS AND DATA

This empirical study was conducted in 2010–2011 among students at a university in South Africa. The institution was sampled from one of the provinces of South Africa, and in the second stage a department was chosen. A total of 200 students were randomly selected from the engineering faculty. They belonged to different years of study at the undergraduate level. A proportionate number of students was randomly selected from each year of study. Students were contacted, not based on the possession of a cell phone, for a face-to-face survey that was administered to them with the help of a structured questionnaire. The interview included sections covering the cost and uses of the cell phone and their views on communication in general and cell phone communication in particular.

Following Abeele, Beullens, and Roe (2013), a number of measures that refer to cell phone use by students were used in this study. They include the number of incoming and outgoing voice calls during the week and the weekend, the duration of these calls, the number of messages they send and receive, and their use of the cell phone for other functionalities. Five key variables thus measured the cell phone usage of the respondents: the frequency of cell phone use, the number of calls made and received in a typical day, the duration of the calls made and received, and the daily use of the cell phone for disparate purposes and for the functions built within the cell phone.

A number of indicators capture the effects of the cell phone use of the students. The students were asked to report on the importance of the cell phone in their life, the number of people they talked to after having owned or were able to use a cell phone, the changes occurring in the personal nature of communication and the usage pattern of the internet since their access to the cell phone. Perception variables used in the study refer to the views of the respondents regarding changes in communication, cost of calls, internet use, and international communication. Consistent with other studies (Karim, Oyebisi, & Mahmud, 2010, for instance), variables such as gender are taken into account in the analysis of the use of the cell phone users.

The uses of the cell phone are measured in terms of the functions of the cell phones as presented in Table 1. In order to understand the extent of communication the cost of communication was also gathered. These measures provide information about the cost involved in cell phone communication which has an effect on their use of the cell phone. In this study we refer to all personal communications of the respondents which are carried out by means of the cell phone. Rather than the content of communication, we are interested in the extent and volume of cell communication that are

Table 1. Respondents' Details.

Details	Boys		Girls		Both	
	N	%	*N*	%	*N*	%
Own and currently use a cell phone	113	96.6	80	100	193	98.0
Use a cell phone but not owned	4	3.4	0	0	4	2.0
Ownership duration of the cell phone						
1–6 months	2	1.7	4	4.9	6	3
6–12 months	2	1.7	1	1.2	3	1.5
1–3 years	12	10.5	5	6.1	17	8.6
Above 3 years	99	86.1	71	86.6	170	86.3
Married	3	2.5	1	1.2	4	2.0
Single	115	97.5	81	98.8	196	98.0
Educational level (completed)						
Secondary school	68	85.0	64	92.8	132	88.6
Diploma	4	5.0	0	0	4	2.7
Bachelors	8	10.0	4	5.8	12	8.1
Masters	0	0	1	1.4	1	0.7
Religion						
Christian	38	50.0	34	52.3	72	51.0
Hindu	12	15.8	10	15.4	22	15.6
African traditional	9	11.8	8	12.3	17	12.1
Islam	6	7.9	4	6.2	10	7.1
Race						
Black	52	44.1	36	43.9	88	44.0
Indian	42	35.6	32	9.0	74	37.0
White	19	16.1	8	9.8	27	13.5
Colored	4	3.4	6	7.3	10	5.0
Other	1	0.8	0	0	1	0.5

measured using the variables of the frequency of the cell phone use for voice calls, the number of calls made and received per day, the duration of the calls, and the time of calls. Communication can be for personal purposes or academic reasons and with any member of their contacts – friends, family, teachers, and classmates. In regard to the perceptions of cell phone communication, we asked the respondents whether the cell phone has increased or decreased their communication, the effects of the cell phone on face-to-face communication, the duration of communication after having a cell phone, the importance of the cell phone in personal communication, the use of the cell phone for international communication, and the cost and benefits of cell phone communication.

ANALYSIS

Respondents: Background Characteristics

Table 1 presents the features of the respondents. As noted above, the sampled respondents were contacted not on the basis of the possession or access to a cell phone. All the students were using a cell phone at the time of interview. All except 2% had a cell phone of their own. By and large, they had been using a cell phone for the last three years or more. Only a few (around 10%) had possessed a cell phone for less than three years, of which only very few had obtained it within the previous year of the study. The split between men and women is 59% and 41% respectively. All except 2% were single. Around 90% of them had completed their secondary school (matric) while 8% had a Bachelor's degree at the time of the survey. Other categories in education were not significant in number. Students belonged to different religious faiths. Half of them were Christians while Hindus and African traditional religion formed the other two major categories of religious groupings. Racially there were three major groups consisting of black African students (44%), Indian students (37%) and white (13.5%) students. The mixed race, Colored[2] students, comprised only 5% of the total sample. The average age of the respondents was 22.7 years (S.D.=2.7) for the whole sample, with men having a higher mean age than women (22.9 and 22.5 respectively) which is not significantly different.

In terms of the educational levels, the split between men and women was not prominent in the Chi-square test. However the majority of women had a matric and their number in the other educational categories of Bachelor's degree and diploma was lower than that of men respondents. Not much variation was found between men and women in the religious background of students. Racially and in relation to men, there were fewer women who were Indian or white. The majority of them were black Africans, for both groups of young men and women.

The cell phone serves a range of functions of both communication and entertainment for the users. The respondents were queried about the major purposes for which they use their cell phones. The listed purposes, as shown in Table 2, differed from sending/receiving short messages to photos. The responses showed that the majority of the uses were confined to sending and receiving short messages (98%), using the cell phone as a watch to check time (94.5%), a calculator (86.5%), and a camera (81%), browsing the internet (76%), playing games (75%), listening to music (73%), and sending and receiving images such as photos. About 40% of them operated their cell phone to send and receive emails, but not as much as they browsed the internet. This indicates the multiple purposes for which the internet was used by these young respondents at the university.

The cell phone is also used as a technological device capable of performing several functionalities. While looking at the differences in the use of cell phones for purposes other than making and receiving voice calls, it was evident that there were statistically significant variations between men and women in the number and types of uses. In the use of sending and receiving emails, playing videos on the cell phone, and in using the gadget as a

Table 2. Major Uses of Cell Phone Other Than for Voice Calls by Students.

Uses	Boys		Girls		Both	
	N	%	*N*	%	*N*	%
Sending and receiving short messages	115	97.5	81	98.8	196	98.0
Sending and receiving email**	44	37.3	43	53.1	87	43.7
Playing games	89	75.4	61	74.4	150	75.0
Playing music	81	69.8	63	76.8	144	72.7
Playing video*	62	53.0	54	65.9	116	58.3
As a watch	110	94.0	78	95.1	188	94.5
As a calculator	101	85.6	72	87.8	173	86.5
As a camera**	89	76.1	73	89.0	162	81.4
Browsing internet	85	72.0	67	81.7	152	76.0
Sending and receiving photos	71	61.2	58	72.5	129	65.8

Note: Chi-square test, Sig: $*p < .1$; $**p < .05$.

camera, the gender difference was clear. More women than men used the cell phone for sending and receiving emails (53% and 37% respectively), playing videos (66% for women and 53% for men), and using the cell phone for its camera features (89% by women and 76% by men). Although it is not statistically significant, more women than men browsed the internet.

Significant differential patterns were observed across different racial categories in the use of the cell phone for playing games and listening to music (Table 3). More Africans and Indians than whites used both these functions.

Like many other modern gadgets, the cell phone has the inbuilt element of planned obsolescence. It is evolving day by day and becomes obsolete soon after its introduction in the market. Often the new features, functions, and uses that are integrated into it make a new model an attractive buy and the existing one redundant and not standard to possess. This necessitates an upgrade of the cell phone at frequent intervals. For the respondents who are currently university students, the upgrade of their cell phones is made normally once in every two years or two to five years (Table 4). About 44% were among this group. A quarter of them had never upgraded their phones ever since they first received one. But another 20% had the chance and the resources to change their phones for a better model once a year. The break-up figures for men and women revealed that there were no obvious differences in the upgrade of the cell phone during the periods of once in between six months and one year, once in every two years,

Table 3. Major Uses of Cell Phone Other Than for Voice Calls by Race.

Uses	African		Indian		White		Others		All	
	N	%	*N*	%	*N*	%	*N*	%	*N*	%
Sending and receiving short messages	87	98.9	73	98.6	26	96.3	10	91	196	98.0
Sending and receiving email	35	40.2	30	40.5	13	48.1	9	82	87	43.7
Playing games**	72	81.8	55	74.3	14	51.9	9	82	150	75.0
Playing music**	69	80.2	54	73.0	13	48.1	8	73	144	72.7
Playing video	51	58.6	41	55.4	16	59.3	8	73	116	58.3
As a watch	80	92.0	73	98.6	25	92.6	10	91	188	94.5
As a calculator	76	86.4	65	87.8	21	77.8	10	91	173	86.5
As a camera	71	81.6	60	81.1	22	81.5	9	82	162	81.4
Browsing internet	66	75.0	57	77.0	20	74.1	10	91	153	76.5
Sending and receiving photos	60	69.8	46	63.9	15	55.6	8	73	129	65.8

Note: Chi-square test, Sig: $^{**}p < .05$.

Table 4. Upgrade, Cost and Payment of Charges.

Costs (Rands)	Boys		Girls		Both	
	No.	%	No.	%	No.	%
Upgrade the cell phone						
Once in 6 months	7	6.0	0	0	7	3.5
Once every year	19	16.2	21	25.6	40	20.1
Once every two years	41	35.0	26	31.7	67	33.7
Once in 2–5 years	21	17.9	17	20.7	38	19.1
Never	29	24.8	17	20.7	46	23.1
Cost	Mean	S.D.	Mean	S.D.	Mean	S.D.
Average cost of the cell phone and connection ($N = 178$)	1,610.1	1,238.7	1,761.1	1,113.5	1,672.9	1,187.4
Average monthly contract fee ($N = 55$)	193.7	115.1	175.8	176.5	187.9	136.8
Averages monthly pay as use fee ($N = 140$)*	117.0	97.0	150.3	128.0	132.5	113.3
Average monthly charge ($N = 170$)	163.2	128.9	163.2	114.9	163.2	122.9
Who Pays the Cell Phone Charges	No.	%	No.	%	No.	%
Self	62	52.5	49	60.5	111	55.8
Parents	51	43.2	28	34.6	79	39.7
Others	5	4.2	4	4.9	9	4.5

Note: Independent t-test. Sig: $*p < .1$.

and once in 2–5 years. As students and who are still receiving support from parents and family members, frequent upgrading is not normally an easy option for them. In terms of race, some differential features are evident. About one-third of Africans and Others have never upgraded their phones, which is higher than the percentage (23%) for the whole sample (Table 5). The percentage for Indians has been the lowest (15%). They are also the ones who upgraded their phones most regularly, more so than any other racial group in the sample. About two-thirds (67%) have upgraded their phones within the last two years.

As seen from the average cost of the phone and connection charges, the cell phones the students had been using were somewhere in the middle range, costing an average of 1,673 rands (11.18 rands is roughly a US dollar at the current exchange rate in November 2014) a piece. On average they paid a sum of 163 rands toward monthly charges. Most of the students had opted not for a fixed monthly contract but for the "pay as use" facility which allows them to budget their usage within their availability of funds.

Table 5. Upgrade, Cost and Payment of Charges by Race.

Costs (Rands)	African		Indian		White		Others		All	
	N	%	*N*	%	*N*	*N*	%	*N*	%	*N*
Upgrade the cell phone										
Once in 6 months	4	4.6	2	2.7	1	3.7	0	0	7	3.5
Once every year	18	20.7	14	18.9	4	14.8	4	40.0	40	20.1
Once every two years	21	24.1	34	45.9	10	37.0	2	20.0	67	33.7
Once in 2–5 years	17	19.5	13	17.6	6	22.2	2	20.0	38	19.1
Never	26	29.9	11	14.9	6	22.2	3	30.0	46	23.1
Cost	Mean	S.D.	Mean	S.D.	Mean	S.D.	Mean	S.D.	Mean	S.D.
Average cost of the cell phone and connection ($N=178$)	1,537.1	1,208.1	1,738.9	1,151.5	1,822.0	1,251.9	1,877.7	1,177.5	1,672.9	1,187.4
Average monthly contract fee ($N=55$)	135.5	73.5	194.3	120.9	255.6	226.6	108.3	20.2	187.5	136.8
Averages monthly pay as use fee ($N=140$)	127.1	108.4	129.8	97.0	183.8	178.1	92.9	55.6	132.5	113.3
Average monthly charge ($N=170$)	154.8	144.6	169.5	105.3	169.6	96.4	176.3	137.7	163.2	122.9
Who Pays the Cell Phone Charges	No.	%	No.	%	No.	%	No.	%	No.	%
Self	61	70.1	31	41.9	14	51.9	5	50.0	111	56.3
Parents	22	25.3	40	54.1	13	48.1	4	40	79	39.7
Others	4	4.5	3	4.1	0	0	1	10.0	8	4.0

A total of 140 students fell into this category, paying an average of 113 rands a month for the services they received from the cell phone service providers. Although there are benefits, the monthly contract fees are more expensive than the "pay as use" plan. Those who were on the monthly contract plan (a quarter of them) were paying an average amount of 302 rands a month, which is more than 127% of the "pay as use" plan. In the case of the monthly charges the students paid for their "pay as use" plan, a significant difference was found in the independent *t*-test between men and women. Here women students spent more than the average for the whole sample and that of the men. No statistically significant variation was found in the ANOVA test among different racial groups in the sample (Table 5). The amount spent by the Other racial category has been lower than that spent by whites, Indians, and Africans.

As regards who make the payments of charges for the services, more than half of them (56%) reported that they were paying themselves and for another 40% it was their parents who foot the bill. Those who reported that they were paying for their cell phone charges, many of them used the money they received from their parents for incidental expenses. Some of them also did small jobs on the campus and outside. For a majority of them this is a priority item of their basic personal expenditure. No significant difference between men and women was found in this variable of who paid the charges. More African students than students from other racial categories spent their own money, and only a quarter of them depended on their parents for this purpose. This is more than the total sample proportion for self-payment and less than for payment by parents (Table 5).

Usage of Cell Phone for Voice Calls

The frequency of cell phone usage (Table 6) supports the conclusion that the device has become an indispensable part of their daily personal communication. A large majority of them made use of their cell phone every day, to make and receive voice calls. If this group is combined with those who used it many times a week, the percentage is 96. Barely 3% of the students were using it once or twice a week for voice calls. The number of calls they made on a typical day ranged from less than two, between three and 10, and more than 10. Of these, the majority (60%) were in the medium range of between three and 10 calls. Close to one-third made less than two calls in a typical day. One-tenth had made more than 10 calls in a day. The call

Table 6. Use of Cell Phone for Voice Calls.

Indicators	Boys		Girls		Both	
	No.	%	No.	%	No.	%
Frequency of cell phone use for calls						
Everyday	94	79.7	73	91.3	167	84.8
Many times a week	18	15.3	6	7.5	24	12.2
Once or twice a week	5	4.2	1	1.3	6	3.0
Number of calls made on a typical day						
Less than two	36	30.8	20	24.4	56	28.1
Between 3 and 10 calls	71	60.7	49	59.8	120	60.3
More than 10	10	8.5	13	15.9	23	11.6
Number of calls received on a typical day						
Less than two	25	21.2	9	11.0	34	17
Between 3 and 10 calls	62	52.5	48	58.5	110	55.0
More than 10	31	26.3	25	30.5	56	28.0
Duration of an average call						
Less than a minute	31	26.5	17	21.3	48	24.4
Two-five minutes	70	59.8	54	67.5	124	62.9
Six-ten minutes	10	8.5	6	7.5	16	8.1
Above ten minutes	6	5.1	3	3.9	9	4.6
Time cell phone usually used						
7–10 am	10	8.5	12	14.6	22	11
10 am–1 pm	6	5.1	6	7.3	12	6
1–5 pm	6	5.1	2	2.4	8	4
5–7 pm	1	0.8	2	2.4	3	1.5
After 7 pm	26	22.0	15	18.3	41	21.5
No specific time	69	58.5	45	54.9	114	57

rates in South Africa are not one of the best for subscribers on the continent. Rather, they are toward the high end of the continuum. Only recently have the rates come down slightly following the strong and determined intervention of the government.

As regards the calls received in a day the pattern appears to be different from that of the calls the respondents made. For 28% of the sample, the received calls were in the region of more than 10 a day. This is more than double the percentage of the calls made by these students from their cell phones. Within the sample no significant variation was evident between men and women. In terms of the percentages women students were more frequent users than men as more than 90% of them had used their cell phones every day. Again, more women had made over 10 calls a day (8.5% for men and 16% for women). In the number of calls received, the percentages did not vary significantly. In relation to African and white students, those in the Other race category are more intense users of their cell phones. They have a higher percentage of users in the category of more than 10 calls a day than the whole sample (Table 7).

When Pearson's correlation test was employed, the correlation between the number of calls made and received by students is quite significant (r = .552; $p < .01$). This implies that the increase in the number of calls made has a corresponding increase in the number of calls received by the respondents. While the receiver does not pay for the calls, the caller pays. Correlation was also seen between the frequency of the use and the calls made a day ($r = -.293$; $p < .01$) and calls received per day ($r = -.309$; $p < .01$). They are negatively correlated as the value of intense use (everyday) was reverse-coded. In other words, the frequency of cell phone use for voice calls produces a corresponding increase in the number of calls made and received in a normal day.

Table 7. Use of Cell Phone for Voice Calls by Race.

Indicators	African		Indian		White		Others		All	
	N	%	*N*	%	*N*	%	*N*	%	*N*	%
Frequency of cell phone use for calls										
Everyday	72	82.8	63	87.5	21	77.8	11	100	167	84.8
Many times a week	13	14.9	7	9.7	4	14.8	0	0	24	12.2
Once or twice a week	2	2.3	2	2.8	2	7.4	0	0	6	3.0
Number of calls made on a typical day										
Less than two	24	27.3	17	23.3	12	44.4	3	27.0	56	28.1
Between 3 and 10 calls	56	63.6	46	63.0	12	44.4	6	55.0	120	60.3
More than 10	8	9.1	10	13.7	3	11.1	2	18.0	23	11.6
Number of calls received on a typical day										
Less than two	16	18.2	10	13.5	6	22.2	2	18.0	34	17.0
Between 3 and 10 calls	53	60.2	42	56.8	11	40.7	4	36.0	110	55.0
More than 10	19	21.6	22	29.7	10	37.0	5	45.0	56	28.0
Duration of an average call										
Less than a minute	19	22.1	18	24.3	9	34.6	2	18.0	48	24.4
2–5 minutes	52	60.5	49	66.2	16	61.5	7	64.0	124	62.9
6–10 minutes	8	9.3	7	9.5	0	0	1	9.0	16	8.1
Above 10 minutes	7	8.1	1	3.8	0	0	1	9.0	9	4.6
Time cell phone usually used										
7–10 am	10	11.4	9	12.2	2	7.4	1	9.0	22	11.0
10 am–1 pm	7	8.0	4	5.4	1	3.7	0	0	12	6.0
1–5 pm	1	1.1	3	4.1	3	11.1	1	9.0	8	4.0
5–7 pm	1	1.1	2	2.7	0	0	0	0	3	1.5
After 7 pm	24	27.3	9	12.2	6	22.2	2	18.0	41	20.5
No specific time	45	51.1	47	63.5	15	55.6	7	64.0	114	57.0

How long does an average call last for these young cell phone consumers? Table 4 provides the details. For a large majority, a call which is made or received usually lasts for less than five minutes. A quarter of the respondents reported that their average calls lasted less than a minute. For another two-thirds the average call duration was in the region of between two and five minutes. Longer calls that lasted beyond six minutes per instance were applicable for only 13% of the whole sample. Both men and women were in the same categories in the duration of calls made and received. Note that an average cost of making a call in South Africa at the time of the study ranged between 2 and 3.5 rands per minute, depending on the time they make the calls. A loaf of bread at the time cost about eight rands. The cost factor determines the frequency and length of communication over the cell phone. This is very relevant for students who do not have a steady income and their communications contacts are mostly friends of the same category. More than half of them revealed that they did not adhere to any specific time to make or receive calls on their cell phone. One-fifth of them preferred to use it after 7 pm when the call rates were slightly lower than that for during the day. There were no gender or racial differences in this measure of the time of the cell phone use for voice calls.

The relation between the measures of the cell phone use and cost (cost of the cell phone and connection, monthly contract charges, monthly charges on the basis of use and monthly charges) was examined (Table 8). In this analysis, the log-transformed measures for the cost variables were used. The measure of the frequency of use was significantly correlated with two cost variables, namely the "pay as use" monthly charges, and average monthly charges. The frequency of use therefore corresponds to the

Table 8. Correlations between Cost and Cell Phone Use Variables.

Measures	Frequency of Use	Calls Made	Calls Received	Call Duration
Cost of phone and connection	−.024	.285**	.239**	.109
Contract fee (monthly)	−.151	.297*	.128	.225
Pay as use fee (monthly)	−.297**	.341**	.279**	.282**
Average monthly charges	−.273**	.279**	.207**	.094

Note: Frequency of use: 1, everyday; 2, many times a week; and 3, once/twice a week; Calls made and received: 1, less than 2, 2 = 3−10 calls, and 3 = more than 10 calls; and Call duration: 1 = less than a minute, 2 = 2−5 minutes, 3 = 6−10 minutes, and 4 = more than 10 minutes.
Sig: $*p < .05$; $**p < .01$.

monthly charges students pay for their calls. The increase in the frequency of cell phone use (note the coding of this variable where the lower value is meant for higher frequency) is dependent on the charges they pay. As causality cannot be determined in a cross-sectional study like this, it can be assumed that the intensity of the use is limited to or controlled by the monthly charges or vice versa. Notably, neither the monthly contract fee nor the cost of the phone and connection affect the frequency of cell phone use. A monthly contract plan comes with a package of a certain number of calls that the subscriber can make from their cell phones. The frequency of the use therefore does not change according to the monthly contract charges students pay.

The number of calls made in a typical day is positively and significantly correlated with all the cost variables employed in the study. As pointed out earlier, the caller pays and not the receiver. The cost of the phone and connection is often tied in with a plan that allows for a certain number of call minutes that the subscriber can make every month. The higher the cost of the phone and the connection, the more the number of minutes allowed to make calls. The same rule applies to other cost measures used in this study. The number of calls received also tends to correlate with the three measures of cost. This is a reciprocal way of communication in which the calls one receives are indirectly related to the calls one makes. Depending on the number of calls one is able to make, subject to cost factors, one can expect more or less the same number of calls from friends and other regular contacts. The only variable associated with call duration is the "pay as use" charges. In other words the length of cell phone conversation, both through incoming and outgoing calls, is not directly related to the cost of the phone and connection, contract fee, or monthly charges students pay. It is clear that if one is able to pay more, the student can have a longer voice call duration.

The correlations presented above underscore the point that cell phone use among students in South Africa is largely contingent on the call rates which the service providers charge. In particular, the number of voice calls the students make hinges on the contract they chose to use and the amount they pay as call charges, monthly or "pay as use."

Effects of Cell Phone Use on Personal Communication

As described in the section on the background of the respondents, all students at the time of data collection had a cell phone for their personal use.

About 95% of them had been using it for more than a year and 86% had it for over three years. This characteristic of the respondents leads to questions as to how the possession and the use of the cell phone influenced their communication patterns. We asked them questions to understand this dimension of their daily communication patterns. The responses are shown in Table 9.

For a large section of respondents (88%), the cell phone is important or very important in their current life stage as youthful students. Of these, about one-third (63%) of them reported that it was really very important in their life as far as their communication needs were concerned. The remaining 12% reported that the cell phone as a communication tool was less important for them at the moment. As we do not find any statistically significant variation between men and women in this category, it is safe to assume that the importance of the cell phone in the life of the respondents is similar, irrespective of gender. In the case of 82% of students, the cell

Table 9. Effects of Cell Phone on Communication.

Communication effects	Boys		Girls		Both	
	No.	%	No.	%	No.	%
Duration of ownership of cell phone						
1–6 months	2	1.7	4	4.9	6	3
6–12 months	2	1.7	1	1.2	3	1.5
1–3 years	12	10.5	5	6.1	17	8.6
More than 3 years	99	86.1	71	86.6	170	86.3
Importance of cell phone in life						
Important	27	22.9	22	26.8	49	24.5
Very important	77	65.3	49	59.8	126	63.0
Somewhat/not important	14	11.9	11	13.5	25	12.5
Use of cell phone on the number of people talked to						
Increased a great deal	58	49.2	43	52.4	101	50.5
Increased somewhat	38	32.2	25	30.5	63	31.5
Remain unchanged	18	15.3	11	13.4	29	14.5
Decreased	4	3.3	3	3.6	7	3.5
Personal nature of communication after using the cell phone						
Became more personal	68	57.6	43	53.1	111	55.5
Unchanged	42	35.6	32	39.5	74	37.2
Less personal	8	6.8	5	6.2	13	6.5
Use of the internet after acquiring the cell phone**						
Frequency increased	73	61.9	59	72.8	132	66.3
Frequency not increased	45	38.1	20	24.7	65	32.7

Chi-square test, Sig: $^{**}p < .05$.

phone has changed their way of communicating with others. This effect was evident in the increased number of people they talked to since they had first started using a cell phone. This applies to both young men and women. More than half of them reported that their use of the cell phone was influenced by the nature of their communication making it more personal. But for another one-third (37%) the use of the cell phone had not changed matters in this regard.

One of the most striking features revealed in the data was how the use of the cell phone had influenced their internet use. Two-thirds of the sample agreed that the availability of the cell phone had caused a definite increase in the frequency of their internet browsing. The gender difference is obvious in this, with a significant Chi-square value for the association of the variables. Positive association is evident between the increased frequency of internet surfing by women. Across all the four racial categories a similar increase in the frequency of using the internet has been observed, and no significant differences between races were evident (Table 10). Women respondents, more so than men, browsed the internet frequently, after having acquired a cell phone for their personal communication needs. As already seen in Table 2 more than three-fourths of the respondents had made use of their cell phone to browse the internet, more by young women than by young men.

The association between the variables reveals the effects of the cell phone on the current communication of students. Table 11 illustrates the interaction between the variables that capture the effects of the cell phone on the communication aspects of students and other relevant cell phone usage variables. The variables of the personal nature of communication after adopting the cell phone and the importance of the cell phone in the day-to-day communication of respondents are significantly correlated. The negative correlation is because of the reverse coding of the variables, as described in the footnotes of Table 11. This confirms that the more the use of the cell phone, the more it contributes to the personal characteristic nature of the communication of the youth. The personal nature of communication (becoming more personal due to the use of the cell phone) is significantly associated with an increase in the number of people the respondents talked to over the cell phone. The greater the number of people the respondents talk to on the cell phone, the more personal is the communication. In comparison to other media such as the landline, the characteristic of the cell phone for the personal communication purposes is obvious in this finding.

Table 10. Effects of Cell Phone on Communication by Race.

Communication Effects	African		Indian		White		Others		All	
	N	%	*N*	%	*N*	%	*N*	%	*N*	%
Duration of ownership of cell** phone										
1–6 months	2	2.3	1	1.4	1	4.0	2	20.0	6	3.0
6–12 months	0	0	2	2.7	1	4.0	0	0	3	1.5
1–3 years	13	14.9	2	2.7	1	4.0	1	10.0	17	8.6
More than 3 years	71	81.6	69	93.2	22	88.0	8	70.0	170	86.3
Importance of cell phone in life										
Important	29	33.0	12	16.2	6	22.2	2	20.0	49	24.5
Very important	50	56.8	51	68.9	17	63.0	8	80.0	126	63.0
Somewhat/not important	8	9.1	8	10.8	1	3.7	1	10.0	18	9.0
Use of cell phone on the number of people talked to										
Increased a great deal	44	50.0	34	45.9	15	55.6	8	80.0	101	50.5
Increased somewhat	24	27.3	29	39.2	8	29.6	2	20.0	63	100
Remain unchanged	16	18.2	10	13.5	3	11.1	0	0	29	14.5
Decreased	4		1	1.4	1	3.7	1	10.0	7	3.5
Personal nature of communication after using the cell phone										
Became more personal	49	55.7	39	53.4	18	66.7	5	50.0	111	55.8
Unchanged	32	36.4	30	41.1	8	29.6	4	40.0	74	37.2
Less personal	6	6.8	4	5.5	1	3.7	2	20.0	13	6.5
Use of the internet after acquiring the cell phone										
Frequency increased	56	64.4	51	70.8	17	63.0	8	80	132	67.0
Frequency not increased	31	35.6	21	29.2	10	37.0	3	20	65	33.0

Note: Chi-square test, Sig: $^{**}p < .05$.

The internet use variables are related (Table 11) to the use of the cell phone for the same purpose. Using the cell phone for sending and receiving email is positively correlated to the use of the cell phone for browsing the internet. Similarly, the same variable of sending and receiving email is associated with the increased use of the internet after the cell phone was made available (negative sign due to the coding pattern). This finding indicates that the cell phone is not only having an effect on the personal communication of students under study but also affects the way they adopt other media such as the internet to substitute their communication needs. The use of the cell phone is therefore affecting the way they make use of internet communication for sending and receiving emails. Accessing the internet on the cell phone makes things easier for the respondents to use it whenever they need to do so. Cell phone service providers charge the subscribers separately for this service or include it as part of the package and plan.

Table 11. Correlation between the Use of Cell Phone and Effects on Communication.

Variables	1	2	3	4	5	6	7
1 Ownership of cell phone							
2 Importance of cell phone in communication	.051						
3 Personal nature of communication after cell phone use	−.014	−.147*					
4 No. of people talked to after cell phone use	.031	−.060	.231**				
5 Cell phone for sending/receiving sms	−.036	.039	−.040	−.043			
6 Cell phone for sending/receiving email	.046	−.008	.013	−.009	−.090		
7 Cell phone for browsing internet	.037	.091	−.043	−.033	.046	.206**	
8 Internet use after cell phone use	.037	−.021	.020	−.069	.055	−.174*	−.132

Notes: (1) 1 = 1–6 months, 2 = 6–12 months, 3 = 1–3 years, and 4 = above 3 years. (2) 1 = not really important, 2 = somewhat important, 3 = important, and 4 = very important. (3) 1 = became more personal, 2 = remain unchanged, and 3 = became less personal. (4) 1 = increased a great deal, 2 = increased somewhat, 3 = unchanged, 4 = decreased somewhat, and 5 = decreased a great deal. (5) 0 = no, 1 = yes. (6) 0 = no, 1 = yes. (7) 0 = no, 1 = yes. (8) 1 = increased a lot, 2 = increased somewhat, 3 = decreased somewhat, 4 = decreased a lot.
Sig: *$p < .05$; **$p < .01$.

Another prominent factor that is obvious in the effects of the cell phone on the communication aspects of the youth is that the length of the ownership of the cell phone does not produce any corresponding increase or decrease in the selected variables that show the effects of communication. This allows us to conclude that it does not take much time for the users of cell phones to have an impact on their communication patterns. Rather, it is instantaneous and fast. Whether someone has been using the cell phone for one month or one year it does not make much difference in the way it affects the communication behavior of the youth.

Perceptions of the Students on Cell Phone Communication

A further series of questions was asked in order to comprehend what the respondents think about the cell phone in the contemporary world. The responses to these were rated on a four-point Likert scale which are presented in Table 12. The maximum possible score for any one item in the Table is 4. The questions were meant to collect the perceptions of the young respondents on the cell phone and communication patterns, and not specifically about their own personal use and communication experience. In particular, the items (total of 14) included in the interview schedule were intended to seek their views and opinions on how the cell phone has affected or influenced the general communication behavior of the students such as face-to-face, personal and international communication, the length of communication, redundancy of the landline, drawing on information, and the cost of communication. The perceptions of the youth on the existing means, tools, and channels of communication and their varied purposes are often based on their current usage patterns and can therefore shed light on certain aspects of their contemporary communication behavior.

Initially, the items that scored above the median score of two should be examined (Table 12). This should also be in line with the question as some questions were stated negatively (such as "the use of cell phone decreased face-to-face communication," and "the cell phone is not very useful for international communication") and therefore the scores can be lower than two. Regardless of gender, there was a majority view that the cell phone has become an essential tool for personal communication (item # 5). Given the higher scores obtained for the first three items (2.99 to 3.68) it is clear that the young respondents believed that the arrival of the cell phone in the country has certainly increased the levels of communication from the existing levels, reduced the way people used to communicate face-to-face, and changed the length of phone communication. In the majority view (item # 4), the increase in the use of cell phone in the country is mainly due to the prevailing quality of the landline and the services. As regards the use of the cell phone for communication beyond the borders of the country, a higher score was obtained for the item (#6) that agrees with the perception that communication with other countries has increased in this cell phone age. This view can be read with the view of the item that the cell phone is not very useful for international communication (item # 7). The lower score for this view implies that the respondents do not agree with the statement and that the cell phone does serve the purpose of international communication. Both young men and women agreed on this as there was no significant difference shown in the independent *t*-test.

There were two statements that pertain to the internet and its relation to cell phone communication. The first statement in this regard (item # 10) is stated negatively. The lower score for this item shows that the respondents

Table 12. Perceptions of Students Regarding Cell Phone Communication.

No.	Perceptions on communication	Boys			Girls			Both		
		N	Mean	S.D.	*N*	Mean	S.D.	*N*	Mean	S.D.
1	Use of cell phone has increased communication	118	3.71	0.56	82	3.63	0.58	200	3.68	0.57
2	Use of cell phone decreased face-to-face communication**	116	3.14	0.89	82	3.23	0.10	198	3.18	0.94
3	Duration of phone communication changed after the advent of cell phone	116	2.94	0.86	82	3.07	0.80	198	2.99	0.84
4	The increased use of the cell phone is due to the quality of existing landline system in the country	111	2.13	0.96	81	2.19	0.87	192	2.15	0.82
5	Cell phone is necessary for personal communication needs	118	3.65	0.53	82	3.63	0.58	200	3.65	0.54
6	Cell phone has increased communication with other countries	114	3.12	0.87	81	3.09	0.83	195	3.11	0.85
7	Cell phone is not very useful for international communication	112	1.90	0.98	82	2.10	1.0	194	1.98	0.99
8	South Africans rarely use cell phone to explore outside world	114	2.24	0.96	81	2.33	0.87	195	2.28	0.92
9	People use cell phone for information from outside world	116	3.04	0.76	82	3.07	0.81	198	2.00	3.06
10	The use of cell phone has decreased the use of the internet**	117	1.64	0.83	82	1.68	0.70	199	1.66	0.77
11	Cell phone takes the money to be used for the internet	112	1.92	0.89	82	2.01	0.85	194	1.96	0.88
12	Cell phone makes people poorer without adequate benefits of use**	113	2.03	0.87	81	2.31	0.90	194	2.14	0.89
13	Cell phone use makes people poorer	114	2.12	0.98	81	2.16	0.90	195	2.14	0.95
14	Government should control the cell phone call rates in the country***	112	2.79	1.19	82	2.94	1.01	194	2.86	1.11

Note: Rated on Likert scale of 1 to 4. 1 = strongly disagree, 2 = disagree, 3 = agree, and 4 = strongly agree. Independent *t*-test, Sig: $^{**}p < .05$; $^{***}p < .01$.

did not believe that the use of the cell phone has not decreased internet use. The second statement in this category is that the cell phone takes up the money that would have otherwise been spent on accessing the internet (item # 11). From the score for this item (1.96) it can be assumed that the respondents did not generally think so. While there is no disagreement between the young men and women in this respect, there was a statistically significant difference in the first internet statement that cell phone use had reduced the use of the internet. Men and women thought differently on this matter.

There were three statements (12–14) that were included to capture the perceptions of the respondents about the cost of the cell phone charges in the country. To begin with, the last item with a score of 2.86 confirms that the prevailing call rates were not acceptable in the view of the respondents. The youth wanted the government to control the call rates as the prevailing rates were unaffordable. This view is supported more by young women than young men. They also felt that cell phone use, in terms of the cost of the calls, makes people poorer (item # 13) which relates to the higher charges that are being levied by the service providers in the country. In a similar vein they argued that, as is clear from statement #12, the cost people pay for the use of cell phone is not commensurate with the benefits people received in return. More women than men think this way. There is thus a significant gender difference as well.

How does racial background affect the perceptions of the respondents on their communication? This is presented in Table 13. In the case of two items (4 and 6) the ANOVA test indicated significant differences among the racial groups in the perception that the increased use of the cell phone is related to the lack of quality of the existing landline system in the country (3, $F = 3.165$, $p < .026$). In this case the whites and Other have higher mean values than the rest of the racial groups. The latter maintained very strongly that it is the failure of the fixed landline system in the country that has led to the increased use of the cell phone. Also, significant differences between racial groups were evident in the perception that the cell phone has clearly increased communication with other countries. Again, as in the previous item, whites and Other racial categories scored higher on this perception than Africans and Indians (3, $F = 2.610$, $p < .053$).

In Table 14 the relationship between the selected perceptions and use of the cell phone is evident. With the exceptions of the sending and receiving of emails on cell phones, the majority are found to be using all the functions of cell phones.

Table 13. Perceptions of Students Regarding Cell Phone Communication by Race.

No.	Perceptions on Communication	African			Indian			White			Others			All		
		N	Mean	S.D.	*N*	Mean	S.D.	*N*	Mean	S.D.	*N*	Mean	S.D.	*N*	Mean	S.D.
1	Use of cell phone has increased communication	88	1.33	0.47	74	1.36	0.71	27	1.19	0.4	11	1.27	0.47	200	1.32	0.57
2	Use of cell phone decreased face-to-face communication	86	1.90	0.92	74	1.82	0.97	27	1.74	0.91	11	1.45	0.93	198	1.82	0.94
3	Duration of phone communication changed after the advent of cell phone	86	1.99	0.91	74	2.05	0.84	27	1.93	0.68	11	2.00	0.63	198	2.01	0.84
4	The increased use of the cell phone is due to the quality of existing landline system in the country**	86	2.85	0.88	71	2.66	0.95	24	3.17	0.87	11	3.36	0.92	192	2.85	0.92
5	Cell phone is necessary for personal communication needs	88	1.40	0.54	74	1.34	0.58	27	1.33	0.56	11	1.18	0.41	200	1.36	0.55
6	Cell phone has increased communication with other countries***	86	1.79	0.78	72	1.86	0.83	27	2.07	1.04	10	2.50	0.85	195	1.89	0.85
7	Cell phone is not very useful for international communication	87	3.01	0.97	71	2.99	1.05	25	3.20	0.82	11	2.82	1.17	194	3.02	0.99
8	South Africans rarely use cell phone to explore outside world	87	2.66	0.87	72	2.82	1.01	25	2.56	0.92	11	3.00	0.63	195	2.72	0.92
9	People use cell phone for information from outside world	86	2.02	0.75	74	1.84	0.74	27	1.89	0.89	11	2.18	0.98	198	1.94	0.78
10	The use of cell phone has decreased the use of the internet	88	3.22	0.81	73	3.37	0.79	27	3.59	0.64	11	3.55	0.52	199	3.34	0.77
11	Cell phone takes the money to be used for the internet	86	2.91	0.92	71	3.14	0.80	27	3.04	0.94	10	3.50	0.71	194	3.04	0.88
12	Cell phone makes people poorer without adequate benefits of use	86	2.92	0.84	73	2.85	0.94	25	2.80	0.87	10	2.50	1.08	194	2.86	0.89
13	Cell phone use makes people poorer	87	2.90	0.93	74	2.92	0.95	24	2.58	0.93	10	2.80	1.13	195	2.86	0.95
14	Government should control the cell phone call rates in the country	85	1.98	1.07	72	2.24	1.09	26	2.50	1.24	11	2.00	1.18	194	2.14	1.12

Note: Rated on Likert scale of 1 to 4. 1 = strongly disagree, 2 = disagree, 3 = agree, and 4 = strongly agree. Independent *t*-test, Sig: $^{**}p < .05$; $^{***}p < .01$.

Table 14. Perceptions of Students Regarding Cell Phone Communication by Selected use of Cell Phone.

No.	Perceptions on communication	SMS		Email		Games		Internet	
		No.	%	No.	%	No.	%	No.	%
1	Cell phone has increased communication with other countries	191	98.5	86	44.3	148	75.9	150	76.9
2	Cell phone is not very useful for international communication	190	97.9	85	44.0	149	76.8	150	77.3
3	South Africans rarely use cell phone to explore outside world	191	97.9	86	44.3	149	76.4	151	77.4
4	People use cell phone for information from outside world	194	97.9	86	43.7	148	74.7	151	76.2

DISCUSSION

This study of a sample of the South African university students illustrates the prevailing trends in the usage patterns of young respondents, the effects of the cell phone use on their personal and private communication, and their perceptions on cell phone communication. The analysis also showed how they were different in these in terms of gender division. The sample students were drawn from diverse racial and cultural backgrounds. A number of factors were found to be prominent in the usage pattern of the students. These factors relate to the major purposes for which the cell phone is used: for communication purposes such as for making and receiving voice calls, sending and receiving short messages, accessing the internet, and for email communication. In some of these communication activities, young men and women differed in their usage pattern. Apart from gender differences that were noticed in some of these aspects examined in this paper, the findings are in agreement with other studies (Axelsson, 2010; Oksman & Turtiainen, 2004). Cotton et al. (2009) referred to a gender divide in both the ownership and the usage of cell phones among middle-school children in the United States but the differences disappeared when affinity and skill were taken into account in the analysis. In the ownership of cell phones, Zainudeen, Iqbal, and Samarajiva (2010) showed a gender divide in countries such as Pakistan and India. But this difference was not as prominent in countries such as Sri Lanka, the Philippines, and Thailand, which were the other countries they selected as their sample. The research conducted among the Swedish youth by Axelsson (2010) showed that the use and attitude toward cell phone communication in a certain life stage tended to exhibit similarities irrespective of the cultural background of the users. Axelsson (2010) acknowledged the additional factors other than the ability to pay the cost and the technological expertise of the young users that determined whether and how the cell phone and other ICTs are used. Two issues are to be considered here, as Axelsson (2010) raises: the use and attitudes are culturally dependent or life-stage dependent. It appears that in the case of the South African students sample, culture does not seem to affect as much as the latter factor of life-stage. The sample of respondents represented a culturally diverse group but exhibited similar use and perception characteristics. In the case of South African students, the use of the cell phone was widespread as all of them had possessed a cell phone for a reasonable period of time. This is therefore more a life-stage factor than a cultural one. The cost of having and maintaining a cell phone service is not cheap in South Africa. But the students have managed in a planned manner to use the cell phone by effectively organizing the device for their voice call needs within their means.

The adoption and ownership of a cell phone has a status dimension as Cawley and Hynes (2010) recognized in their study of Irish teenagers. They noted that in the case of their sample, for participants the ownership of a cell phone is something which is taken for granted, and does not confer any social status on the users. The cell phone for the Finnish teenagers is an organic part of their life rather than a symbol of status (Oksman & Turtiainen, 2004). If we consider this dimension of social status as one that motivates South African students to possess a cell phone and shapes their communication needs accordingly a similar picture emerges. Nearly two-thirds of our respondents, regardless of the split between young men and women, viewed the ownership of a cell phone as a symbol of social status. Although all the respondents owned or used a cell phone the status dimension of it is acknowledged and accepted as a determining factor for cell phone usage. It is not yet a gadget the possession of which is largely taken for granted as part of their life style. Ling (2008) also thought that social status is a driving force behind owning a cell phone.

The usage pattern of the young South African engineering students has similarities with their counterparts elsewhere. It has been reported that a high rate of penetration of the cell phone has occurred among the Finnish youth and the appliance has become an inevitable part of their life (Oksman & Turtiainen, 2004). Our data also showed a similar high penetration rate in the sample and their regular use for the various functionalities of the cell phone. Swedish students, for instance, have shown similar trends in terms of the number of voice calls they make and receive a day and in terms of the duration of these voice calls (Axelsson, 2010). In comparison to Irish teenagers (Cawley & Hynes, 2010) a great majority of South African students tend to make a somewhat greater number of calls. These usage patterns present comparable trends. But in the case of their American counterparts there is no comparison with the usage of South African students. In a study of American adolescents, Weisskirch (2008) reported that the respondents made and received an average of 16 calls per day. Does this signify that the youngsters have the same mindset with regard to mobile technology?

Understanding the usage patterns of the users in relation to cell phone technology is the first step toward an understanding of their communication patterns and its social consequences. In this sense, this South African study offers the basis for the analysis of the communication patterns and social consequences that have become necessary as a result of the widespread adoption of the technology by the youth. As researchers such as Cawley and Hynes (2010) have argued, the communication patterns of youngsters are constantly evolving and are therefore quite complex to comprehend. But knowledge of their usage patterns and perceptions of communication will facilitate a greater understanding of this complexity.

NOTES

1. Psychological neighborhoods, according to Wurtzel and Turner (1977, cited in Wei & Lo, 2006, p. 57) refer to a supportive community to provide ready connectedness and support immediate interaction.
2. Colored means a mixed race of blacks, white, or Indians.

REFERENCES

Abeele, M. V., Beullens, K., & Roe, K. (2013). Measuring mobile phone use: Gender, age and real usage level in relation to the accuracy and validity of self-reported mobile phone use. *Mobile Media and Communication, 1*, 213–236.

Aker, J. C., & Mbiti, I. M. (2010). Mobile phones and economic development in Africa. *The Journal of Economic Perspectives, 24*, 207–232.

Atkin, D. J. (1993). Adoption of cable amidst a multi-media environment. *Telematics and Informatics, 10*, 51–58.

Axelsson, A.-S. (2010). Perpetual and personal: Swedish young adults and their use of mobile phones. *New Media and Society, 12*, 35–54.

Bond, E. (2010). Managing mobile relationships: Children's perceptions of the impact of the mobile phone on relationships in their everyday lives. *Childhood, 17*, 514–529.

Bratton, M. (2013). Citizens and cell phones in Africa. *African Affairs*, *112*, 304–319.

Campbell, S. W., & Kwak, N. (2010). Mobile communication and civic life: Linking patterns of use to civic and political engagement. *Journal of Communication Inquiry*, *60*, 536–555.

Castells, M., Fernández-Ardèvol, M., Qiu, J. L., & Sey, A. (2007). *Mobile communication and society: A global perspective*. Cambridge, MA: The MIT Press.

Cawley, A., & Hynes, D. (2010). Evolving mobile communication practices of Irish teenagers. *Aslib Proceedings*, *62*, 29–45.

Cohen, A. A., & Lemish, D. (2003). Real time and recall measures of mobile phone use: Some methodological concerns and empirical applications. *New Media and Society*, *5* 167–183.

Cotten, S. R., Anderson, W. A., & Tufekci, Z. (2009). Old wine in a new technology, or a different type of digital divide? *New Media and Society*, *11*, 1163–1186.

Drotner, K. (2000). Difference and diversity trends in young Danes' media uses. *Media, Culture and Society*, *22*, 149–166.

Fortunati, L. (2002). The mobile phone: Towards new categories and social relations. *Information, Communication and Society*, *5*, 513–528.

Goggin, G. (2013). Youth culture and mobiles. *Mobile Media and Communication*, *1*, 83–88.

Harvey, J., & Sturges, P. (2010). The cell phone as appropriate information technology: Evidence from The Gambia. *Information Development*, *26*, 148–159.

Hutchby, I., & Barnett, S. (2005). Aspects of the sequential organization of mobile phone conversation. *Discourse Studies*, *7*, 147–171.

Igarashi, T., Takai, J., & Yoshida, T. (2005). Gender differences in social network development via mobile phone text messages: A longitudinal study. *Journal of Social and Personal Relationships*, *22*, 691–713.

Ilahiane, H., & Sherry, J. (2009). Economic and social effects of mobile phone use in Morocco. *Ethnology*, *48*, 85–98.

ITU. (2014). Retrieved from http://www.itu.int/en/ITU-D/Statistics/Pages/stat/default.aspx

Karim, N. S. A., Oyebisi, I. O., & Mahmud, M. (2010). Mobile phone appropriation of students and staff at an institution of higher learning. *Campus-Wide Information Systems*, *27*, 263–276.

Leung, L., & Wei, R. (1999). Who are the mobile phone have-nots? *New Media and Society*, *1*, 209–226.

Leung, L., & Wei, R. (2000). More than just talk on the move: A use-and-gratification study of the cellular phone. *Journalism and Mass Communication Quarterly*, *77*, 308–320.

Licoppe, C. (2004). 'Connected' presence: The emergence of a new repertoire for managing social relationships in a changing communication technoscope. *Environment and Planning D: Society and Space*, *22*, 135–156.

Ling, R. (2008). *New tech, new ties: How mobile communication is reshaping social cohesion*. Cambridge, MA: MIT Press.

Mihailidis, P. (2014). A tethered generation: Exploring the role of mobile phones in the daily life of young people. *Mobile Media and Communication*, *3*, 58–72.

Morimoto, S. A., & Friedland, L. A. (2011). The Lifeworld of youth in the information society. *Youth and Society*, *43*, 549–567.

Mutula, S. (2002). The cellular phone economy in the SADC region: Implications for libraries. *Online Information Review*, *26*, 79–92.

Napolitano, C. M. (2010). "MXing it up": How African adolescents may affect social change through mobile phone use. In M. U. Bers & G. G. Joam (Eds.), *New media and technology: New directions for youth development* (Vol. 128, pp. 105–113). London: Wiley.

Oksman, V., & Turtiainen, J. (2004). Mobile communication as a social stage. *New Media and Society*, *6*, 319–339.

Özcan, Y. Z., & Koçak, A. (2003). A need or a status symbol? Use of cellular telephones in Turkey. *European Journal of Communication*, *18*, 241–254.

Peters, O., & Allouch, S. b. (2005). Always connected: A longitudinal field study of mobile communication. *Telematics and Informatics*, *22*, 239–256.

Rice, R. E., & Katz, J. E. (2003). Internet and mobile phone digital divides. *Telecommunications Policy*, *27*, 597–623.

Rivera-Sánchez, M. M., & Walton, M. (2013). Making sense of life's transitions mobile phones and the creation of alternative spaces by South African youths. *Journal of Creative Communications*, *8*, 177–191.

Robinson, S., & Stubberud, H. A. (2012). Communications preferences among university students. *Academy of Educational Leadership Journal*, *16*, 105–113.

Rogers, E. M. (2003). *Diffusion of innovations* (5th ed.). New York, NY: Free Press.

Thulin, E., & Vilhelmson, B. (2007). Mobiles everywhere: Youth, the mobile phone, and changes in everyday practice. *Young: Nordic Journal of Youth Research*, 15, 235–253.

Thurlow, C., & McKay, S. (2003). Profiling "New" communication technologies in adolescence. *Journal of Language and Social Psychology*, *22*, 94–103.

Vishwanath, A., & Goldhaber, G. M. (2003). An examination of the factors contributing to adoption decisions among late-diffused technology products. *New Media and Society*, *5*, 547–572.

Wei, R. (2006). Lifestyles and new media: Adoption and use of wireless communication technologies in China. *New Media & Society*, *8*(6), 991–1008.

Wei, R., & Lo, V.-H. (2006). Staying connected while on the move: Cell phone use and social connectedness. *New Media and Society*, *8*, 53–72.

Wei, R., Lo, V.-h., Xu, X., Chen, Y.-N. K., & Zhang, G. (2014). Predicting mobile news use among college students: The role of press freedom in four Asian cities. *New Media & Society*, *16*, 637–654.

Weisskirch, R. S. (2008). Parenting by cell phone: Parental monitoring of adolescents and family relations. *Journal of Youth and Adolescence*, *38*, 1123–1139.

Wurtzel, A. H., & Turner, C. (1977). Latent function of the telephone: What missing the extension means. In I. Pool (Ed.), *The social impact of the telephone* (pp. 246–261). Cambridge, MA: The MIT Press.

Zainudeen, A., Iqbal, T., & Samarajiva, R. (2010). Who's got the phone? Gender and the use of the telephone at the bottom of the pyramid. *New Media and Society*, *12*, 549–566.

ADOLESCENTS' SELF-DEFINING INTERNET EXPERIENCES

Åsa Andersson, Margareta Bohlin, Linda Lundin and Emma Sorbring

ABSTRACT

Purpose – *The purpose of this study was to investigate how young women and men perceive the Internet as a phenomenon and what role and meaning they ascribe to the Internet as an arena for defining themselves and for shaping their identity.*

Methodology/approach – *The empirical data consist of narratives written by Swedish adolescents. Using content analysis the analysis was carried out in three steps: (1) finding categories and themes, (2) calculation of statistical differences in category frequencies, (3) a theoretically informed interpretation of central themes, using Bourdieu's concept of different forms of capital, and Giddens' concept of "pure relations."*

Findings – *The narratives exemplify how computer literacy and technological competence can be converted into social, cultural, and symbolic capital. Gender differences occur both in statistical differences between category frequencies in girls' and boys' narratives and in the interpretation of central themes. But there are also several examples that show*

Technology and Youth: Growing Up in a Digital World
Sociological Studies of Children and Youth, Volume 19, 105–132

ISSN: 1537-4661/doi:10.1108/S1537-466120150000019004

more complex and contradictory tendencies, exceeding or transformative of gender differences and hierarchy.

Originality/value – *This study considers adolescents' own perspectives on an arena of great importance. The analyses have been performed both qualitatively and quantitatively, which gives a nuanced picture of young people's self-defining experiences on the Internet.*

Keywords: Adolescent; internet; identity; gender; narratives; self-defining context

INTRODUCTION

Adolescents explore their personal and social sense of their selves in many different arenas in society: in peer groups, in school, and in their families. Depending on the social context, children and adolescents commit to different aspects of identity. For example, some children and adolescents may learn that physical appearance is particularly important and valuable for a positive personal and social sense of self; whereas others may learn that school achievement and careers are particularly important (e.g., Grogan, 2007). Crucial for every aspect of identity is, however, the relational and social character: how the self becomes mirrored and confirmed by other individuals and groups.

For young adolescents today the Internet has always existed, but even in a short historical perspective the Internet must be considered a new kind of social environment providing what seem to be endless opportunities for seeking out, initiating, and maintaining social contacts with peers as well as strangers of different ages and cultural backgrounds from around the world. Most of the 9–16-year-olds in the EU's 25 countries have a profile on a social network where they can socialize as well as express their thoughts and feelings (Alexandersson & Hansson, 2011). In Sweden, seven out of 10 Internet users participate in social networks/media and 95% of youths between 16 and 25 years of age visit Facebook, 81% of them daily. The most popular digital buys among adolescents (ages 12–15) in Sweden are music and apps. At the same time 80% use Internet to seek facts for schoolwork (Findahl, 2014).

An important and interesting area of research has thus concerned what types of activities adolescents engage in on the Internet and what type of environments they are part of and create there. However, the main question in the present study considered how young women and men in Sweden perceive the Internet as a phenomenon and what role and meaning they ascribe the Internet as an arena for defining themselves and for shaping their identity. Since gender socialization is such a crucial adjustment to societal demands, and since these adjustments are integral parts in the formation of identity, thereby strongly affecting the individual's self-perception (Connell, 2009), a gender perspective is an important approach to consider using in this kind of study.

Alternative Arena for Identity Experimentation

The Internet offers a new culture of simulation where human nature is reconstructed by allowing individuals to experiment with imaginative creations of the self beyond real-life limits. On the Internet the self is no longer a biologically bound individual. Instead the user can participate detached from his or her physical body (Zhao, Grasmuch, & Martin, 2008). The Internet has changed the traditional conditions of identity construction, where people can project an identity that is more socially desirable than the "real" offline identity, allowing adolescents to portray what they see as their alternative or even true self without having to worry about their own or other people's external attributes (Bargh, McKenna, & Fitzsimons, 2002; Larraín, Zegers, & Alejandro, 2007). Online young people can experiment with different roles, fantasies, ideas, actions, and creations of self-presentations and identities (Rosenmann & Safir, 2006; Suler, 2002; Zhao et al., 2008). The Internet thereby offers an alternative arena for identity experimentation with a higher degree of freedom from social pressures, norms, and expectations (Schmitt, Dayanim, & Matthias, 2008). The extent of identity experimentation seems though to depend on whether the online setting is anonymous or non-anonymous (Cinnirella & Green, 2007; Zhao et al., 2008). People are more likely to construct online identities that correspond to offline identities in non-anonymous online settings, such as Facebook, compared to anonymous environments, such as chat rooms and bulletin boards (Cinnirella & Green, 2007; Zhao et al., 2008). Nevertheless, regardless of the extent of anonymity, several studies indicate that it is fairly uncommon for adolescents to construct online identities that diverge greatly from their offline identity, involving for example changes of gender and sexuality (e.g., Larraín et al., 2007; Schmitt et al., 2008). Instead, it is more common to present an online identity that is very similar to the offline identity. Internet should thus be seen as a context, among others, where the individual's social identity is expressed, and where the construction of identity takes place. For example, the constructed online identities tend to reproduce offline class, ethnicity, gender, and culturally typical patterns of social role behaviors (Tynes, 2007; Valentine & Holloway, 2002).

Transformation of the Social World

Previous research suggests that one of the most important aspects of adolescents' use of the Internet is the way it transforms their social world by influencing how they communicate, socialize, and experiment with self-presentation (Gross, 2004; Mazur & Kozarian, 2009; Wolak, Mitchell, & Finkelor, 2002). Internet communication is regarded as a complement to offline social interaction (Pempek, Yermolayeva, & Calvert, 2009). Young people regard a phenomenon such as social media-sites not only as fun, but also as a controlled means for representing themselves (Bargh et al., 2002). Previous research shows that adolescents on social networking sites tend to strategically choose layout designs and texts that are meant to reflect their ideal selves and personalities (Livingstone, 2008). For example, photos displayed online tend to be particularly flattering and carefully chosen to express the individual as desirable (Gibbs, Ellison, & Heino, 2006; Mazur & Kozarian, 2009; Tynes, 2007; Valentine & Holloway, 2002; Zhao et al., 2008).

The anonymity of the Internet allows individuals also to experiment with facets of their selves that they find uncomfortable expressing offline (Bargh et al., 2002; Rosenmann & Safir, 2006; Suler, 2002). Studies on self-presentation and interaction in blogs (Mazur & Kozarian, 2009) and in online teen chat rooms (Subrahmanyam, Smahel, & Greenfield, 2006) have shown how young people self-disclose and freely discuss private ideas and concerns related to sexuality, relationships, and drug use in these arenas. The anonymity of the Internet reduces the risks of placing oneself in a vulnerable and potentially awkward situation (Tidwell & Walther, 2002).

As shown in Valkenburg and Peter's (2009) study on adolescents' online communication, the Internet may thus increase intimacy and formation of close friendships by fostering self-disclosure (see also Bane, Cornish, Erspamer, & Kampman, 2010). As shown in studies done by Schmitt et al. (2008) and Valkenburg, Schouten, and Peter (2005), there are age-differing dimensions where preadolescents tend to use personal homepages and networking sites to write about their skills and accomplishments; whereas older adolescents tend to write about values, social relationships, and personal characteristics. Self-expression and peer interaction on blogs become increasingly advanced over the course of adolescence and into emerging adulthood (Davis, 2010). However, sharing and connecting with others online is more common among older adolescents than among adults, and they are also more willing to open up to others in these contexts (Christofides, Muise, & Desmarais, 2012; Hollenbaugh & Ferris, 2014; Walrave, Vanwesenbeeck, & Heirman, 2012).

Paradoxically, as a social arena, the Internet can increase offline isolation while at the same time enhance social relationships online (Lanthier & Windham, 2004; Wolak et al., 2002; Ybarra, Alexander, & Mitchell, 2005). For example, adolescents who feel socially anxious report that they perceive the Internet as particularly valuable for socializing and for disclosing intimate information about themselves (Bargh & McKenna, 2004; Peter & Valkenburg, 2006; Valkenburg & Peter, 2007). In this case, the Internet offers an alternative social arena for individuals who otherwise have difficulties in forming and maintaining social relationships. Online social networking sites can then be important tools for creating and maintaining social relationships and can also be a mobilizing tool for voicing subjects of concern (Ackland & O'Neil, 2011; Shapiro & Margolin, 2013). However, social media also provide platforms for repeating and disseminating negative rumors to a large audience and can be a platform for bullying (Bolinder, Chenik, & Juth, 2011; Landstedt & Persson, 2014). Research has also revealed that text messaging for sexual purposes, for example, sexting, is associated with symptoms of depression, impulsivity, and substance abuse (Temple et al., 2014). The evidence for teenagers in general exposing themselves to higher risks online than offline is not conclusive (Reich, Subrahmanyam, & Espinoza, 2012; Vallor, 2011). For example, a recent review indicates an overall positive association between the use of social networking sites and sense of social connectivity (Shapiro & Margolin, 2013).

Gender Online

Since our characteristics and expectations of behavior are related to the differing conditions that women and men live under, one of the most important factors involved in forming an identity is gender. Most aspects of social life are influenced by the ways in which gender is performed and perceived. Gender structures are created and repeated throughout life in the family, at school, among peers, at the work place, during leisure time, and by authorities etc. (Connell, 2009; Giddens, 2009). Gender socialization is an adjustment to societal demands, but at the same time it shapes the identity and affects the individual's self-perception of what is important. Previous studies have shown that when it comes to differences in usage, girls tend to use the Internet more for social interaction and educational purposes; whereas boys tend to use the Internet more for playing computer games and for other forms of entertainment and leisure activities (e.g., Colley & Maltby, 2008; Davis, 2010; Willoughby, 2008).

Women tend to author blogs to a greater extent than men. However, Herring and Paolillo (2006) argue, in their study on gender and genre variation in blogs, that those who tend to be seen as the most influential bloggers usually are men. In the use of blogs, studies show that both adolescent boys and girls tend to use blogs primarily to self-present themselves in a self-focused manner. However Mazur and Kozarian (2009) have shown a gender difference in their study of adolescents' blogs, where girls tend to use personal homepages mainly to convey emotions and to write about and discuss relationships with their friends. Boys, on the other hand, tend to use personal homepages mainly to write about societal issues, meet new friends online, and experiment with their identities (Mazur & Kozarian, 2009; Schmitt et al., 2008). Furthermore, young female bloggers were found to self-disclose and discuss more private and intimate concerns than young male bloggers (Bortree, 2005; Hollenbaugh, 2010). However, young men have been found to disclose phone numbers and home addresses on social networking profiles to a greater extent than young women (Fogel & Nehmad, 2009). For both girls and boys, an increase in the complexity of personal homepage content can be seen with increasing age (Pfeil, Arjan, & Zaphiris, 2009).

THEORETICAL FRAMEWORK

With the aforementioned studies in mind it may be possible to relate the Internet as a social arena to the sociological theories of Bourdieu (1984, 1986, 1998) and Giddens (2009). They have in different ways conceptualized the social world as comprising systems of relations affected by authority and power structures, which in turn means that identity formation to a high degree involves handling, re-creating and sometimes changing social positions of varying types. Bourdieu (1984) has conceptualized the processes where social relations are formed and developed as depending upon the assets of varying types of capital: economic, cultural, social, and symbolic. This capital is, according to Bourdieu, related to and defined by the objective structures of social fields (1998). The three latter forms of capital mentioned will be relevant for the context of the present study, since striving for acknowledgment, confirmation and sometimes status appear to be central in identity formation. Cultural capital involves competence, skills, and knowledge; social capital involves an individual's social networks; and finally, symbolic capital comprises the acknowledgment of the other forms of capital resulting in prestige and eminence. Together these forms of capital influence which social positions individuals can have access to and move between, and can be related to skill development and access to social positions online. Consequently, previous research indicates that the Internet contributes to new ways of increasing social capital by enhancing relationship building, maintaining pre-existing relationships, and fostering fast and efficient social interaction (Pfeil et al., 2009; Young, 2005).

The increase of symbolic and cultural capital gained by the Internet is closely connected to globalization. The Internet has contributed to the radicalization of some processes associated with late modernity, conceptualized in the social sciences as medialization, technologization, and the virtual quality of time and space. These processes, referred to by Giddens (2009) as the time-space compression and the disembedding of social systems, signify the lifting of social relationships from local contexts where the articulation of social life is altered as time and space are reorganized. The Internet has opened new channels of communication and interaction between individuals where collaboration and interaction become possible without regard to geographic location. This allows people from different cultural, religious, and ethnic backgrounds to meet, interact, and have discussions (Tynes, 2007). Access to an unlimited global audience has also rendered it possible to increase a person's symbolic capital due to the opportunity to form status and even to gain fame on and through the Internet (Reysen, Lloyd, Katzarska-Miller, Lemker, & Foss, 2010). The globalization that the Internet reflects and provides also makes it possible to increase one's cultural capital by searching for information and improving one's knowledge, skills, and competence from a wide range of international sources (Meyen, Pfaff-Rudiger, Dudenhöffer, & Huss, 2010). This radically changes the conditions for how relations can be established. In contrast to traditional contexts – where relations have been closely connected to personal ties in external social and economic conditions – the contexts found in the era that in sociological terms is referred to as late modernity offers other types of conditions for relationship building. Especially intimate relationships are now, as Giddens has stated, assumed to be "pure," that is, unaffected by external conditions such as the economy and collectivistic social regulations. These pure relations are thereby believed to exist only based on what the relationship itself can offer the individual involved in it (Giddens, 1991). The conception and maintenance of pure relations involve an increased focus on individualization and emotional aspects in relationship building.

Bourdieu's concepts of different forms of capital and Giddens' concept of pure relations both deal with relational aspects of societal structures and the social life of individuals, while a combination of the concepts will be used in the current study to focus on and analyze different tendencies in young people's perceptions of the Internet as a social arena for self-definition.

THE CURRENT STUDY

Much of the previous research presented above examines how adolescents use the Internet and what type of activities they engage in online. Although, different aspects of young men's and women's Internet use have been illuminated by research, there are not many studies in which the voices of the young people themselves are heard. Research that considers adolescents' own perspectives is however needed for a deeper and more nuanced understanding of young people's everyday lives. Hence, the aim of the present study was to investigate how young women and men in Sweden perceive the Internet as a phenomenon and what role and meaning they ascribe to the Internet as an arena for defining themselves and for shaping and developing their identity. We sought to investigate these research questions by asking pupils in upper secondary schools to write narratives about the Internet and the influence it has had on their lives. Considering the fact that almost all Swedish teenagers in this age group are frequent users of the Internet and can be referred to as so-called "Internet natives," it is supposed that most teenagers would have some and probably much experience from the Internet.

METHODS

Informants

The number of informants was 121 (59 girls, 62 boys), aged 16–18, and recruited from two upper secondary schools in western Sweden. In the first school (3,466 pupils whereof 51% girls) 41% of the pupils' parents had higher education and 11% were pupils with immigrant background. In the second school (1,296 pupils whereof 46% girls), the percentages were 43% and 12%, respectively. In upper secondary schools nationwide the percentages were 46% of the pupils had parents with higher education and 19% were pupils with immigrant background. Representatives for the schools were asked to select classes that together represented a diverse socioeconomic background as well as equal numbers of boys and girls. The pupils studied different programs: Technology, Social Sciences, Electronics and Telecommunications, Science, Catering, Art, and Child Care. Since the assignments were distributed and collected at the same time, there were no external dropouts. However, two assignments were incomplete and were therefore excluded, resulting in a sample of 119 (58 girls and 61 boys). All of the participants used the Internet. About half of the sample (37% girls, 62% boys) used the Internet more than 3 hours per day, and a little bit less than half (54% girls, 33% boys) used it everyday, but less than 3 hours; a smaller proportion of the sample (9% girls, 5% boys) used the Internet only a couple of times a week. It should be noted that the narratives are retrospective, which means, for example, that the amount of time the adolescents spent on the Internet might not have been the same if the narratives had been written during the period they wrote about.

Procedure and Instrument

Teachers in each school were contacted, and the aim and research questions were explained. The researchers visited the classes during their ordinary lessons, when the teacher was present, but kept a very low profile, in order to let the participants write freely. The pupils were informed that their participation was voluntary and that their stories would be handled confidentially. The purpose and significance of the study was explained a couple of days before the actual writing session in order to enable the pupils to prepare themselves for the task of writing about the Internet and its significance for their self-definition. A document with background questions and pieces of blank paper were distributed and the pupils wrote down their narratives during an ordinary 45-minute lesson.

The study focused on the participants' own written narratives. Other than that there were only background questions on gender and age. The pupils were subsequently asked to write a narrative. To facilitate the exercise, the informants were given two vignettes. The two examples presented in the vignettes were an intense interest in music with powerful psychological and social experiences related to music, and an intense experience of and insight into social problems and injustices triggered by a TV-series (appendix). The pupils were then asked to write their narratives about how they experience and perceive the Internet as a factor in their self-definition.

Procedure of Analysis

The handwritten narratives varied widely in length. The shortest were about two or three sentences and the longest were two and half pages, with an average of one page. Altogether the 119 pupils wrote 70 full handwritten pages, which equal approximately 15,400 words. The narratives were analyzed for content. Content analysis is often used to analyze documents and texts, either by quantifying or by qualifying the data. The aim of the study and the research questions are important tools in the coding of the material and in the analysis. Questions like "What is discussed in the text?" and "How is the subject described?" are asked and analyzed throughout the reading of the texts (Watt Boolsen, 2007). In the quantitative content analysis, categories are coded and statistically examined, while the qualitative content analysis results in theoretically informed descriptions of phenomena important for understanding the dynamics in the texts.

Accordingly, the data in this study was analyzed both quantitatively and qualitatively with content analysis. The first step in the analysis was qualitative in that the focus was to discover categories in the data. In order to accomplish this, all narratives were read through by the researchers, and words and sentences about perceptions, experiences, and beliefs regarding the Internet were marked in the text. The second step was to quantify qualitative data. The words and sentences were summed up into 47 categories, which in turn were coded into 14 broader categories and this was done in collaboration between the researchers. The 119 narratives were split into four bundles – one for each researcher – and each text was coded according to the 14 categories. In this process one new category was found, and after consultation within the research group the categories were increased to 15. These were used to code the data. The researchers then swapped bundles and the plausibility of encodings was controlled and sometimes changed after a discussion within the research group. The encodings were compiled in two stages. The number of times each category occurred in the texts (the individuals' narratives) was calculated (at the most 119 times) as well as the number of times a category was found in the whole material (one category could occur several times in each narrative).

In the next step of the analysis the 15 categories were compiled into six more condensed themes. To further deepen the analysis, three themes of the total 12 themes were focused on in a qualitative content analysis. The categorization in the qualitative content analysis is to some extent similar to the quantitative content analysis. The coding of the texts is done in several steps, where categories and themes are found. In this analysis, the qualitative analysis was present in the categorization process, which formed the basis for the statistics from the beginning. However, to further develop the qualitative analysis, the three chosen themes are inspired by and make use of Bourdieu's theoretical concepts describing how social relations and thereby identities are constituted (1984). The themes further penetrated in the narrative of each informant were thus the following: cultural capital, social capital, and symbolic capital. These themes were selected since they refer to conceptualized experiences in several identity-forming contexts but, at the same time, are relatively unexplored in the context of the Internet. Bourdieu's theory of the different types of capital has previously been used in media research for the purpose of describing and theorizing a typology of Internet users with respect to the kinds of capital Internet users gain online and which factors influence the patterns of usage (Meyen et al., 2010). Our use of the different types of capital, however, derives from content analysis of young peoples' own narratives of the Internet as a factor in their construction of identity.

RESULTS

In accordance with the process described in the method section, the results and findings will be presented in three steps: first, a presentation of the themes that appear in the material, second, a calculation of statistical differences between category frequencies in girls' and boys' narratives, and finally, a theoretically informed interpretation of central themes.

First Step: Themes Appearing in the Material

Fifteen categories appeared in the material and were organized into six different themes. The basis for these six themes will be described briefly below.

Three of the themes below (numbers 1, 2, and 5) will be focused on further in the qualitative analysis that appears later on in the results.

1. *Social capital*: This theme consists of categories that describe what it is like to search for experiences in different social arenas, where social contacts between individuals and groups, as well as their interaction, are created and maintained. Social capital in Bourdieu's (1984) theory represents the aggregate of the actual or potential resources that are linked to possession of a network providing mutual acquaintance and recognition (Bourdieu, 1986). The participants' narratives in this study express experiences of relations that have been established in the past, relations in the present, and possible future relations, the way these relations are created and maintained through the Internet, and what significance they are given.
2. *Cultural capital*: The categories that make up this theme are based upon the adolescents' experiences of the possibility they have to further increase their own development and knowledge. Bourdieu (1984) stated that cultural capital consists of competency, skills, and knowledge that people create in relation to their surroundings and especially in relation to the legitimate culture. As a consequence of cultural globalization the definition and role of the legitimate culture are somewhat diffused, although not absent. Activities that are already known to the adolescents are perceived as possible to expand upon within the increased availability of a new arena. There are experiences of personal development, improvement of their own capacity, and amplified knowledge.
3. *Realities in interaction*: This is a theme that comprises categories of feelings that the new arena – and one's own existence there – are infinite. The connectability offered generates possibilities but can sometimes result in an overwhelming feeling of not being in control of time and space. These processes are similar to the ones referred to by Giddens (2009) as a time-space distinction and the disembedding of social systems. The social relations from local contexts are altered and time and space are reorganized. The Internet provides a new social environment for adolescents with almost endless opportunities for contacts with peers and strangers of different ages and cultural backgrounds from around the world. The participants describe experiences of interaction between a virtual arena and everyday life as having different paces and to a certain extent as different realities. These two realities have a floating connection between feelings of anonymity and freedom and feelings of

rationality and practicality. The Internet has thereby contributed to radicalizing some processes in late modernity.

4. *Creation of lifestyles*: In this theme categories of the different ways one creates lifestyles are included. Among these are experiences of leisure time activities that enrich everyday life, making it more fulfilling, offline and online. The Internet provides a multitude of role models, which can be important agents in the process of consolidating input from the social environment that serves to shape an individual's beliefs, attitudes, and identity. For the adolescents in this study there are new opportunities to give vent to their own creativity, performances, culture, and esthetics. Also, there are expanded opportunities for admiration and inspiration from significant others that can bring meaning to their own lives.
5. *Symbolic capital*: According to Bourdieu (1984), symbolic capital comprises the acknowledgment of the other forms of capital (economic, social, and cultural) resulting in prestige and eminence. In the narratives this theme is based on categories that reflect the adolescents' need for confirmation and also how they can gain achievements. The Internet is in this case an environment where what several participants call their "real selves" may expand through acknowledgment by others. At the same time, it is a place that craves performances and a demand to maintain prestige.
6. *Milieu of uncertainty*: This is a theme where the participants express feelings and experiences of being uncertain of oneself as well as of other people and their intentions. This may also involve feelings of lost self-control. Reflections about the risk of being explored and abused both emotionally and in practice are experienced as unpleasant uncertainties.

Second Step: Category and Theme Frequencies in Boys' and Girls' Narratives

Table 1 presents the number of times each category was mentioned in the adolescents' narratives. This means that one category could have been mentioned by the same individual on several occasions. Hence the number of times a theme was mentioned is the sum of the number of times the categories within that theme were mentioned. Reports by boys and girls are shown both together and separately. The binominal test revealed that girls described the Internet as a tool for control and support to a greater extent than boys ($z = 2.01$, $p = .022$) Furthermore, boys reported using the

Internet for lifestyle development to a greater extent than girls ($z = 3.06$, $p = .001$).

The number of individuals mentioning a given category is also presented in Table 1. In this presentation a category cannot be considered more than once for each individual. The categories were incorporated in the same six themes as mentioned above. The number presented for each theme does not equal the sum of the number of categories mentioned within a given theme, but rather the number of times a given individual mentioned that particular theme. Reports by boys and girls are shown both together and separately. The results for how many individuals mentioned a given category are in line with the result above, indicating how many times each

Table 1. The Number of Times a Category Was Mentioned in the Material and by How Many Individuals (in Total and Separately for Women and Men, $F = 58$; $M = 61$).

Themes and Categories	Number of Times Categories Were Mentioned in the Material			Number of Individuals Mentioning a Given Category		
	Total	Women	Men	Total	Women	Men
Social capital	*152*	*79*	*73*	*82*	*42*	*40*
Searching in a social arena	63	33	30	42	21	21
Establishing social relationships	89	46	43	63	35	28
Cultural capital	*144*	*72*	*72*	*83*	*41*	*42*
Deepened development	73	30	43	57	24	33
Insight and values	41	21	20	23	13	10
Tools for control and support	30	21[a]	9[a]	22	16[a]	6[a]
Realities in interaction	*121*	*59*	*62*	*71*	*33*	*38*
Erased boundaries (vastness)	25	8	17	21	7	14
Reciprocal action Internet and IRL	53	31	22	38	23	15
Media-specific	43	20	23	33	15	18
Creation of lifestyles	*90*	*30*[b]	*60*[b]	*59*	*19*[a]	*40*[a]
Symbolic capital	*87*	*50*	*37*	*43*	*25*	*18*
Imaginary solution	58	34	24	33	18	15
Achievement	14	6	8	8	2	8
Confirmation	15	10	5	10	6	4
Milieu of uncertainty	*47*	*21*	*26*	*36*	*16*	*18*
Vulnerability and risk	18	10	8	13	8	5
Media takes over	29	11	18	20	6	14

[a]Difference significant at the .05 level.
[b]Difference significant at the .01 level.

category was mentioned in total. Girls hence reported using the Internet more for control and support than boys ($z = 1.92$, $p = .027$); whereas boys reported using the Internet more for lifestyle development than girls ($z = 1.84$, $p = .033$).

Previous research indicates that girls tend to use personal homepages in order to convey emotions and discuss intimate concerns and relationships with their friends to a greater extent than boys (e.g., Hollenbaugh, 2010; Mazur & Kozarian, 2009). In line with the study by Colley and Maltby (2008), the boys in our study tended to use the Internet for entertainment and leisure to a greater extent than the girls. Altogether these results indicate that gender-stereotypical behaviors that are perceived in offline contexts, for example that women tend to focus on relationships (e.g., Skeggs, 1997), are also found on the Internet. Girls and women tend to have access to more social settings for support, and to discuss emotions and feelings, than men and boys do, and it is difficult to alter one's identity between stereotypically feminine and masculine expressions (Whitley & Aegisdottir, 2000). Accordingly, this seems to be valid also for the Internet.

Third step: Interpretation of Central Themes

In general the narratives reflect the Internet as an open, self-defining arena that is predominantly perceived as positive and as offering almost endless possibilities to experiment with, try out and develop social competence. Three of the six themes mentioned previously in the quantitative analyses seem particularly relevant for an in-depth qualitative analysis.

The sample drawn from these three themes has been based on the question in focus in the present study – what role and what meaning do adolescents ascribe to the Internet as an arena for defining themselves and for shaping and developing their identity?

It turned out that the themes described with the terminology influenced by Bourdieu were particularly complex with regard to defining and creating identity as they partly relate to each other but not in a totally coherent way, and they do not only involve aspects that are media-specific. These three themes also comprised interesting gender dimensions. However, the expressions of social, cultural, and symbolic capital in the adolescents' narratives indicate that Internet arenas are not free from social differentiating, positioning, and hierarchies, which in turn reflects an ambivalence regarding possibilities on the one hand and limitations on the other.

The results will be illustrated through quotations from the interviews. To enlighten the reader about the person behind the quotation, information about the individual – gender, age, educational program, and Internet use – is presented. The educational programs are represented by a letter: Technology (T), Social Sciences (SS), Electronics and Telecommunications (ET), Science (S), Business (B), Art (A), Catering (C), and Child Care (CC). Internet use is represented by a digit: More than 3 hours a day (>3) and everyday, but less than 3 hours (<3).

Social Capital

The Internet is described in the narratives as a social arena radically extended over time and space, enabling new ways of communicating and socializing. Internet-related activities are thereby perceived as unique in terms of speed, availability of information, and opportunities to initiate and form social contacts of varying kinds, as has also been indicated by previous studies (e.g., Colleyey & Maltby, 2008; Pfeil et al., 2009; Tynes, 2007; Young, 2005). The extraordinary speed and availability provide individuals with possibilities to seek out people of interest, instead of only having social interactions with people in their immediate surroundings, which indirectly means that young people can communicate and share aspects of their identity with a larger crowd of like-minded people. The significant increase in pace and quantity thereby involves broadened opportunities with regards to many different types of alternative arenas and contacts, as exemplified in following statement from one of the narratives:

> I am only a few clicks from joining a deep philosophical discussion or a conversation with a computer expert that can help me with various problems.
>
> [...] Many of the Internet forums I have visited have been international and I have communicated with people from many different countries, which is something I would never have done if the Internet did not exist. [...] and the spontaneous co-operation I have had with strangers in online games, something that does not happen very often in real life. This has definitely improved my ability to co-operate with others. (male, aged 18 S, >3)

By facilitating access to a wide social network, the Internet provides a unique arena perceived as enriching personal development. The Internet is also described as providing an opportunity to maintain contact with friends and distant family members due to its convenience and disengagement from time and space. For example, a 17-year-old girl (B, >3) proposes that Facebook has allowed her to have better and closer contact with her father since her parents' divorce than she would otherwise have had.

The meaning of intimacy here is transformed into an emotional experience rather than a spatially anchored experience. Furthermore, many narratives contain stories about relationships, both with friends and romantic partners that only occur online. In some narratives these disembodied relationships are described as preferable to relationships "in real life." There are also examples of narratives where the adolescents describe romantic relationships that have been going on for years without them actually having met the other person in real life. Especially some of the girls portray these types of relationships as preferable to romantic relationships "in real life." The adolescents also describe strong friendships and exemplify the openness and closeness that the disembodiment – a characteristic of online meetings – allows:

> He has been somebody that I have been able to open up to and knowing that we will perhaps never meet gives me a sense of security. I think that a lot of people feel the way I do, that it is easier to open up to somebody that is "at a safe distance." (male, aged 16 ET, >3)

> You become more self-confident. You dare to do way more on the Internet. You do not have to look at people and you do not have to see their reactions. It is so much easier. (female, aged 17, SS, >3)

There are also examples of how both boys and girls describe romantic relationships that after a while take place both on the Internet and in other contexts, as a "result" of the initial Internet communication. It is also evident from the narratives that the Internet is perceived as an alternative arena allowing freedom, with reduced risks of rejection and humiliation. These findings are in line with those of Schmitt et al. (2008) in their description of the Internet as an arena that offers identity experimentation with a higher degree of freedom from social pressure and limiting expectations:

> I did not think at first that one could develop feelings for another person on the Internet, but I have realized that you can. On the Internet you get to know a person in a different way, it is not someone's physical characteristics you first see, but who they are. [...] everyone gets a chance to get to know one another. (female, aged 17 CC, >3)

The descriptions of the relationships initiated on the Internet, and how they are valued, are in many ways in line with what Giddens has conceptualized as pure relations; a social relation that is internally referential and fundamentally dependent only on the satisfaction or benefit which is gained by the relationship itself (Giddens, 1991). Pure relations strive toward individualism and freedom, in contrast to the collective, traditionally anchored and socially positioned patterns of relations. In other words, pure relations should be achieved as the result of voluntarily chosen relations only, based on quality and reciprocity, and free from social positioning. Further, radicalization of different patterns of relations in late modernity are reflected by the descriptions of the Internet as offering the opportunity to really get to know somebody and to be able to evaluate whether one could have an authentic exchange and benefit from engaging in a relationship with that person.

Some narratives also indicate that the personal self and opportunities for development of the personal self are perceived as less constricted by time and space, resulting in an extended sense of the self. Such possibilities are in turn perceived as allowing for deliberating, authentic, and constructive potential for "one to be anyone" or "to be who I really am." Despite expressions of these potentials as contemporary and natural phenomenon, there are also critical reflections around the authenticity of the extended selves appearing on the Internet:

> You only show one side of it. You never write about how boring life is sometimes, how angry you are or how annoying something is. You only write about the positive. [...] On the Internet and particularly on blogs you can expose only one side of yourself. This means you can build an identity that is not for real, a life that is not for real and I think that is wrong. That can make young girls look up to bloggers who are actually writing about a life they do not live. (female, aged 17, SS, <3)

Such critical evaluations stand in contrast to perceptions of the "Self" appearing on the Internet as more authentic. As the girl quoted above writes later in her narrative: "You need reality too."

Two opposite trends thus emerge with potential for radical freedom, individuation, and authenticity on the one hand, and risks of emptiness, senselessness, and falseness on the other.

Cultural Capital

In many of the narratives written by the adolescents, the Internet is described as an arena and facility for development and education. What differentiates this arena from others are qualities similar to the ones just described as enabling social relations. Experiences that have contributed to forming insights and fundamental values are also characterized by the disembodiment of time and space, which is perceived as a determinant for receiving information and knowledge one could not have received in any other way. Education is perceived as taking the form of both increased awareness of oneself and in societal matters, as a result of broader perspectives. However, many narratives indicate that this process of increased awareness is not so easy. The unlimited availability of information is in some narratives described as overwhelming and, in turn, demands the development of strategies to distinguish, select, and critically evaluate information. The strategies described vary as well with regard to the way information and knowledge are perceived, as in how they are seen as resulting in increased personal competence. In line with patterns of relations in other arenas, skills and competence are perceived as the means to mark status and differentiate the self from unwanted others (Bourdieu, 1984). In some narratives, skills and competence achieved via the Internet are also described as strong and gratifying assets for the individual in contexts outside the Internet:

> The Internet has meant a lot to me and I have learned a lot that I would otherwise not have learned if I had not spent so much time on the Internet. [...] have had so many nice moments with file sharing and file sharing has probably, in one way or the other, increased my interest in networks and other things related to computers, and I have done much better than some other people in school as a result of this. (male, aged 16, T, >3)

> The music online has influenced me a lot. It has made me interested in musical instruments, for example drums and guitar, later this helped me get into a band [...] it made my lifestyle into something different than it used to be. (male, aged 16, TE, <3)

In several narratives the Internet is also described as an arena where one can discover creative abilities and progressively develop competencies that strengthen one's self-esteem and social positions.

> I want to write about how the Internet has helped me find a part of my personality that I did not know that I had, or had not found yet: my creativity. [...] I have been blogging for four years now, on and off, but during the past year I have become very serious about it and I have created a (according to myself) "killer" design. [...]. Today my blog is my life. I write about anything I can think of and try to be as active in the blog world as possible in order to strive forward. (female, aged 17 SS, >3)

The phenomenon of blogging is one of the media-specific activities that appear in the narratives. The Internet is here described as a forum where creative abilities can flourish. By discovering expressive forms and genres, learning about them, and after a while, producing them by oneself, the Internet is seen an effective instrument for developing and expanding one's cultural capital.

> YouTube is a page I often visit and this is where my interest in animation and making video clips started. [...] Suddenly my vivid and exaggerated imagination has become an instrument, and the same goes for my interest in writing stories. (female, aged 16, T, >3)

As an arena for development and education, the Internet seems in some ways to exceed the often polarized gender patterns that occur in other contexts, since the Internet to some extent can abolish social determinations where gender identity is often central. However, the Internet does not exist in isolation from other arenas surrounding individuals. With regard to differences in social interaction patterns online Colley and Maltby (2008) have found that women more often focus on maintaining their existing friendships whereas men are more interested in finding people with similar interests and in forming new social relationships. Hence, from these tendencies we can deduce that women seem more oriented toward increasing their bonding and maintaining social capital; whereas, men seem more oriented toward bridging social capital by seeking and forming new social relationships. This means that the use of the Internet reflects, or at least, relates to the conditions and structures in society in general. Previous studies have confirmed that gender patterns observed in offline settings also exist in online settings. For example, women more often use homepages to convey emotions and discuss relationships; whereas men more often write about societal issues, meet friends, and experiment with their identities (Mazur & Kozarian, 2009). But, as will be described in relation to symbolic capital, some transformative developments can also be perceived on the Internet when it comes to attitudes and hierarchies relating to gender.

Symbolic Capital

The Internet, the communication and the activities taking place there, comprises an arena where special skills can be practiced, developed, and confirmed. Therefore, different types of assets and statuses are relevant for self-definition and the construction of identity in the form of cultural and symbolic capital that can be accumulated or lost. Cultural capital can involve more or less symbolic capital depending upon how it is classified and valued by others; hence, what function it has and whether it is worth something is ascribed to social networks (social capital).

Gender patterns are revealed which indicate in some respects different types of strategies used by girls and boys, but also in some respects strategies that may be seen as exceeding or transformative of the gender hierarchy. Boys' narratives tend to focus mainly on achievements where the Internet is portrayed as an arena for demonstrating skillfulness and in some cases an arena where fame can be achieved, for example as a successful player of different types of online games. However, there is also the possibility to gain fame using skills not only practiced on the Internet, but mainly communicated there:

> I discovered many possibilities to become famous via the Internet, I started producing music and spread it via different homepages, such as for example Myspace. As a result I have admirers and followers all over Sweden. […]. Without the Internet I do not know what my life would have been like or where I would have been today. (male, aged 16 ET, >3)

Similar types of confirmation and acknowledgment of prestige can be seen in the narratives written by girls. However, these are described somewhat differently, by the girls as enhancing, for example, positive comments on blogs or YouTube-movies. Characteristic for this arena though, in contrast to traditional everyday arenas, seems to be that young women's achievements can be ascribed status and fame among both boys and girls. Examples are given of how girls become central figures and role models as a result of skillfulness, a feature not so common in traditional arenas.

However, something in our material that is mentioned exclusively in the narratives written by girls is how specific sites and blogs can be used as a resource to help and support the handling of various problems. For example, some narratives describe how Internet contacts, and in some cases even community-like networks, have been formed around experiences of eating disorders and how these arenas have constituted help and support in the process toward recovery. This confirms previous research findings about how young women have been found to self-disclose and discuss private and intimate concerns online (Bortree, 2005; Hollenbaugh, 2010). But in addition to that, these different kinds of more or less specialized networks also provide possibilities for recognition of a person's inside knowledge, which is converted into social and symbolic capital.

The symbolic capital also partly comprises the way the social and cultural forms of capital together constitute the means of achieving prestige and fame, resulting in securing or risking an individual's status in relation to other people. Symbolic capital is therefore strongly dependent on power structures and social hierarchies. However, the examples mentioned above also indicate possibilities for alternative types of confirmation in Internet-related arenas that may indicate a slight change in gender identities and their possible statuses.

DISCUSSION AND CONCLUSION

Adolescents in modern societies are living in a rapidly changing world that is associated with new ways of living, such as new ways of interacting socially, new ways of disseminating and manipulating information, which in turn can be associated with changes in cultural, economic, and political activity. New technology, such as the development of the Internet, partly alters how people interact and communicate and can influence how individuals self-present themselves and how individuals experiment with and form their identities. Adolescents in contemporary Western societies constitute the first generation growing up in a computer culture with the Internet as an integrated everyday arena.

The Internet adds new ways of increasing social capital by enhancing the building and maintaining of relationships and it makes social interaction fast and efficient (Pfeil et al., 2009; Young, 2005). For the participants in this study Internet literacy seems to be perceived as a multidimensional competence, increasing cultural capital by, for example, developing abilities to access, analyze, evaluate, and create online content. The Internet is thereby also defined as an arena for enhancing skills, and achieving status – symbolic capital.

For self-definition and relation building on the Internet there is, however, an important tension that appears in the narratives. As asserted by Bargh et al. (2002) and Larraín et al. (2007) and also confirmed by the current study, young people can use strategies to project an online identity that is more socially desirable than their offline identity. They can experiment with different roles, fantasies, ideas, and actions, which can represent their identities (Rosenmann & Safir, 2006; Suler, 2002; Zhao et al., 2008). In the narratives in our material, these tendencies are most often valued as positive and as offering individual freedom and creativity. There is however a crucial ambivalence here when it comes to the meaning of authenticity. On the one hand, the experienced freedom in the Internet arena is granted a kind of utopian quality, where individuals can come out as their "real selves" and establish pure relations with other individuals' "real selves." The Internet media may in this respect appear to offer the independence and freedom from social determinations that the desired pure relations require. Individuals can choose who they want to be or at least which aspects of the self that will be put forward in interactions with others. The availability, the range, the speed, and the spatial and physical independence in the social interactions taking place on the Internet may also radicalize the opportunities for pure relations. On the other hand though, doubtful remarks are seen in some of the narratives, which challenge this authenticity by contrasting it to "reality" – which in these cases refers to the world outside of the Internet. This kind of skepticism instead defines the online context as a smokescreen for the real self, and the conditions for establishing pure relations are then not seen as the most promising, but rather in need of a reality check.

Regarding the gender perspective, according to a study by Colley and Maltby (2008) on male and female personal perspectives with reference to the impact of the Internet, there is a gender difference when it comes to social capital, where women tend to be more oriented toward increasing their bonding and maintaining social capital; whereas men tend to be more oriented toward bridging social capital by seeking and forming new social relationships. Similar conclusions can be drawn from the present study, as has been shown both by the statistical differences in category frequencies in girls' and boys' narratives and in the interpretation of central themes in the narratives. But there are also several examples that show more complex and contradictory tendencies concerning gender. Primarily the boys in the study describe how they use the Internet as an arena for improving capabilities, which in turn can improve status and reputability. These tendencies can however also be seen in some of the girls' narratives, and there are also examples of boys who ascribe this kind of status, reputability, and role model-function to girls and their creative skills. Though a tendency only visible in narratives written by girls, in the present study, is the description of the Internet as an arena also used for help and support in relation to the experience of challenges and difficulties such as eating disorders. In relation to this finding it is however important also to reflect upon the methodological implications of using narratives. Gender norms will influence not only behaviors but also what is liable to be narrated depending on the gender of the participant. To express vulnerability and exposedness is for example not in line with norms of masculinity and is thereby probably more difficult to state in narratives written by boys.

In this explorative study the purpose was to examine how adolescents perceive the Internet as a phenomenon and what role and meaning they ascribe to the Internet as an arena for defining themselves and for shaping their identity. The results have contributed to an increased and complex understanding regarding how adolescents perceive their offline worlds as incorporated into their online worlds and vice versa.

In relation to the question about the Internet as a self-defining arena, the main results indicate that the adolescents participating in the present study primarily used the Internet in order to communicate and interact in various social arenas. In these arenas, relations were created, maintained, and sometimes deepened. Furthermore, the adolescents described how the social interaction and communication on the Internet offers a potential for interacting with a wide social network and becoming a part of various social groups with various social identities linked to them. The Internet provides the opportunity to self-present in a relatively controlled way by carefully choosing what aspects of the self to reveal and enhance and what aspects to exclude. Adolescents can thereby use the online arenas to increase their self-esteem by presenting themselves in a way that is expected to be perceived positively by others. Furthermore, young men and women can more specifically choose to interact socially in certain contexts where they can more easily self-verify and where their ideal self is presumed to be accepted; hence, the Internet is a most crucial arena for self-definition and self-presentation. As expressed by one of the young men:

> You could say that it is thanks to the Internet that I am the person I am today. (male aged 16, ET, <3)

REFERENCES

Ackland, R., & O'Neil, M. (2011). Online collective identity: The case of the environmental movement. *Social Networks*, *33*(3), 177–190.

Alexandersson, M., & Hansson, T. (Eds.) (2011). Unga nätmiljöer: Nya villkor för samarbete och lärande *[Young network environments: new conditions for learning and collaboration]* (1st ed.). Lund: Studentlitteratur.

Bane, C. M. H., Cornish, M., Erspamer, N., & Kampman, L. (2010). Self-disclosure through weblogs and perceptions of online and 'real-life' friendships among female bloggers. *CyberPsychology, Behavior and Social Networking*, *13*, 131–139.

Bargh, J. A., & McKenna, K. Y. A. (2004). The internet and social life. *Annual Review of Psychology*, *55*, 573–590.

Bargh, J. A., McKenna, K. Y. A., & Fitzsimons, G. M. (2002). Can you see the real me? Activation and expression of the "true self" on the internet. *Journal of Social Issues*, *58*, 33–48.

Bolinder, G., Chenik, M., & Juth, N. (2011). Sociala medier förändrar relationen mellan patient och vårdgivare [Social media is changing the relationship between patient and caregiver]. *Läkartidningen*, *13*, 727–729.

Bortree, D. S. (2005). Presentation of self on the web: An ethnographic study of teenage girls' weblogs. *Education, Communication and Information*, *5*, 25–39.

Bourdieu, P. (1984). *Distinction: A social critique of the judgment of taste*. Cambridge, MA: Harvard University Press.

Bourdieu, P. (1986). The forms of capital. In J. Richardson (Ed.), *Handbook of theory and research for the sociology of education*. New York, NY: Greenwood.

Bourdieu, P. (1998). *Practical reason. On the theory of action*. Stanford, CA: Stanford University Press.

Christofides, E., Muise, A., & Desmarais, S. (2012). Hey mom, what's on your Facebook? Comparing Facebook disclosure and privacy in adolescents and adults. *Social Psychological and Personality Science*, *3*, 48–54.

Cinnirella, M., & Green, B. (2007). Does "cyber conformity" vary cross culturally? Exploring the effect of culture and communication medium on social conformity. *Computers in Human Behavior*, *23*, 2011–2025.

Colley, A., & Maltby, J. (2008). Impact of the Internet on our lives: Male and female personal perspectives. *Computers in Human Behavior*, *24*, 2005–2013.

Connell, R. (2009). *Gender in a world perspective*. Cambridge: Polity Press.

Davis, K. (2010). Coming of age online: The developmental underpinnings of girls' blogs. *Journal of Adolescent Research*, *25*, 145–171.

Findahl, O. (2014). *Svenskarna och internet 2014* [The Swedes and the Internet]. Advance online publication. Retrieved from www.iis.se/docs/SOI2014.pdf

Fogel, J., & Nehmad, E. (2009). Internet social network communities: Risk taking, trust and privacy concerns. *Computers in Human Behavior*, *25*, 153–160.

Gibbs, J. L., Ellison, N. B., & Heino, R. D. (2006). Self-presentation in online personals: The role of anticipated future interaction, self-disclosure, and perceived success in Internet dating. *Communication Research*, *33*, 152–177.

Giddens, A. (1991). *Modernity and self-identity. Self and society in the late modern age*. Cambridge: Polity Press.

Giddens, A. (2009). *Sociology*. Cambridge: Polity.

Grogan, S. (2007). *Body image: Understanding body dissatisfaction in men, women and children*. London: Routledge.

Gross, E. F. (2004). Adolescent internet use: What we expect, what teens report. *Journal of Applied Developmental Psychology*, *25*, 635–649.

Herring, S. C., & Paolillo, J. C. (2006). Gender and genre variation in weblogs. *Journal of Sociolinguistics*, *10*, 439–459.

Hollenbaugh, E. E. (2010). Personal journal bloggers: Profiles of disclosiveness. *Computers in Human Behavior*, *26*, 1657–1666.

Hollenbaugh, E. E., & Ferris, A. L. (2014). Facebook self-disclosure: Examining the role of traits, social cohesion, and motives. *Computers in Human Behavior*, *30*, 50–58.

Landstedt, E., & Persson, S. (2014). Bullying, cyberbullying, and mental health in young people. *Scandinavian Journal of Public Health*, *42*, 393–399.

Lanthier, R. P., & Windham, R. C. (2004). Internet use and college adjustment: The moderating effect of gender. *Computers in Human Behavior*, *20*, 591–606.

Larraín, M. E., Zegers, B., & Alejandro, T. (2007). Do adolescents compromise their identity when chatting on the Internet? *International Journal of Applied Psychoanalytic Studies*, *4*, 91–109.

Livingstone, S. (2008). Taking risky opportunities in youthful content creation: Teenagers' use of social networking sites for intimacy, privacy and self-expression. *New Media & Society*, *10*, 393–411.

Mazur, E., & Kozarian, L. (2009). Self-presentation and interaction in blogs of adolescents and young emerging adults. *Journal of Adolescent Research*, *25*, 124–144.

Meyen, M., Pfaff-Rudiger, S., Dudenhöffer, K., & Huss, J. (2010). The internet in everyday life: A typology of Internet users. *Media, Culture & Society*, *32*(5), 873–882.

Pempek, T. A., Yermolayeva, Y. A., & Calvert, S. L. (2009). College students' social networking experiences on Facebook. *Journal of Applied Developmental Psychology*, *30*, 227–238.

Peter, J., & Valkenburg, P. M. (2006). Research note: Individual differences in perceptions of internet communication. *European Journal of Internet Communication*, *21*, 213–226.

Pfeil, U., Arjan, R., & Zaphiris, P. (2009). Age differences in online social networking. A study of user profiles and the social capital divide among teenagers and older users in MySpace. *Computers in Human Behavior*, *25*, 643–654.

Reich, S. M., Subrahmanyam, K., & Espinoza, G. (2012). Friending, IMing, and hanging out face-to-face: Overlap in adolescents' online and offline social networks. *Developmental Psychology*, *48*(2), 356–368.

Reysen, S., Lloyd, J. D., Katzarska-Miller, I., Lemker, B. M., & Foss, R. (2010). Intragroup status and social presence in online fan groups. *Computers in Human Behavior*, *26*, 1314–1317.

Rosenmann, A., & Safir, M. P. (2006). Forced online: Push factors of Internet sexuality: A preliminary study of online paraphilic empowerment. *Journal of Homosexuality*, *51*, 71–92.

Schmitt, K. L., Dayanim, S., & Matthias, S. (2008). Personal homepage construction as an expression of social development. *Developmental Psychology*, *44*, 496–506.

Shapiro, L. A. S., & Margolin, G. (2013). Growing up wired: Social networking sites and adolescent psychosocial development. *Clinical Child and Family Psychology Review*, *17*(1), 1–18. Advance online publication. doi:10.1007/s10567-013-0135-1

Skeggs, B. (1997). *Formations of class and gender. Becoming respectable*. London: Sage.

Subrahmanyam, K., Smahel, D., & Greenfield, P. (2006). Connecting developmental constructions to the internet: Identity presentation and sexual exploration in online teen chat rooms. *Developmental Psychology*, *42*, 395–406.

Suler, J. R. (2002). Identity management in cyberspace. *Journal of Applied Psychoanalytic Studies*, *4*, 455–459.

Temple, J. R., Donna Le, V., van den Berg, P., Ling, Y., Paul, A. J., & Temple, B. W. (2014). Brief report: Teen sexting and psychosocial health. *Journal of Adolescence*, *37*(1), 33–36.

Tidwell, L. C., & Walther, J. B. (2002). Computer-mediated communication effects on disclosure, impressions and interpersonal evaluations. *Human Communication Research*, *28*, 317–348.

Tynes, B. M. (2007). Role taking in online "classrooms": What adolescents are learning about race and ethnicity? *Developmental Psychology*, *43*, 1312–1320.

Valentine, G., & Holloway, S. L. (2002). Cyberkids? Exploring children's identities and social networks in on-line and off-line worlds. *Annals of the Association of American Geographers*, *92*, 302–319.

Valkenburg, P. M., & Peter, J. (2007). Preadolescents' and adolescents' online communication and their closeness to friends. *Developmental Psychology*, *43*, 267–277.

Valkenburg, P. M., & Peter, J. (2009). Social consequences of the Internet for adolescents: A decade of research. *Current Directions in Psychological Science*, *18*, 1–5.

Valkenburg, P. M., Schouten, A. P., & Peter, J. (2005). Adolescents' identity experiments on the Internet. *New Media & Society*, *7*, 383–402.

Vallor, S. (2011). Flourishing on Facebook: Virtue friendship & new social media [Original paper]. *Ethics and Information Technology*, *14*(3), 185–199. doi: 10.1007/s10676-010-9262-2

Walrave, M., Vanwesenbeeck, I., & Heirman, W. (2012). Connecting and protecting? Comparing predictors of self-disclosure and privacy settings use between adolescents and adults. *Cyberpsychology: Journal of Psychosocial Research on Cyberspace*, *6*(1). doi:10.5817/CP2012-1-3

Watt Boolsen, M. (2007). Kvalitativa analyser *[Qualitative analyzes]*. Göteborg: Gleerups förlag.

Whitley, M., & Aegisdottir, A. (2000). The gender belief system. Authoritarianism, social dominance orientation, and heterosexuals' attitudes. *Sex Roles*, *42*(11–12), 947–967.

Willoughby, T. (2008). A short-term longitudinal study of Internet and computer game use by adolescent boys and girls: Prevalence, frequency of use, and psychosocial predictors. *Developmental Psychology*, *44*, 195–204.

Wolak, J., Mitchell, K. J., & Finkelor, D. (2002). Close online relationships in a national sample of adolescents: A description. *Adolescence*, *37*, 441–456.

Ybarra, M. L., Alexander, C., & Mitchell, K. J. (2005). Depressive symptomatology, youth Internet use, and online interactions: A national survey. *Journal of Adolescent Health*, *36*, 9–18.

Young, B. (2005). A study on the effect of Internet use and social capital on the academic performance. *Development and Society*, *35*, 107–123.

Zhao, S., Grasmuch, S., & Martin, J. (2008). Identity construction on Facebook. Digital empowerment in anchored relationships. *Computers in Human Behavior*, *24*, 1816–1836.

APPENDIX: VIGNETTES

Something That Has Influenced Who I am

Sometimes you can realize that you have had experiences, or been in situations or with people that have contributed to shaping who you are as a person. When I was your age there were plenty of experiences and situations that meant a great deal to me and that I think have influenced who I now am.

I particularly recall experiences involving music. I have always been interested in music and listened a lot to music both at school and in my spare time. I have been to concerts by my favorite band and I have spent time with friends who have had the same interest. I have also been in different bands as a singer. During my teenage years music was a way of life and I spent lots of time listening to music with several of my friends. I believe this has shaped me into the person I am today and that it has influenced how I see myself.

One strong memory that has influenced me is when I saw the TV series *Roots*. The series depicted how people from the African continent were shipped as slaves to the United States and followed subsequent generations all the way up to the 1970s. Being able to identify with characters in history that so grippingly described an inhumane and deeply unjust system had a strong and long-lasting impact on me and has contributed to how my view of the world was formed. I thought a great deal about issues concerning power, oppression, and racism, and how this affects the relationships of people and groups.

I would now like you to describe a similar experience that you have had and that you believe has or will shape you and the way that you see yourself. When I was young the Internet did not exist – but today the Internet is important for many people. *Choose, for example, a situation, an experience, a personal contact, an activity or such that you have experienced on the Internet. It is completely up to you to choose the experience and location.* It could be something that you do both on the Internet and outside the Internet, either now or at some time in the past. The main thing is that it is something that you believe has shaped you as a person. *Tell us what happened, what you did, what you thought and what you felt. Write down your story and if you need more paper just let us know.*

Thank you!

YOUNG COLLEGIANS: BETWEEN THE PHYSICAL AND NUMERICAL TERRITORIES, DIFFERENT PROCESSES OF INDIVIDUATION

Elmir de Almeida, Marilena Nakano,
Maria Elena Villar e Villar and Vanderlei Mariano

ABSTRACT

Purpose – *The paper presents final results of a comparative research on young Brazilian collegians in the 18–24-year-old age bracket. The objective was to understand the interactions and ways in which they transit within the physical and digital Web spaces and, within their transits, set up "circulatory territories," deepen and enrich their secondary socialization setups and sociability, as well as their processes of individuation within the historic condition in which youth lives.*

Methodology/approach – *The study is supported by conceptual contributions offered by the sociology of youth, circulatory territory, socialization, sociability, and individuation. Research was carried out with students of two different universities: a public/state one and a public/municipal/foundational one situated in different urban centers of the*

Technology and Youth: Growing Up in a Digital World
Sociological Studies of Children and Youth, Volume 19, 133–162

ISSN: 1537-4661/doi:10.1108/S1537-466120150000019005

southeastern region of the country. Procedures were qualitative and quantitative – closed questions forms, interviews, registries in field note-books, etc.

Findings – *The results of the investigation demonstrate: (a) that the representatives of the two student collectives studied circulate in physical and digital territories, setting up circulatory territories; (b) there are also different youth lifestyles, either due to social positions and the fact that they possess social capital, or because of differences and inequalities referring to gender, race/ethnicity (whites and non-whites), living situations.*

Originality/value – *In this manner, the study indicates the importance of questioning the homogenized image of connected youth since some collegians' lives are limited due to their condition as young workers, while others live their youth condition as a social moratorium, being able to produce other manners of being and of living in the world.*

Keywords: Circulative processes; individuation; to experience and young college students

INTRODUCTION

This study was carried out in Brazil with young collegians from two types of public universities that are especially characteristic of this country. It approaches youth in their relation to the Internet in such a manner as to undo certain homogenized views that work with images of a global subject, for example that of wired youth, without polemicizing the differences and inequalities among the groups that circulate in the virtual world. It is important to consider and collect the existing diversity and inequality among youth groups that circulate within this territory as a result of their condition, the interest groups to which they approach, their life stories, and the different contexts that depend on the country in which they live.

The specificities that are part of the Brazilian case could possibly permit problematizing certain categories that have been a part of studies about youth in different countries. For example, as indicated by Nakano (2013), the notion of precariousness, produced in France and widely used in Brazilian studies to deal with the relationship between youth and the social structure and their relation with the State, loses its analytical potential if not relativized in the case of this country since the welfare state did not establish itself as it did in France (Castel, 1995; Paugam, 2003).

The same can be said of the notion of "emergent adulthood" according to Arnett's (2004) proposal with reference to the insertion of youth in the working world, in a world that has become more complex, introducing new elements in their identity-building processes, as well as adding new challenges for youth to reach adulthood. Also, in this case, in order for this notion to explicit its analytical potency, it is necessary that in each country this be relativized in accordance with the existing reality, as suggested by Brandão, Saraiva, and Matos (2012) for the Portuguese and Brazilian cases.

In the Brazilian case, as in other countries of South America, the expansion of basic schooling and higher learning are very recent educational and social changes, which began in the final decades of the 20th century, and, among us, it is very common to find a large number of young collegians who are associated, concomitantly, with student and worker statutes, concretely establishing the idea that in Brazil, besides schooling, "work also forms youth" (Sposito, 2005).

If young Brazilians are giving evident signs that they are immersed in globalization movements and advanced modernity, promoting significant changes in their identity-building resources and individuation (Martuccelli, 2012), on the other hand, Brazilian society presents socioeconomic and cultural peculiarities that suggest care in using the theses of prolonging studies, of tardy and difficult insertion in the labor market, and of the unchallenged existence of a phase in the life cycle understood as "emergent adulthood" (Arnett, 2004; Brandão et al., 2012).

CONTEXTUALIZING THIS STUDY AND THEORETICAL INPUTS

In the case of the young Brazilian collegians, they participate in a system of higher learning that is quite singular when compared to the systems implemented in different countries in the world. More than 70% of Brazilian collegians study in higher schooling institutes offered by the private sector – entrepreneurial or denominational, especially youth from the lower strata of the population, student-workers, in this way demonstrating the importance of not approaching the universe of young collegians as a homogenous group, given that internally there is an enormous diversification, inasmuch as it includes youth who access and frequent public universities, enjoying a free higher schooling and belonging to the middle class, with higher incomes and higher cultural capital (Bourdieu, 2007), and youth that accesses paid higher schooling, in private institutions or foundations set up by municipal governments.

The effort undertaken in this research, the results of which are here reported, is part of an academic-scientific undertaking that is being carried out by Brazilian researchers since the 1990s. Differing from what happened in Human and Social Sciences in North America and in Western Europe, as well as in Spanish America, it was only toward the end of the 1980s that academic researchers of these areas recommitted themselves to thematize more regularly the phase of life identified as youth, the condition of being young, and the individual youths as testified in the assessment on post graduate studies on youth within the fields of Education and the Social Sciences, coordinated by the Brazilian researcher Sposito (2002, 2009).

A movement similar to that carried out by the academic world in the field of studies on youth was also verified within the settings of state institutions dedicated to study and research (Brasil. Instituto de Pesquisa Econômica Aplicada [IPEA], 2009), of non-governmental organizations (*Ibase/Polis*, 2005; *Instituto Cidadania*, 2003), as well as of multilateral organisms associated with Brazilian institutions (*UNESCO*, 2006, 2007).

The results of this state-of-the art type research, coordinated by M. P. Sposito, referring to Brazilian studies carried out in the field of studies on youth, in those of Education and the Social Sciences, within the 1999–2000 period, demonstrate that one of the themes that deserved considerable attention from the academic researchers was that of contemporary relations of Brazilian youth with the higher schooling universe (Carrano, 2009; Sposito, 2009).

Carrano (2009), on examining the production that has been dedicated to the theme of "young collegians," assessed that the studies carried out in that period portrayed a "very broad and diversified ambit of works that, in their heterogeneity of research problems, theoretical-methodological guidelines, as well as their levels of deeper study [permitted] a significant apprehension of the situation of collegians in Brazil (Carrano, 2009, p. 182). This is because the researches problematized, with greater or lesser emphasis, issues such as access to – and conditions for – permanence of youth at this level of schooling, the schooling trajectories and permanence of the collegians from the popular classes, the choice of careers, the formation and professional insertion, and also, the opinions, interests, and experiences of the university students. However, Carrano (2009) also observed that the academic work evaluated by him presented gaps and proposed new challenges to investigation about young Brazilian collegians.

For Carrano (2009), a challenge posed to the Brazilian researchers that worked on scrutinizing the relations that youngsters establish with the universe of higher schooling would be that of "making an inventory of the multiplicity of the variables that configure the condition of being a student," as well as "how do the experiences of experimentation of university life happen after their ingress" to this level of schooling. About this last aspect, such a perspective of investigation would apply to youth situated in diverse positions of the Brazilian social structure, whether their origins are urban or rural, "of the popular classes, of the middle classes and those from the economic elites," since one of the keynotes of the studies examined by Carrano (2009) was that which privileged "the analysis of the life of the young student from an institutional perspective and from the unilateral condition of student, disconsidering other existential and biographical variables of the young students" (Carrano, 2009, p. 181).

Within Brazilian reality, the warnings posed by Carrano (2009) should not be disdained because data from the Higher Schooling Census in Brazil (Brasil. Inep, 2011, 2013) demonstrates that only very recently have enrollments in higher schooling presented considerable increase, presenting a socioeconomic and ethno-racial difference in the total number of students that have succeeded in accessing and attending higher schooling in the country. According to the Ministry of Education, and Inep – the Brasil. Inep (2013), from approximately 1.3 million enrollments in 1980 to 7,526,681 in 2013. Of these, 5,373,450 are in the private sector (76% of the total).

Studies such as those of Pinto (2004), Catani and Hey (2007), Gomes and Moraes (2009), Kauling et al. (2010), Schwartzman (2012), and Neves (2012), among others, dedicated to the analyses of the meanings contained in the process of expansion of opportunities to access higher schooling in the country, in the transition from the 20th to the 21st century, demonstrated that the phenomenon of the recent expansion of vacancy availability at this level of education occurred through the participation of the public sector – at federal, state, and municipal levels, as well as by the private sector. All of them, however, point out the strong presence and participation of private initiative in this process, ever since the 1960s.

Besides the situations mentioned, those authors also highlighted the persistence of regional imbalances in the availability of vacancies and the differences in the quality of academic and scientific work of the institutes of higher schooling in the country. They also highlighted some of the mechanisms put into effect by the Brazilian State in order to minimize long-standing socioeconomic and ethno-racial inequalities that characterize our higher schooling, on implementing policies such as (FIES), *Student Financing Fund (Fundo de Financiamento Estudantil* – FIES*)*, the *University for All Program* (ProUni), the social programs of affirmative action for youngsters from low income families, non-white youngsters, the "*permanence grant*," among others. (de Almeida, 2014). With these actions, the different governments of the Brazilian State seek to promote some kind of differentiation within the universe of the youngsters that have accessed the institutions of higher schooling in the country.

Neves (2012), in her appraisal of the expansion of enrollments in Brazilian higher schooling in the 1998–2010 period, registered that the access to this level of schooling still figures as a usufruct right predominantly enjoyed by white youth that belong to middle and higher classes. However, on analyzing data referent to family income of the new generation of students, the author also admitted the occurrence of "some changes in the social composition" within the universe of collegians and identified a small alteration in the percentage of non-white youth that has been accessing the higher schooling courses (Neves, 2012, pp. 10–11). In this manner, even though this level of schooling still maintains the marks of its atavistic elitism, it is inevitable to admit that it is going through a rejuvenating process in its population, and has been accessed by youth situated in positions that are close to the base of the social pyramid, and its ethno-racial composition already shows signs of internal differentiation.

The above-mentioned references lead us to refer to Carrano's (2009) advertence when he suggests that the researchers who want to study young Brazilian collegians should take into consideration "the new makeup of the university public, their diversity," which is a result of the expansion of enrollments as well as of the institution of certain policies, in association with "the permanence of the historic conditions of inequality that are associated to the inequalities that result from contemporary contexts." For Carrano (2009, pp. 180–181), we must observe and carefully examine the "existential and biographical variables" of the new students, since these could reveal manners that are singular to certain youth segments in experiencing their youth.

The systematic study of the variables highlighted by Carrano (2009) becomes more pressing insofar as it is ever more evident that it is the new generations and the young collegians that access and manage the new information and communication technologies – the NICTs, as demonstrated by the study of Gatti and Sá Barreto (2009) and the results of the inquiries promoted by the *Comitê Gestor da Internet no Brasil* – CGIB (2010, 2011, 2012). The results of these studies reveal that the individuals between 15 and 24 as well as those between 25 and 34 years of age, with higher and secondary schooling levels, from different social strata, situated in distinct regions within the country are those who most access and use the computer, the internet, and mobile phones in the country. In this manner, it is not possible to disagree that, as in regular schooling, the new technologies of information and communication transverse the quotidian and the condition of being young of these players, make their experiences unique and differentiate them, historically and culturally, from those who integrated previous generations of Brazilian collegians.

OBJECTIVES OF OUR STUDIES AND CONCEPTUAL CONTRIBUTIONS

In this text our proposal is to face the new epistemological and methodological challenges in the field of study about youth, by means of developing *a comparative study* on the interactions and the manner in which young collegians in urban centers transit through the physical and digital territories – the Web – in order to understand the circulatory practices *in* and *among* the physical and virtual territories and how these practices currently inform and conform their processes of individuation.

In our study we operate with the classic concept of socialization established by social theory that defines it as being the process "by which a society supplies itself with members and, at the same time, the process whereby an individual becomes a member of a society" (Martuccelli, 2012, p. 69). We also consider, with Berger and Luckmann (2001), that the process of socialization that constructs the individual and assures the objectivity of a corporate order should be understood considering the spaces and times in which this is materialized as its starting point. In this manner, in the life of individuals we can distinguish two types of socializing arrangements: those that come from the circles of primary social interactions – the family – "primary socialization," and those that expand this experience of the individuals and that derive from new social circles – formal schooling, the working world, the spaces and times of the neighborhood where the individuals live, getting married, processes of mobilization or socio-spatial circulation, etc. (Berger & Luckmann, 2001, pp. 164–184).

By a process of individuation we understand a dynamic process, undoubtedly implicated with processes of socialization, which reveals itself in a certain historic moment, when a society structurally fabricates a type of individual. According to Martuccelli (2007, 2012), the process of individuation combines "a diachronic axis with a synchronic one, so as to interpret the horizon of a life – or of a generation, and the consequences of the great historic transformations" (Martuccelli, 2007, p.15).

In order to understand the practices that the collegians undertake within the physical and digital territories, our principal reference source is the contribution of Tarrius (2000, 2002), since in his studies about "globalization from below" he alerts us to the fact that the processes of circulation or mobility include spatial elements – not only the points of departure and of arrival, but also the multiple temporalities since the subjects carry with themselves the history of their place of origin, a history that interferes in the socializing processes imposed and experienced within the places of arrival.

For Tarrius (2000), it is necessary to consider that space and territory do not have an unequivocal identity considering that, if all space is circulatory, not every space can be considered configured as territory, seen that the idea of territory includes the socialization of the spaces based on differential mobility or circulation logics. Besides this, for the French anthropologist, the notion of circulatory territory demands a double rupture in the notions of territory and circulation insofar as it suggests that the social order that emerges from sedentary lifestyles is not a *sine qua non* for the emergence of territories, therefore demanding a rupture with logistical concepts of circulation and of flows in order to give social meaning to spatial mobility. In this manner, a circulatory territory leads us to refer to the spaces that serve as support to displacements within it; the individuals invest and recognize themselves within a shared history of migration, generator of an original social bond. (Tarrius, 2000, pp. 55–56, 2002, n/p., free translation).

Having these as conceptual references, for over two years we have dialogued with young collegians, situated in different urban centers, from diverse courses and university careers, and connected to two higher schooling institutions: a public state one and the other, a public municipal/foundational one. The public state one is within the city of Ribeirão Preto while the public municipal/foundational one is in the city of Santo André, both in the State of São Paulo.

Both the city of Santo André and that of Ribeirão Preto represent the processes of change that have marked capitalism and the Brazilian socio-spatial urban formation in the transition from the 20th to the 21st century. The first one is part of the metropolitan region of Greater São Paulo and, as part of its history, has housed the automotive industry since 1950. However, as of the 1980s, it has been experiencing the vicissitudes of the deindustrialization process, the advance of automation in the productive processes and in labor management, as well as the expansion of outsourcing. All these processes have provoked tangible changes with regard to work, employment, and unemployment and that affect more acutely the young segments of the population living there. In this city our interaction occurs with students from the *Centro Universitário Fundação Santo André – CUFSA*, a public municipal foundation under private law, which permits it to charge monthly fees for the courses it offers.

Ribeirão Preto, on the other hand, is an urban non-metropolitan locality, situated in the northeast of the State of São Paulo and represents another facet of the growth and modernization of Brazilian capitalism. Since the 1980s, Ribeirão Preto is a part of what some economists have called the region of a *Brazilian or Paulista California*, considering the processes of modernization and of endogenous and exogenous "monetizing" of its economic sector, the expansion of the transformation industry, agro industry, and agribusiness as well as the services sector (Jannuzzi, 1995). In that city our interaction has occurred with the *Universidade de São Paulo – campus Ribeirão Preto – USP-RP* a public autarchy set up by the government of the State of São Paulo.

Within the demographic contingents of the two cities and in the cities of the respective surroundings there is a strong representation of the heterogeneity that is present in the Brazilian middle class, as there are also representatives of the new segments of workers that, according to the denomination given by Souza (2010) are the "Brazilian battlers." These are so called because, ever since the 1990s, even though they repositioned themselves within the sphere of labor and formal employment and ascended economically, they still face constraints in the access and usufruct of the products of culture, as well as of the indistinctly discerned rights of social welfare public policies currently in force in Brazil in the areas of education, health, and social security, among others (Pochmann, 2012; Souza, 2010; Souza & Lamounier, 2010).

In this manner, the present text addresses themes that originate from a more ample research, focalizing the transits within the physical and virtual territories, that young collegians connected to these two universities carry out, marked by the phase of life in which they are and by the place that they occupy in the socioeconomic structure of Brazilian society, seeking to extract elements in order to understand their socialization and individuation proccsses.

ADOPTED METHODOLOGICAL PROCEDURES

Our study was developed in two stages. In the first one, we developed a quantitative research with a sample group, having the standard error rate set at 0.05% and the level of reliability rate at 95%. During this phase of the investigation our objective was to produce a profile of the students presently at the university institutions already mentioned, so as to understand the socioeconomic and socio-demographic characteristics of these students within the social and ethnic rejuvenation of the youth from urban centers who reached higher schooling levels this decade.

During the second stage, we endeavored to understand the values espoused by the collegians, their participation in movements and social groups, their principal groups for secondary socialization, as well as their sociability within and among the physical and virtual territories. For this undertaking we used instrumental qualitative research tools – registries in field diaries and interviews.

The students who dialogued with us were selected using the following criteria: (a) acceptance of the invitation to contribute with the study, based on a position manifested during the implementation of the first stage of the research; (b) be in the 18–24-year-old age bracket; (c) be enrolled and frequenting courses in different careers of undergraduate study.

The set of quantitative data collected during the initial phase of the research permitted us to perceive some logic in the circulation of students in the physical territories having as a differential element between the two universities studied the fact of whether the student works or not or if they incorporated "temporary migratory" practices in their experiences as youths. It was evident that the young student-workers lived as "imprisoned" in the home-work-school triangle. On the other hand, those who only studied revealed multiple possibilities of circulation in the physical milieu.

Eighty students participated in the qualitative research, 40 from each university. It was demanded from each one of them that they produce a diary during one month and that, in it, they register the following elements: (a) date, time, and place of entry on the Internet; (b) equipment used; (c) reasons for entering the virtual world; (d) report on the ways they coursed within the virtual world, themes considered, interlocutors, interests, and objectives; (e) description of the problems or facilities to enter the virtual world and circulate in it.

Based on the quantitative data that we collected, we presented a socioeconomic, socio-demographic characterization of the young people in both universities – CUFSA and USP-RP. In the analytical part we will give emphasis to similarities, differences, and inequalities identified between the two groups of students; we will also discourse about the circulatory practices that they engage in, as students, in the physical territories.

Following this, our focus will come back to circulatory practices that the subjects of the research undertake within the circulatory territory of the Internet, once again with the intention of highlighting similarities, differences, and inequalities in the practices that they formalize in the digital world, distinguishing them as instrumental or meaningful practices.

THE YOUNG COLLEGIANS WHO PARTICIPATED IN THE RESEARCH: AGE GROUP, ETHNIC, AND GENDER DIFFERENCES

Data collected with the students of both Institutions reveal a process of expansion, social rejuvenation, and differentiation that higher schooling in Brazil has been undergoing over the last decades, and the ages of the collegians of CUFSA and USP-RP are expressions of this process in the urban milieu as can be seen in Fig. 1.

However, the rejuvenation observed among the subjects of our research manifests itself differently in the two Institutions: while at CUFSA the proportion of youngsters aged 17–24 is 80%, at USP-RP that proportion rises to 92%. The distinguishing data about the ages of the youngsters

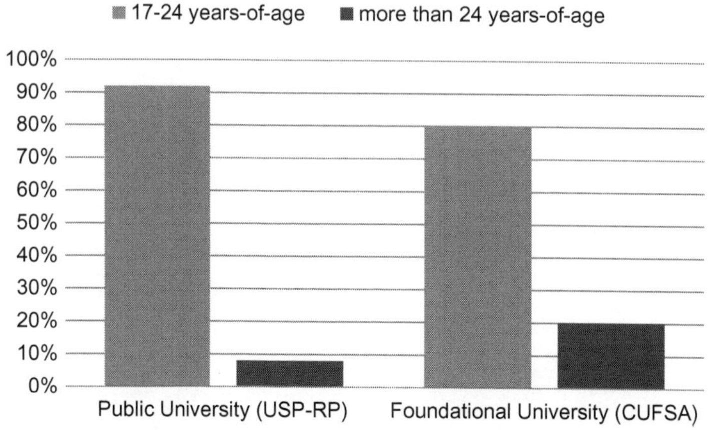

Fig. 1. Collegians according to Age Group, by University, 2011 (%).

researched, in the Brazilian case, cannot be analyzed in an isolated manner, since when these are associated to those related to the self-representation that they have about "color of the skin" the differences increase: if in both groups of collegians white youngsters are predominant, those belonging to the public state university – USP-RP – are "whiter" (84%) than those of the municipal/foundational one – CUFSA – (72%), according to the result indicated in Fig. 2.

With regard to gender, we have inverse tendencies in each of the two universities (Fig. 3), even though the differences between them are small: at CUFSA the men predominate (51%) and at USP-RP, the presence of young women is slightly superior (54%) to their male peers (46%).

BRAZILIAN YOUNG COLLEGIANS: THE DIFFERENCES AND INEQUALITIES OF THEIR RELATIONS WITH THE WORKING WORLD

As we advance in the reading of the data we collected, we observe that in the small differences – age, skin color, and gender, the percentages lead us to some of the longer lasting inequalities that mark Brazilian society and, within this, significant contingents of its youth. Henriques (2001), on analyzing the history of racial inequality in Brazil, ascertains that "poverty is strongly concentrated in early childhood and youth, but, more categorically, among the blacks of these age groups" (Henriques, 2001, p. 14), and

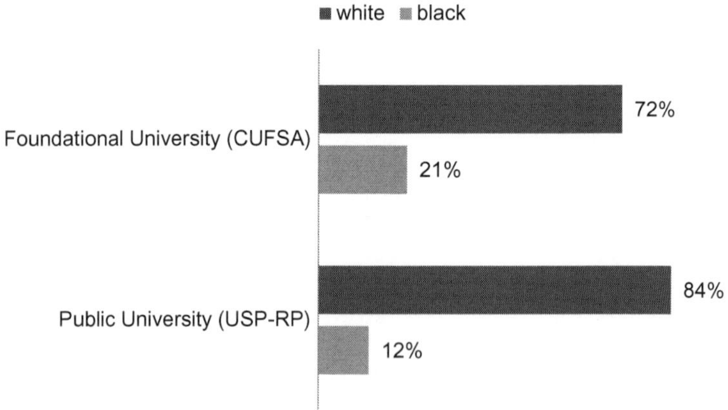

Fig. 2. Collegians according to "Skin Color," by University, 2011 (%).

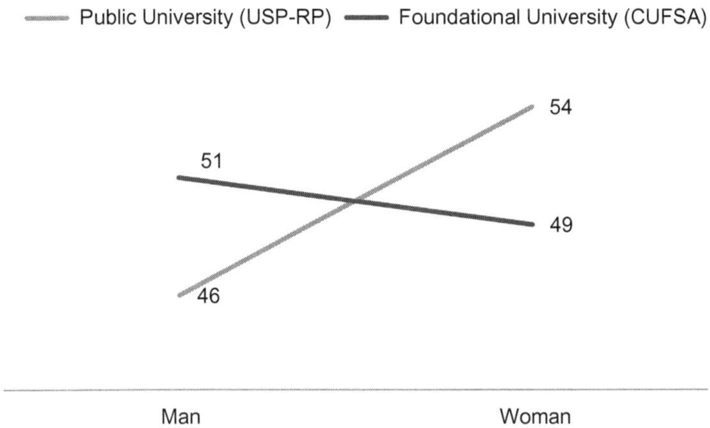

Fig. 3. Male and Female Presence in the Two Universities Researched – 2011 (%).

to this he adds the historic inequality that involves the women in their relation with the men. This phenomenon appears in Santo Andre, but not in Ribeirão Preto.

The numbers that refer to insertion in the work sphere, family income, and parents' schooling, as well as taking into account the maximum level of schooling achieved by fathers and mothers of students researched, characterizes decisively some inequalities among the youth of both universities. If it is a blatant fact that both Institutions have always received more students from the different strata of the Brazilian middle class – from the well-to-do middle class as well as children of the working class that have climbed economically in the country – the "Brazilian battlers," we verify that of the total of students aged 17–24 at CUFSA (the foundational university), 94% are worker-students, whereas at USP-RP (the public university) the percentage of youngsters in the same situation is diametrically inverse, reaching only 8% (Fig. 4).

Besides this, among the CUFSA students that combine the pace of going to school and working, 45% of them work 40 or more hours/week and 20% of them between 20 and 40 hours/week. Of the 94% of collegian-workers from CUFSA, 55% had formal employment contracts, but 45% of them worked in precarious conditions, especially when employed in the service sector. This batch of data leads us to consider that while the USP-RP collegians live their young condition as a kind of "social moratorium"

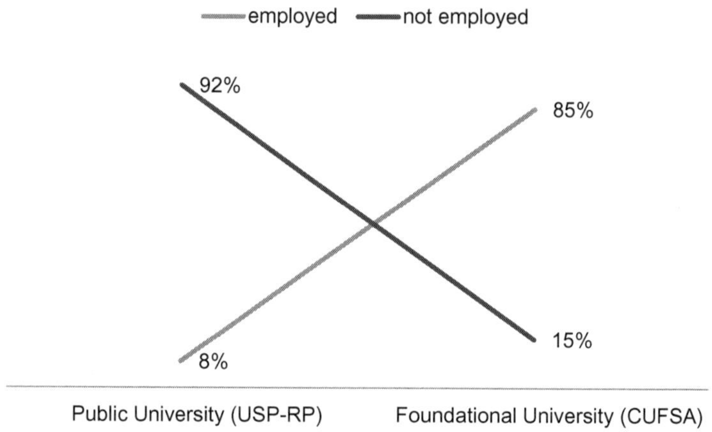

Fig. 4. Collegians, according to Employment, by University – 2011 (%).

the same does not occur with the greater part of the CUFSA students, that is, these do not succeed in living their youth as a socially conceded time of selective tolerance, while permitting certain youth segments to dedicate themselves to schooling and non-schooling formation, shaping experimental socio-cultural practices, revert from already made choices, and postpone assuming commitments and responsibilities associated to the entry into adult life, such as working, setting up a family, and having children. (Margulis & Urresti, 1998, free translation).

The CUFSA students, on the other hand, experience a "confluence of several socializing processes in their experience of being young" insofar as in their daily life they experience, at the same time, socializing processes established by the family, the school, and the job, leading them to believe that, in Brazil, "work is also part of being young and it becomes increasingly complex to establish the socio-cultural profile of the youth category in our country without effectively and symbolically mediating it" (Sposito, 2005, p. 226). But here it is necessary to define clearly that work produces certain youth segments as for example the CUFSA students, and not other, as the collegians at USP-RP. In this case, the data we have collected and analyzed substantiates a very clear opposition between them.

Still within the field of differences observed between the two collegian groups, data presented in Fig. 5, concerning the monthly income of the family nucleus of origin, does not reveal any other dissimilarity in the indicators. However, in the case of the CUFSA students, the monthly salary

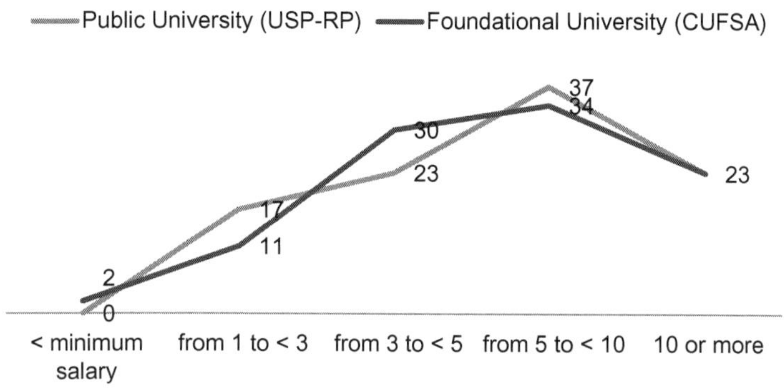

Fig. 5. Collegians according to Family Income, by University – 2011 (%).

that they receive is a part of the family income transforming their situation into being quite different from that of the USP-RP students.

The indicators of the cultural heritage of the fathers and mothers of the collegians researched also differentiate the groups of collegians: the data attests that the parents of the USP-RP students achieved a greater number of schooling years than those of the worker-students at CUFSA, when considering those that accessed secondary school, higher schooling and post graduate studies (Fig. 6).

In this dimension of family life of the young students researched, it can also be verified that each one of them presents tendencies that are different from the national ones. Nationally, it is the women who have achieved a greater number of years of study than the men, according to what the schooling indicators of the Brazilian demographic totals demonstrate over the last decades, seeming to indicate that the existing historic inequalities in the attainment of the right to education have affected more the men than the women, because it is the women that have longer schooling trajectories than the men (Brasil. Instituto Brasileiro de Geografia e Estatística [IBGE], 2012). However, this national reality was verified only at USP-RP. At CUFSA, the mothers' schooling is only higher than the fathers' until secondary school. In this manner, within the composition of these differences, the configuration of inequality can be perceived. An inequality that must be understood within the present state of the debate on the issue of the Brazilian middle class's composition. For this reason it is important to discuss the very understanding of what the middle class is, even if only in a quick and brief manner, in order to handle an understanding of the students in both universities, and not fall into an empty debate on this issue.

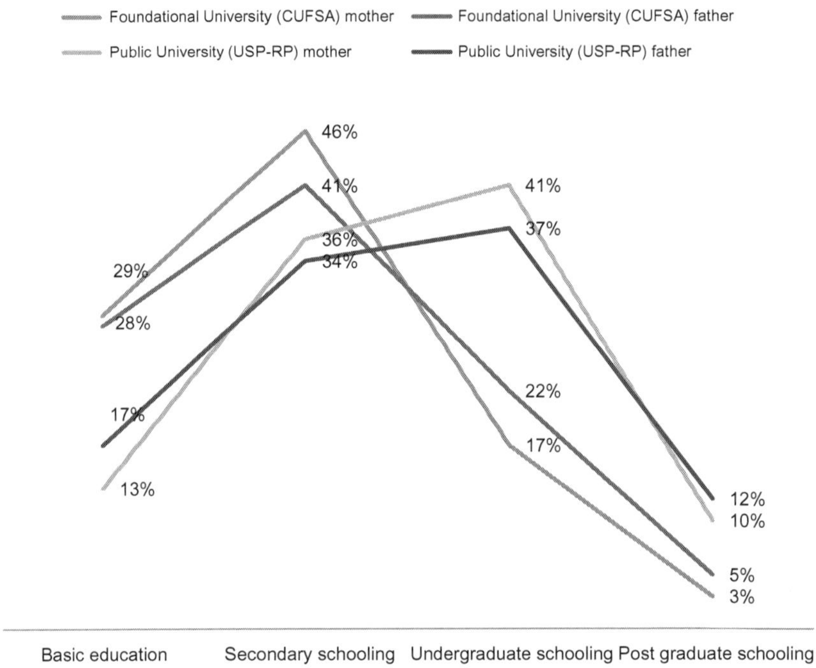

Fig. 6. Schooling of Father and Mother – CUFSA and USP-RP and Linear Tendency of Parental Schooling of Parents in Both Schools (%).

The two situations described above impact unequally, not only because of the young age of the collegians we are studying, but also because of their school life, as well as the circulatory practices that they undertake in the real world and in the virtual world, imposing efforts and differentiated manners to constitute themselves as players and to enjoy and take advantage of artifacts, products, and cultural processes that are at their disposal.

YOUNG COLLEGIANS FROM URBAN AREAS AND DISSIMILAR EXPERIENCES OF CIRCULATORY PRACTICES IN THE PHYSICAL SPHERE

Based on the concept of circulatory territory (Tarrius, 2000, 2002), it was possible to verify similarities as well as differences and inequalities in the circulatory processes of the collegians researched. In the case of the

CUFSA students, 87% of them live and reside in the region in which the university is installed. It is possible that, for them, the choice of the university they are attending is based on the relative ease in the physical access to it, but it could also be due to the difficulties they encounter in *home-work-school* commuting.

Even if they live in the same city in which the university is situated, or in neighboring cities, the students are obliged to circulate, in order to study and to go home. This demands certain knowledge as to how to circulate, involving different means of transportation: buses, vans (people carriers), cars, getting a ride, or even walking. For the students at CUFSA this occurs principally at night, when a greater number of them go to the university. It also involves different forms of circulating, depending on the kind of transportation used.

According to the CUFSA students, among the means of transportation used, the two most common ones are buses and vans. In general, those who circulate by bus, commute anonymously, creating mechanisms for not speaking to anyone beside them as, for example, pretending they are sleeping, putting on earphones to hear music; looking at their cell phone even though there is nothing new to see. This happens especially when they are going home, toward night's end, when they are exhausted by a long day's work with more hours added in night study. The great trial for them is to study and work at the same time, something that only about 15% of young Brazilians do. The youngsters hardly ever establish relations with other people on the bus, except for the friends they have made in other places, but not on the bus. These are usually from the same neighborhood or the university.

Contrarily, the scene in the vans is different. Each van is registered at the city hall in order to be legalized, and attends the public of a certain region of the city. However, there are illegal vans that emerge from relations established with the drivers of the legalized vans. In this means of transportation, students talk about everything, from personal issues, such as sexual relations, to more general issues such as those about classes they attend at the university. The van seems to be free "territory" where one can talk about anything.

When they leave the bus or the van the students enter the university building and in this space they meet and speak very little. They act, almost always, without any kind of intimacy. In spite of the exceptions, polished and distant greetings predominate, as if they hardly know each other. The common spaces of the university building become spaces for circulation. The common spaces of the university building hardly ever are transformed into a place for meeting. They go directly to the classroom, where stronger ties will be established, some of which long-lasting. It is the case of privileging the *locus* from which they will be able to obtain a certificate.

Students that live in many places of the Region where CUFSA is situated, come across each other at the university, but they only come across each other and do not relate. The common spaces of the university seem to be constituted as spaces for circulating and do not transform themselves into territories. While on their itineraries, circulating in buses and vans, the territory is not constituted. It is quite true that they talk about themselves and about many other issues, but these do not establish more profound social relations that would produce a territory.

For the CUFSA students, the home/work/school triangulation has made clear that these are the fixed points of circulation within the physical space of the city of Santo Andre, or in the spaces of localities in the vicinity. These points function as anchors that fasten them to the contours of those territories, spaces that stamp their socializing marks on the young students.

Besides this, as is the case of the USP-RP students, they choose their courses mainly based on what they like, but the choice is also guided by the gamble they make on getting new opportunities for work. They work in a conspiracy with time because, if in the present they act based on their desires, they also have their eyes focused on the future at this time, a look that is marked by the need to have work, to take on the responsibilities that are socially sanctioned for adults.

On the other hand, among the USP-RP students, only 39% live and reside in Ribeirão Preto or in the cities in the surrounding area; 60.7% of them come from different localities within all the quadrants of the State of São Paulo and of the country. This situation demonstrates that these collegians shape circulatory practices that differ from those that mark the everyday routine of the young-worker-students of CUFSA. (Fig. 7).

YOUNG COLLEGIANS FROM USP-RP: "TEMPORARY MIGRATIONS" AND HOME-SCHOOL-HOME TRANSITS

Among the students of USP-RP that do not work, their movement has a different kind of complexity since the fixed points tend to be the home/school/home spaces. On the other hand, for those who have had to move from their original family nucleus, situated in other parts of the state or of

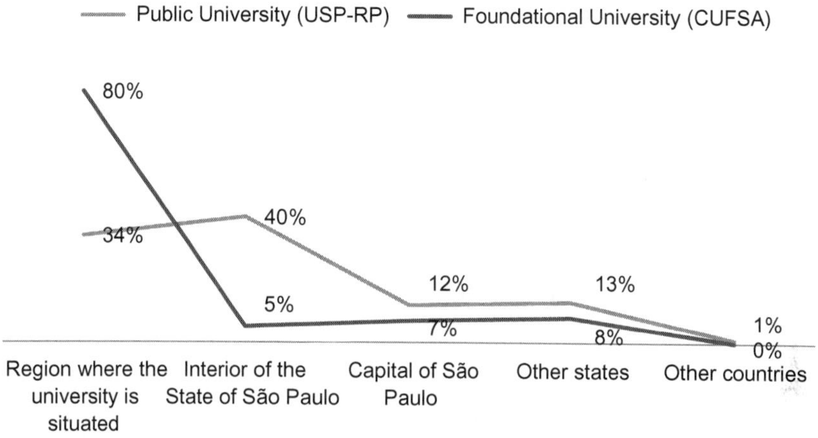

Fig. 7. Collegians according to Place of Birth, according to the Universities Researched – 2011 (%).

the country, the circulation practices are determined by other fixed points, and these necessarily incorporate the socio-cultural experiences of having migrated and of having become a migrant – whether transitory or not. In this case, it must be admitted that, differently from the CUFSA students, these youngsters insert themselves in another universe of circulatory experiences, of socializing and individuation arrangements.

For the youngsters who experience such situations, the advantage represented by the access to public higher schooling imposes considerable alterations in their daily life and some losses on the level of sociation – secondary socialization and sociability (Berger & Luckmann, 2001; Simmel, 2002). They need to intensify their processes of autonomy, since they feel the need for different "supports" – material and symbolical, to sustain themselves to face the world of the university that they have chosen, so as to fill in for what is missing – temporarily or not – of the resources, affection and sensibilities that are innate to the family circle, and also of the social relations they left in the cities in which they were born and in which they lived during childhood and adolescence.

On moving from their cities to a place where their university is situated, the first challenge is to find housing, be it a boarding house, a "student republic" – *fraternity*, a house, an apartment, a room at a friend's or family member's house, or even a studio in the student housing buildings, etc. In this process they avail themselves of the material and immaterial support that is made available by their own families and to a much lesser degree to the state social ones or to the public state university institution that has received them.

Some arrive with their parents and these interfere in choosing the place. About this kind of situation a student reported that when he arrived in the city, he went with his parents to visit a "boarding house" and decided that it was best not to stay there because the students who were already installed there "smoked marijuana." This interference of the parents annoys many students. Many boarding houses or republics end up being just a place to sleep because the place to meet and to establish relations is the campus of the university itself. There they study, eat, practice sports, establish interaction, and establish social networks, especially with their peers in the same course. In this manner, university life, beyond school work, installs itself for these students: if they free themselves partially from parental tutelage and protection, they must also seek help and support among their peers, or student assistance so as to withstand university life and the situation of circulating subjects. The experience of migrating faced by them, when it doesn't pass the tests of uprooting and significant ruptures, it imposes the need for new learning and living experiences in socializing processes that would enable them to integrate and adapt to new spaces and times in which they circulate (Martins, 1997).

YOUNG COLLEGIANS FROM USP-RP: SOCIALIZING EXPERIENCES AFFORDED BY "TEMPORARY MIGRATION"

In this manner, many youngsters who study at USP-RP are temporarily from Ribeirão Preto, but, at the same time they are not, even if the visits to their parents' home becomes ever rarer. They interfere and reconfigure certain spaces in the city with their youthfulness, their demands, needs, juvenile desires – material and subjective. Then they leave and new youngsters will come, eternally renewing the city that modified its features, not only by the form in which it was economically established, but also by producing new social relations that involve a "circulating knowledge." In this manner circulatory territories are configured, which begin with the decision to live in the city and find a place to live, frequently within relationship networks. They exchange furniture, so that part of the initial costs of installing themselves can be recovered. They exchange knowledge: local ones for use of leisure time, even though these compete with study hours, such as bars to frequent, places to eat. The coming and going from their cities to Ribeirão Preto. Knowledge of how to live in movement not connected to local dwellers with their fixed rules and ways of acting. A typical movement in certain European countries, through the *Erasmus*[1] program, guarantees the circulation of their students, contributing towards knowing how to circulate, beyond that of fixed places. For that group of young students at USP-RP that are living the experience of a temporary migration, it is from this condition of migrant-collegian that they also grow, socialize, acquire new living perspectives for youth, "broaden their worldview, their social relations, their awareness of themselves and of others" (Martins, 1997, p. 44).

Other differences also mark the life of the students investigated. When asked about how many weekly class/hours they have, it is perceived that the USP-RP students have less weekly class/hours than do the CUFSA students. But on the other hand, the former spend more time studying outside class time, which requires knowledge of how to "circulate" seeking books in an autonomous manner, which is different from the latter that have the classroom as the most important space to access knowledge, under tutelage of the professor.

This autonomous circulating knowledge also appears in the possibility to study abroad. Even though the exchange experience that has been given incentive by the Brazilian state is recent, it has been experienced by more students from public universities than by students from institutions such as CUFSA from where, until this date, only seven students have gone abroad.

Whichever the group to which they belong, living a situation as a student under "moratorium," as a worker-student who complements family income, or as a migrant-student, these young players need support since, as Martuccelli (2012, p. 36, free translation) reminds us: "the idea of an individual that supports himself alone, only by inner force is a heroic image that is not backed by objective reality." This is also the case for the young collegians at CUFSA and USP-RP. However, the data that we have at present indicates the fact that the "supports" that they demand or activate are not necessarily identical or, if they are, they are not activated or used with the same purpose or to achieve the same end. An evident case of this is seen in the circulatory territories of the physical space in which they move around, and also those relative to the use of collective public transport, of student housing or "social rental." These are public social programs that contribute towards a less tense and conflictive integration to the higher schooling sphere and that imply the access to subsidies for nourishment, for collective public transport, both municipal and interstate, as for the access and fruition of exchange programs abroad, scholarships, etc.

It is necessary to point out that the different practices within the circulatory spaces of CUFSA and USP-RP, permit us to retain for further analysis, although in a preliminary manner, the fact that they play different games in the processes of establishing identity. If for the great majority of the worker-students of CUFSA their socializing and self-defining processes are firmly anchored in the triad that establishes the fixed points previously mentioned, for the USP-RP collegians, in this case, it is necessary to observe that there are differences since, for almost half of these it is necessary to consider their social and cultural experience resulting from their migration and from having assumed the state of transitory migrant within their student experience, as one of the possibilities of experiencing their youth and establishing themselves as individuals.

But the data we have collected to-date indicate that our interlocutors do not reduce their movements and processes of sociability (Simmel, 2002) only by the circulatory territories of the physical world, they also inform us that, similarly, they circulate, socialize, construct sociability and subjectivity processes within the "numerical territory," the virtual territory of the Internet, and report having also shaped different socio-cultural lifestyles. In this manner, they inform us that they increment and diversify their circulatory practices and their processes of individuation.

YOUNG COLLEGIANS CONNECTED: SIMILARITIES, DIFFERENCES, AND INEQUALITIES

At CUFSA, 96% of the students informed us that they accessed the Internet during the 3 (three) months previous to filling out the research questionnaire, and 42% of them affirmed that it is with the Internet that they spend more of the little free time that they have. At USP-RP, the proportion is almost identical, since 95% of the students inform that they carry out the same practice everyday. In this manner, it is unequivocal that the collegians researched incorporate the Internet and the circulation through its *auto routes* to their quotidian (Boullier, 2010).

Based on a diary where the students registered their quotidian, giving special attention to the circulation they set up in the virtual world, the tools they used and the content involved, the distinctive circulatory processes in the *auto routes* can be better understood.

It was verified that the circulatory practices that the two groups establish in the "numerical territory"[2] start up at spaces such as housing (boarding house, fraternity-republic, or student housing), the school, their parents' home, their friends' homes. Among the CUFSA students, the place of work was also a possibility for transiting through the virtual world. To do this, they use a desktop computer (PC), notebook/laptop or mobile telephone. However, for the CUFSA students, different from the USP-RP students, the equipment is not necessarily individual: in order to use them they need to do some negotiation, which they undertake sometimes with siblings in the same house, with work colleagues, and their working time as well. For the two groups of collegians only the mobile phone is for individual use, an apparatus that they use independently from the fixed point in which they are in the physical world, which makes a "same time" experience possible.

Still regarding the access to and use of those apparatuses it was possible to verify that such practices do not occur on the fringe of the institutions where they organize their daily life, but from them (Winocur, 2006, p. 553). They are simultaneously spatial and temporal markers in the uses they make of the apparatuses. The use of expressions such as "before" – arriving at the university, "after" – leaving work, classes, "during the interval" – of classes, "during" the day, and night, etc. One exception to these ways of circulating occurs with regard to the mobile phone, because in this case they affirm that the use occurs "all the time." In this manner, these technological supports must be "understood in their imbrications with other spheres of social life (Sorj & Martuccelli, 2008, p. 74).

YOUNG COLLEGIANS: CIRCULATORY PRACTICES IN THE DIGITAL TERRITORY WITH INSTRUMENTAL AND MEANINGFUL OBJECTIVES

In a preliminary manner, we consider it pertinent to characterize circulatory practices that the two university groups have as their objective in their "numeric territory" as being *instrumental/practical* or *meaningful.* Based on this scheme we could identify similarities, differences, and inequalities, whether appearing in comparing the two groups, or be it within the same group.

From an instrumental point of view, the exchange of electronic correspondence is common to both groups – send/receive e-mails; circulating electronic pages of renown newspapers of the state of São Paulo or of regional papers; the search for public utility – e-banking, e-government – state or local; "tutorials" for cultural production; consulting doctors' offices; carrying out e-commerce – in large retail stores, for acquiring non-durable goods and services – shoes, air tickets, clothing, costume jewelry; and in large bookshops, etc. The circulation of magazines focused on specific public, such as variety shows, serial story leaflets, or sports activities are also common in both groups. In this manner, for example, the young women students affirmed: (a) access "Facebook to check messages, sites of magazines such as *Superinterssante*, *Veja*, and the *Folha de São Paulo* newspaper to see the news"; (b) "read the daily news, especially that of the election of the new Pope (Francis I)"; or "get information about a book I bought on the Internet"; or "find out if the book had already been posted." The male collegians (a) accessed "the site of the City Hall to check out information on public service exams"; (b) used "the Internet of the desktop computer to register a bank account"; (c) "navigated on the Banco do Brasil site to check out the balance on my account."

From the point of view of the meaningful function of circulation within the "numeric territory" they go in search of cultural mediations that make leisure experiences possible, such as: listening to music, watching musical videos, series, and television programs, watching films – Brazilian or foreign, searching for electronic games, and fun events of gambling, etc. And last, but not less important, they informed about their constant circulations in the virtual social networks – Facebook, YouTube, SKYPE, to hear music and watch movies. In this manner you can find in the diaries of students statements as follows: (a) "I accessed the first site to check e-mails; on site 2 I looked for music I could listen to while I was reading"; (b) "The laptop was connected continuously, even when I wasn't using it. As to the desktop computer, I used the Internet to listen to operas while I was resting in my room"; (c) "Navigated the YouTube to listen to music"; (d) "Accessed YouTube to see a movie; fell asleep watching it…"; (e) Accessed *Youporn* to watch pornographic videos."

However, on analyzing data of the circulatory practices, by Institution and by gender, we verify, for example, that among the CUFSA students, it is the females who most use and circulate within the Internet, both with instrumental and meaningful objectives. With regard to the USP-RP students, the use and circulation shows itself more balanced between genders in achieving their objectives, both instrumental/practical as in those of expression. In this manner it was possible to learn that both men and women make use of mediations that permit "videoconferences" (*Skype or MSN*) in order to establish dialogues in real time with friends, boy or girl friend and family. In this group of collegians, some young women "entered Facebook to read publications, chat with university friends or with my boyfriend and family," "navigated in Facebook to chat with friends, check messages and answer emails." In this group of collegians, some young men also informed us of the use and circulation via Internet aimed at finding partners for the practice of sexual activities or, also, to transit in sites with pornographic material or practice virtual sex with virtual partners: "Accessed Skype with a fake name to look for sex." This difference among men and women deserves to be studied more deeply since it could also indicate how the virtual territory is influenced by the existing culture in the physical territory in which women tend to have fewer rights to a free sexual life than do the men.

Both groups behave in the virtual territory in very similar ways, however, with nuances that differentiate them. In this manner, the CUFSA students, due to the little time that they have for relating outside of the university and of their workplace, it is the internet that permits them to maintain, strengthen, and create new social relations. "At home, at the university, and on the street when going somewhere"; "I use the cell phone as soon as I wake up. When I am going to work, at lunch time, in the car, going to the university, and at the university." They make more virtual connection with the family, even though living with them, because their life experience is that of little time, rushed, and with innumerable activities. They open themselves more to friendships in the virtual world than do the USP-RP students. They also make use of this tool for professional contacts and to study, while those at USP-RP valorize email contacts more with those people with whom they originally related to in the physical territory.

In this manner, the most significant difference that we observed in the circulatory practices within the virtual territory, regards interactions that the collegians carry out with collectives of an ethic-political nature; in this case, more were from USP-RP than from CUFSA, that indicated their circulation in blogs, debate groups, or electronic pages of identities committed to themes such as "human rights," "feminist activism," "collegian movement," that is, they circulate in virtual territories in which certain collective players simultaneously defend the right of those who are different, recognize similarities, and stand for the value of equality.

And the differences between the two groups of collegians are multiplied when we asked some of their representatives about the ways in which they informed and engaged themselves in the "Brazilian June" manifestations. Both the representatives of CUFSA as those of USP-RP related that they sought information about the movements of "Brazilian June" especially in Facebook or blogs of collectives that were visibly committed to the manifestations in different cities in the country, as for example the blog of *Movimento Passe Livre – MPL*[3] – or of *Anonymous-Brasil*. However, it was the young students of USP-RP, given the condition of social moratorium that they experience, who made their self-convocation and went out to the streets, whether to expand the fight for reducing collective transport fares, or of urban mobility, or for the right to education, or for women's rights or of the LGBT segments. In this manner they transit through virtual territories in which some collective players – either of the young category or not – defend and spell out other ways and possibilities of doing politics and/or suggest alternatives for political participation distinct from the experiences offered by the traditional political system (de Souza, 2003; Freire Filho, 2005; Martuccelli, 2012; Sorj & Martuccelli, 2008).

The circulation in the virtual world of both groups appears to configure spaces that are, at the same time, similar and disparate. All of them circulate in the virtual world in innumerous situations, as has already been mentioned, producing relations with a certain permanency and density in the movement that they carry out, configuring circulatory territories with different nuances: the CUFSA students, focused on work and in carrying out their schooling activities, seeking to maintain and broaden the number of friends. Even though the USP-RP students also carry out activities in this line, they tend more toward those contacts that would intensify certain relations that have their origins in the physical world, so as to constitute themselves as players, to potentiate their actions in the physical territory, while diversifying their processes of "subjectification" and individuation.

These differences, although anchored in their social positions, in the living conditions and circulatory experiences that make their existence singular, they are also differentiated by the ways in which the students researched intervene in these realities, live the youthful phase of their lives, and the by-laws of being a collegian.

To live this youthful phase as a moratorium, without having to work, introduces possibilities of circulating in the physical and virtual world that do not exist for those who are worker-students. These, in their turn, seem to use the circulation that they carry out in the virtual world, to fill in the gaps that exist in the physical world as, for example, that of friendship and of work. In this manner, even though the collegians of both universities circulate in the virtual world, inequality occurs when certain supports of the physical world are present and certain differences become visible.

FINAL CONSIDERATIONS

Nowadays youth is no longer spoken of as a homogenous group and, for this reason, the term "youths" has been adopted. But when the use of the internet and the circulation within its territories is dealt with, there is a consensus that these subjects, more than others that circulate in the virtual world, have tended to erase differences that exist among individuals and youth groups. In the Brazilian case, this is aggravated when dealing with thinking about the so-called middle class. Generalizations have tended to overshadow the existing differences and inequalities among youth and among young collegians who have contemporaneously reached higher schooling, even among those who succeed in accessing public universities.

The results of the exploratory studies here presented, on discussing certain elements of the biographies of the students studied, as well as their socialization, circulation, and individuation processes, also demonstrates the importance of recovering the transversalities, the differences, and inequalities that mark the young urban area collegians.

To be a young worker or non-worker collegian, even when the monthly family income of the groups seems similar, impacts decisively on the circulation in the virtual world and the possibilities, within this territory, to set themselves up as players and occupy hybrid spaces. That is to say, connectivity is not necessarily a synonym of interactivity (Garcia Canclini, 2008). Depending on the condition that is being lived, the inequalities are multiplied going from color/race to participating/not participating in the makeup of the family income, schooling of parents, gender, and even in the ways in which they move, as social players, in the circulatory territories within the internet.

Data from our study permits us to posit the hypothesis that if the researched collegians circulate in distinct "circulatory territories" of the urban areas they are certainly conditioned by the class origins of the subjects. And by the cultural capital that they detain, they also indicate that they are living their youth in different processes of "secondary socialization" and individuation. This differentiation permits us to perceive, as well, some of their actions on the conditions that singularize their youth living experiences, and how they plot the "intercultural" meanings of "the times" (Garcia Canclini, 2004).

NOTES

1. Student mobility program created and financed by the European Union (EU), cf: http://eacea.ec.europa.eu/erasmus_mundus/funding/scholarships_students_academics_en.php. Accessed on 31.05.2014.

2. Data obtained from interviews and from the diaries written by the CUFSA and USP-RP students, during the first and second semester of 2013.

3. Brazilian social movement having youth and student interests as their basis and that fights for urban students in the country to have the right to free public transport – bus, train, and metro – besides demanding public discussion on urban reform as well as the individual and collective ways of circulation in the metropolitan areas.

REFERENCES

Arnett, J. J. (2004). *Emerging adulthood.* New York, NY: Oxford University Press.

Berger, P. L., & Luckmann, T. (2001). *La construcción social de la realidade.* Buenos Aires: Amorrortu.

Boullier, D. (2010). Au-delà des territoires numériques en dix thèses. *Urbe - Revista Brasileira de Gestão Urbana*, Pontifícia Universidade Católica do Paraná, *2*(1), 13–27.

Bourdieu, P. (2007). *Campo de poder y reproducción social: Elementos para un análisis de la dinâmica de las classes* (pp. 195–202). Cordoba: Ferreyra Editor.

Brandão, T., Saraiva, L., & Matos, P. M. (2012). O prolongamento da transição para a idade adulta e o conceito de adultez emergente: Especificidades do contexto português e brasileiro. *Análise Psicológica*, ISPA, Instituto Universitário, *30*(3), 301–313.

Brasil. Inep – Instituto Nacional de Estudos e Pesquisas Educacionais. (2013). *Censo da educação superior: 2011 – Resumo técnico.* Brasília, DF: INEP.

Brasil. Instituto de Pesquisa Econômica Aplicada & Sociedade Brasileira de Sociologia. (2012). *Estudo comparado sobre a juventude brasileira e chinesa: Dados preliminares do Brasil* (p. 22). Brasília: IPEA.

Brasil. Instituto de Pesquisa Econômica Aplicada. (2009). *Juventude e políticas sociais no Brasil* (p. 303). Brasília: Ipea.

Carrano, P. (2009). Jovens universitários: Acesso, formação, experiências e inserção profissional. In M. P. Sposito (Ed.), Estado da arte sobre juventude na pós-graduação brasileira: Educação, ciências sociais e serviço social *(1999–2006)* (Vol. 1). Belo Horizonte – ME: Argvmentvum.

Castel, R. (1995). *Les métamorphoses de la question sociale, chronique du salariat.* Paris: Fayard.

Catani, A., & Hey, A. P. (2007). A educação superior no Brasil e as tendências das políticas de ampliação do acesso. *Atos de Pesquisa em Educação*, PPGE/ME FURB, *2*(3), 414–429.

Comitê Gestor da Internet no Brasil. (2010). *Pesquisa sobre o uso das tecnologias da informação e da comunicação no Brasil: 2005–2009*. São Paulo: CGIB.

Comitê Gestor da Internet no Brasil. Núcleo de Informação e Coordenação do Ponto BR. (2011). *Pesquisa sobre o uso das tecnologias de informação e comunicação no Brasil: Tic domicílios e tic empresas 2010*. São Paulo: CGIB.

Comitê Gestor da Internet no Brasil. (2012). *Pesquisa sobre o uso das tecnologias de Informação e Comunicação no Brasil: Tic domicílios 2012*. São Paulo: CGIB.

de Almeida, W. M. (2014). *ProUni e o ensino superior privado lucrativo em São Paulo – Uma análise sociológica*. São Paulo: Musa: FAPESP.

de Souza, J. T. P. (2003). As insurgências juvenis e as novas narrativas políticas contra o instituído. *Cadernos de Pesquisa*, PPGSP/UFSC, nº 32, fevereiro.

Freire Filho, J. (2005). Das subculturas às pós-subculturas juvenis: Música, estilo e ativismo político. *Contemporânea*, *3*(1), 138–166.

Garcia Canclini, N. (2004). *Diferentes, desiguales y desconectados: Mapas de la interculturalidade* (pp. 167–173). Barcelona: Gedisa.

Garcia Canclini, N. (2008). *Leitores, espectadores e internautas*. São Paulo, SP: Iluminuras.

Gatti, B., & Sá Barreto, E. S. (2009). *Professores do Brasil: Impasses e desafios*. Brasília, DF: UNESCO.

Gomes, A. M., & Moraes, K. N. (2009). A expansão da educação superior no Brasil contemporâneo: Questões para o debate. *Anais da 32ª Reunião Anual da ANPEd*. Caxambu, MG, Rio de Janeiro: ANPEd.

Henriques, R. (2001). *Desigualdade racial no Brasil: Evolução das condições de vida na década de 90*. Rio de Janeiro, RJ: IPEA. (texto para discussão nº 807).

Ibase/Polis. (2005). *Pesquisa Juventude Brasileira e Democracia*. Rio de Janeiro/São Paulo: Ibase/Polis.

Instituto Cidadania. (2003). *Perfil da juventude brasileira*. São Paulo: Fundação Perseu Abramo.

Jannuzzi, P. M. (1995). Distribuição de renda: O caso da Califórnia brasileira. *São Paulo em Perspectiva*, *9*(3), 54–61.

Kauling, F. A. S., Moré, R. P. O., Costa, A. M., Bunn, D. A., Pereira, M., & Azevedo, P. (2010). *Acesso à educação superior no Brasil: O desafio da inclusão dos jovens brasileiros segundo o PNE. X Colóquio Internacional sobre* Gestión Universitaria en América del Sur. Mar del Plata Argentina: Universidade Federal de Santa Catarina. Retrieved from http://www.inpeau.ufsc.br/wp/wp-content/BD_documentos/coloquio10/162.pdf

Margulis, M. Y., & Urresti, M. (1998). *La construcción social de la condición de juventude. Viviendo a toda: Jóvenes, territórios culturales y nuevas sensibilidades*. Bogotá, DC: Siglo del Hombre Editores (Departamento de Investigaciones Universidad Central).

Martins, J. S. (1997). Migrações temporárias, problema para quem? In *Exclusão social e a nova desigualdade*. São Paulo: Paulus.

Martuccelli, D. (2007). *Cambio de rumbo: La sociedade a scala del individuo*. Santiago: LOM Ediciones.

Martuccelli, D. (2012). *Lecciones de sociologia del individuo*. Retrieved from http://departamento.pucp.edu.pe/ciencias-sociales/files/2012/06/

Nakano, M. (2013). Entre violence et mouvement social, deux mondes d'une favela, une jeunesse introuvable? Défi méthodologique pour une étude de terrain en milieu sensib. In G. Leturcq, F. Louault, & T. C. S. Marques (Eds.), *Le Brésil: Un laboratoire pour les Sciences Sociales* (pp. 39–51). Paris: L'Harmatan.

Neves, C. E. B. (2012). *Ensino Superior no Brasil: Expansão, diversificação e inclusão. Congresso da Associação de Estudos Latino Americanos – LASA*. San Francisco, CA: LASA. Retrieved from http://www.ufrgs.br/geu/Artigos%202012/Clarissa%20Baeta%20Neves.pdf

Paugam, S. (2003). *La société démocratique face aux inégalités: Les nouvelles négalités entre salariés* (pp. 11–17). *LesCahiers Français*. Paris: La documentation française.

Pinto, J. M. R. (2004). O acesso à educação superior no Brasil. *Educação & Sociedade*, *25*(88), 727–756.

Pochmann, M. (2012). *Nova classe média? O trabalho na base da pirâmide brasileira*. São Paulo, SP: Bomtempo.

Schwartzman, S. (2012). *A educação de nível superior no Censo de 2010*. Rio de Janeiro: Instituto de Estudos do Trabalho e Sociedade. Retrieved from http://www.schwartzman.org.br/simon/2012censosup.pdf

Simmel, G. (2002). *Cuestiones fundamentales de sociologia*. Barcelona, Espanha: Gedisa.

Sorj, B., & Martuccelli, D. (2008). *O* desafio latino-americano*:* Coesão social e democracia. Rio de Janeiro: Civilização Brasileira.

Souza, A., & Lamounier, B. (2010). *A classe média brasileira: Ambições, valores e projetos de sociedade*. Rio de Janeiro: Elsevier.

Souza, J. (2010). *A ralé brasileira: Quem é e como vive*. Belo Horizonte: Editora UFMG.

Sposito, M. P. (Ed.). (2002). Juventude e escolarização *(1980–1998)* (p. 317). Brasília: INEP/MEC. (Série Estado do Conhecimento nº 7).

Sposito, M. P. (2005). Indagacões sobre as relacões entre juventude e a escola no Brasil: Institucionalizacão tradicional e novos significados. *JOVENes Revista de Estudios sobre Juventud*, *22*, 220–245.

Sposito, M. P. (Ed.). (2009). Estado da arte sobre juventude na pós-graduação brasileira*: Educação, ciências sociais e serviço social (1999–2006)* (Vol. 1). Belo Horizonte: Argvmentvum.

Tarrius, A. (2000). Leer, describir, interpretar las circulaciones migratórias: Conveniência de la noción de "territorio circulatório" – Los nuevos hábitos de la identidade. *Relaciones*, *21*(83), 37–66.

Tarrius, A. (2002). *La mondialisation par le bas: Les nouveaux nomades de l'économie Souterraine*. Paris: Ed. Balland.

UNESCO. (2006). *Juventude, juventudes: O que une e o que separa*. Brasília: Unesco.

UNESCO. (2007). *Juventudes: Outros olhares sobre a diversidade*. Brasília: Ministério da Educação; Unesco.

Winocur, R. (2006). Internet en la vida cotidiana de los jóvenes. *Revista Mexicana de Sociología*, *68*(3), 551–580.

KEEP CALM: YOUTH NAVIGATING ADULT AUTHORITY ACROSS NETWORKED PUBLICS

Ana Campos-Holland, Brooke Dinsmore, Gina Pol and Kevin Zevallos

ABSTRACT

Purpose – *Rooted in adult fear, adult authority aims to protect and control youth (Gannon, 2008; Valentine, 1997). Continuously negotiating for freedom, youth search for adult-free public spaces and are therefore extremely attracted to social networking sites (boyd, 2007, 2014). However, a significant portion of youth now includes adult authorities within their Facebook networks (Madden et al., 2013). Thus, this study explores how youth navigate familial- and educational-adult authorities across social networking sites in relation to their local peer culture.*

Methodology/approach – *Through semi-structured interviews, including youth-centered and participant-driven social media tours, 82 youth from the Northeast region of the United States of America (9–17 years of age; 43 females and 39 males) shared their lived experiences and perspectives about social media during the summer of 2013.*

Technology and Youth: Growing Up in a Digital World
Sociological Studies of Children and Youth, Volume 19, 163–211

ISSN: 1537-4661/doi:10.1108/S1537-466120150000019006

Findings – *In their everyday lives, youth are subjected to the normative expectations emerging from peer culture, school, and family life. Within these different and at times conflicting normative schemas, youth's social media use is subject to adult authority. In response, youth develop intricate ways to navigate adult authority across social networking sites.*

Originality/value – *Adult fear is powerful, but fragile to youth's interpretation; networked publics are now regulated and youth's ability to navigate then is based on their social location; and youth's social media use must be contextualized to be holistically understood.*

Keywords: Social media; youth; family; school; peers; adult authority

"They can never, ever, ever, ever, ever, ever find me," 14-year-old Vanessa[1] explained while showing how she has her two older brothers, grandmother, great aunt, three uncles, two aunts, ex-friend, and brother's friend all blocked from viewing her Facebook profile. Why does she block her extended family from viewing information that her 1,704 friends can see? It is because, as she says, "They're nosy." For Vanessa's peers, however, being friends with adult family members on Facebook is the new normal. Not only are 7 in 10 teens Facebook *friends* with their parents, but 91% have *friended* members of their extended family and 30% have *friended* their teachers or coaches (Madden et al., 2013). Children and adolescents are increasingly navigating around-the-clock adult presence and surveillance in their digital worlds (Davies, 2011; Livingstone & Bober, 2013; Madden et al., 2013). However, parental presence within teenage social media networks is primarily concentrated on Facebook (Gao, 2015). Thus, the purpose of this qualitative study is to explore how youth perceive and experience familial and educational-adult authority across social networking sites in relation to their local peer culture.

During the summer of 2013, 82 youth in the Northeast region of the United States of America (USA) shared their lived experiences and perspectives about social media. Within this local peer culture, youth's social media use varies across age, peer groups and social networking sites, and adults in youth's lives dominate Facebook, but hesitate to use other social networking sites. In this dynamic, youth experience adults' control and protection, with adult authorities in the familial and educational contexts prohibiting, guiding, incapacitating and/or surveilling youth's social media use. Thus, it is vital to rethink how youth navigate adult authority across social networking sites. Before exploring this process in detail, it is important to ground this study in the relevant literature and introduce our research design.

BACKGROUND

Scholars are taking an interdisciplinary approach to the study of social media (Gleason, 2013; Lingel, 2013; Marwick, 2012; Matthews-Juarez, Juarez, & Faulkner, 2013; Sumiala & Tikka, 2011; Van Dijck, 2013). In this literature, adult fear about online risk dominates the study of youth and social media use (Allen, 2012; Brown, Keller, & Stern, 2009). However, recent scholarship is beginning to explore social media in relation to youth's lived experiences within networked publics (boyd, 2007, 2014; De Ridder & Van Bauwel, 2013; Mazzarella, 2005; Reich, 2010; Wohn, Ellison, Khan, Fewins-Bliss, & Gray, 2013). Contributing to the latter, this study takes a youth-centered approach to explore how youth navigate adult authority in networked publics. In the following sections, we establish the conceptualization of childhood and adolescence in relation to adults' fear and youth's lack of access to public spaces, explore online risks, parental practices, and youth's resistance, and draw upon Corsaro's interpretative reproduction (2014) to assist in this analysis.

Youth, Adult Fear, and Access to Public Spaces

Western societies temporarily segregate children from adults and define childhood as a time of innocence, vulnerability, irrationality, and dependency on parents to protect children from the all-surrounding risks (Valentine, 1997). Within this conceptual framework, parental authority is governed by fear and sets "boundaries, supervision/care and discipline" to protect children (Valentine, 1997, p. 75). While these characteristics are transferred onto adolescence, some features are unique to the social construction of adolescence.

Adolescence involves the transition from "childhood dependence to adult independence" (Coleman, Catan, & Dennison, 2004, p. 227). This transition is complicated by an ambiguous beginning and end, multiple institutional transitions, adolescents' status ambiguity and continuous "lack of power to influence events and to take control" of their lives, and earlier transitions and later departures in today's society (Coleman et al., 2004, pp. 227–228). The implementation of adult authority thus becomes more complicated, with adolescents engaging in intense negotiations for freedom (Coleman et al., 2004; Davidson, 2011).

Beyond these adultist conceptualizations, childhood and adolescence interact with other systems of inequality to shape children and adolescents' lived experiences, such as that of race/ethnicity, class, sexuality, and gender (Davidson, 2011; Katz, 2014; Pascoe, 2011; Rios, 2011; Thorne, 2015). Thus, the fear-governing parenthood expands and parents of minority children and adolescents confront the intensified possibility of society transforming their children from 'at risk' to 'the risk' (Irwin, Davidson, & Hall-Sanchez, 2013; Pascoe, 2011; Rios, 2011; Wridt, 2004). Thus, it is important to keep minority children and adolescents' experiences with adult authority in context. Childhood (and adolescence) is a permanent segment of the stratified social structure, which children and adolescents navigate with agency (James, 2009; Qvortrup, 2009).[2]

Youth engage their agency in a continuous negotiation with adult authorities for personhood, especially in regards to access to public spaces (Coleman et al., 2004; Corsaro, 2014; James, 2009; Valentine, 1997). At an early age, youth begin to explore adult authority through play and navigate adult directives that aim to control their daily lives (Cobb-Moore, 2012; Kent, 2012). In doing so, youth are not naïve about risk; rather, youth are fearful for their personal safety and continuously work toward enhancing their risk-management competence while resisting adult control (Pini, 2004; Valentine, 1997). In this process, youth actively negotiate for access to public spaces and freedom of expression, a negotiation that expands into networked publics (boyd, 2007, 2014; Valentine, 1997; Wridt, 2004).

Youth's increased use of digital technologies is connected to larger changes in childhood and adolescence as a social structure (Poyntz & Hoechsmann, 2011). In western societies, adolescence is getting longer, while paradoxically *adult* realities such as sexuality and substance use are increasingly becoming part of childhood earlier (Coleman et al., 2004; Poyntz & Hoechsmann, 2011). Consequently, a societal tension emerges that places youth in need of protection and control (Gannon, 2008; Valentine & Holloway, 2001). In response, youth's access to public spaces has been increasingly limited, and youth are increasingly subjected to adult surveillance within the remaining public spaces (boyd, 2014; Valentine & Holloway, 2001). In this context, social media emerges as a public space in which youth can "participate in unregulated publics while located in adult-regulated physical spaces such as homes and schools" (boyd, 2007, p. 136). Youth construct identities, maintain friendships, and participate in local peer cultures in this new public space (boyd, 2014; Itō et al., 2010; Livingstone, 2008).

Unlike publics within socio-spatial environments, social media involves a networked public where the network "mediates the interaction between members" (boyd, 2007, p. 125). A networked public is characterized by the persistence of media that enables asynchronous communication, *searchability* of individuals' digital bodies, replicability, and invisible audiences (boyd, 2007, 2014). Although this phenomenon is not subject to time and space, youth's membership and social media performance is peer-driven (boyd, 2007, 2014).

Thus, it is important to explore how youth perceive and experience *adult authority* across networked publics.

Risk, Adult Power, and Youth's Resistance

Emerging from an adult perspective, the study of youth's Internet use places exposure to risk at the center of inquiry (Ahn, 2011; Brown & Bobkowski, 2011; Davidson & Martellozzo, 2013; Hinduja & Patchin, 2008; Mitchell, Finkelhor, & Wolak, 2001; Vandebosch & Van Cleemput, 2009). Scholars have identified several predictors of victimization, such as time spent online, information sharing, and chat room use (Davidson & Martellozzo, 2013; Mitchell et al., 2001; Mitchell, Finkelhor, & Wolak, 2007; Staksrud, Ólafsson, & Livingstone, 2013; Vandoninck, d'Haenens, De Cock, & Donoso, 2012; Wolak, Finkelhor, & Mitchell, 2004; Ybarra & Mitchell, 2008). Although a contested finding, children are more likely to talk to strangers online and to be the victims of Internet-initiated sex crimes, while adolescents are more likely to meet strangers offline and to be the targets of unwanted sexual solicitation (Mesch, 2009; Mitchell et al., 2001; Peter, Valkenburg, & Schouten, 2006; Staksrud et al., 2013; Vandoninck et al., 2012; Wolak et al., 2004; contested by Ybarra & Mitchell, 2008). Similarly, as age increases, the risk of cyberbullying increases (Kowalski & Limber, 2007; Navarro & Jasinski, 2012). Now, unpacking the complexities of online victimization, scholars in this area are beginning to re-conceptualize victimization.

More specifically, scholars are now exploring the complicated realities of cyberbullying (Pyżalski, 2012; Vandebosch & Van Cleemput, 2009; Wolak, Mitchell, & Finkelhor, 2007). For example, perpetrators and victims tend to know each other, there is an overlap between being a victim and a perpetrator, and online aggression between those who know and do not know each other offline differs (Erdur-Baker, 2010; Kowalski & Limber, 2007; Vandoninck et al., 2012; Werner, Bumpus, & Rock, 2010; Wolak et al., 2007; Ybarra & Mitchell, 2008). In light of these findings, it is important to limit the application of the *cyberbullying* label to instances that fit traditional bullying characteristics: imbalance of power, perpetrator's negative intentions, and the repeated targeting of the victim (Pyżalski, 2012).

Within this literature, scholars also explore adult responses to online risks within the familial and educational contexts, primarily focusing on parental practices. Interestingly, most parents are aware of the Internet's benefits and opportunities for their children; however, they experience anxiety and concern about their children's Internet and social media use (Álvarez, Torres, Rodríguez, Padilla, & Rodrigo, 2013; Livingstone, 2007; Madden et al., 2013; Mitchell, Finkelhor, & Wolak, 2005). In response to these anxieties and concerns, parents use either control or warmth strategies with their children (Livingstone, 2007). Control strategies involve regulation, restriction, or surveillance, such as rules about time or site usage, and the use of filters or monitoring software (Álvarez et al., 2013; Valcke, Bonte, De Wever, & Rots, 2010). Warmth strategies are based on active communication between parent and child, such as talking about risks and experiences, and spending time on the Internet together (Álvarez et al., 2013; Valcke et al., 2010). The most common practices involve communication and rule-setting about usage, and only a minority of parents use filtering and blocking software (Álvarez et al., 2013; Eastin, Greenberg, & Hofschire, 2006; Livingstone & Helpsper, 2007; Mitchell et al., 2005; Valcke et al., 2010).

In exploring parental practices, it is important to consider differences across parent groups. Parents of higher socio-economic status and those with high levels of Internet use are more likely to mediate their children's Internet use through communication and rule-setting, and less likely to use restrictive monitoring or technological means (Álvarez et al., 2013; Livingstone & Helpsper, 2007; Padilla-Walker, Coyne, Fraser, Dyer, & Yorgason, 2012; Valcke et al., 2010). The parents of younger children are far more likely to mediate, while the parents of boys are more likely to use direct supervision and time-or-content restrictions (Álvarez et al., 2013; Eastin et al., 2006; Lee, 2013; Livingstone & Helpsper, 2007; Mitchell et al., 2005; Padilla-Walker & Coyne, 2011; Padilla-Walker et al., 2012). Most of these strategies have been found ineffective; however, parental monitoring can reduce risk of being exposed to strangers online and being the target of unwanted sexual solicitation, and in some instances the use of filtering software is effective in reducing the risk of online aggressive behaviors (Livingstone & Helpsper, 2007; Mesch, 2009; Mitchell et al., 2001; Navarro & Jasinski, 2012; Shin, Huh, & Faber, 2012; Williams & Merten, 2011).

Similar to parents, school personnel conceptualize youth's online access both as a possible risk and as a pedagogical resource (Greenhow, 2011; Notley, 2008). Pedagogically, social networking sites can be used for peer-to-peer knowledge sharing and instilling civic engagement (Greenhow, 2011; Itō, 2006). However, the risk perspective dominates educational policy about social networking sites. From the educational perspective, social media can perpetuate and promote bullying, sexual victimization, and the unwanted spread of pornography and violence (Bhat, 2008; Notley, 2008). In public schools, browser filtering keeps students from accessing web-based emails, social networking sites, and instant-sharing sites (Meeder, 2005; Notley, 2008). Especially concerned with school-based cyberbullying, schools attempt to control the *impulsive* cyberbully and prevent victimization (Bhat, 2008). However, research in this area is extremely limited and in much need of development (Pascoe, 2011).

In discussing adults' attempts to control youth's social media use, it is important to note the difference between adults and youth's perceptions. For instance, Livingstone and Bober (2013) find that while 86% of parents ($n = 677$) said they did not allow their children to give out information on the Internet, only 49% of their children ($n = 1{,}257$) confirmed that their parents restricted the information they could share. More broadly, youth view parental attempts to monitor their usage as curtailing their freedom and actively resist (Davies, 2011; Livingstone & Bober, 2013). While younger children tend to be compliant with parents' opinions and rules due to their own fears and uncertainties about the Internet, adolescents actively resist, giving the appearance of compliance while in actuality disobeying parental rules (Davies, 2011).

Resistance to adult control is part of youth's online practices. Youth delete emails and web browser histories, minimize windows when someone walks in, and hide or mislabel files (Livingstone & Bober, 2013). Although social networking sites are public, youth build barriers to prevent the infiltration of adult authority, achieving privacy in public (boyd, 2007, 2014). Youth report low levels of concern about third parties accessing their social media data; and instead are in search of privacy from adults who have power over their lives (boyd, 2014; Madden et al., 2013). Privacy is not in opposition to public, but must be continuously achieved (boyd, 2014). Thus, "[p]rivacy doesn't just depend on agency; being able to achieve privacy is an expression of agency" (boyd, 2014, p. 76). Beyond technological maneuvering, youth achieve privacy through controlling the situation, focusing on omission, encoding messages, and adapting to ongoing adult surveillance (boyd, 2014; Madden et al., 2013).

Interestingly, today's social world is no longer online or offline. Instead, it is a hybrid online-offline social world where youth see social media use as an "extension of offline interactions" (Madden et al., 2013, p. 6). In this context, parents and school staff are very present in youth's digital lives. With a nationally representative sample ($N = 802$ teens), Madden and colleagues (2013) describe a very family-friendly Facebook, with 91% of youth *friending* extended family, 76% *friending* siblings, and 70% *friending* parents. Although less prevalent, 30% of youth *friend* their teachers or coaches, with older youth (14–17 years old) and girls *friending* school staff at higher rates than their counterparts. While adult presence dominates Facebook, other social networking sites are attracting youth and youth's enthusiasm for Facebook is "waning" (Gao, 2015; Lenhart et al., 2015; Madden et al., 2013, p. 7).

Thus, it is important to explore how youth perceive and experience *familial* and *educational-adult authorities* across networked publics.

Social Media, Interpretative Reproduction, and Peer Culture

Sociologists focus on the study of youth peer cultures (Adler & Adler, 1998; Eder, 1995; Pascoe, 2011; Thorne, 2015). According to interpretative reproduction, youth collectively aim to interpret the world around them and consequently "produce their own peer worlds and cultures" (Corsaro, 2014, p. 23). While constrained by social structure and societal reproduction, youth create and participate in a series of peer cultures across the life course and institutional contexts (Corsaro, 2014). As a peer-driven social phenomenon, social media use is largely connected to local peer cultures (boyd, 2007, 2014).

The mass movement from MySpace to Facebook in 2005–2007 reflects the significance of local peer cultures in youth's social media use (boyd, 2014; Livingstone, 2008). While MySpace and Facebook were both created around the same time (2003 and 2004, respectively), Facebook did not open usage to the high school population until 2006, at which time MySpace had been firmly established as the go-to youth social networking site (boyd, 2014). Between 2005 and 2007, youth could choose between two social networking sites of fairly equal popularity (boyd, 2014). However, the choice to move (or not move) from MySpace to Facebook was rooted in how the sites were understood within local peer cultures (boyd, 2013, 2014; Livingstone, 2008).

For Livingstone's British respondents, choice of site was understood as a marker of relative maturity within local peer culture, with older teens moving to Facebook away from the less mature MySpace used by younger teens (2008). boyd (2013, 2014) shows that in some American local peer cultures, MySpace and Facebook became defined in racial and class terms. White teens began leaving MySpace for Facebook, as MySpace became associated with blackness and assigned the cultural markers of *ghetto*, *hip hop*, and *low brow*, while Facebook became associated with whiteness and assigned the cultural markers of *classy*, *trendy*, and *highbrow* (boyd, 2013, 2014). The movement of white teens from MySpace to Facebook did not create new racial divisions, but was rather the reflection of previously existing racial divisions (boyd, 2013, 2014). Both boyd (2014) and Livingstone (2008) illustrate how youth's choices about social networking sites occurred within the context of local peer cultures and how the meanings local peer cultures assigned to social media are related to structural divisions and concerns within those peer cultures.

Thus, it is important to explore how youth perceive and experience familial and educational-adult authorities across networked publics in relation to *their local peer culture*.

RESEARCH METHODS

In rejecting top-down approaches, sociology of childhood scholars embrace qualitative methods, using ethnographic approaches to study youth's peer cultures and interviews to uncover youth's perspectives and lived experiences (Greig, Taylor, & MacKay, 2007; Lange & Mirendoff, 2009). In practice, researchers must negotiate their status as adults, use age-appropriate language and methods, and respect youth's agency (Corsaro & Molinari, 2008; Engel, 2005; Greig et al., 2007; Mayall, 2008; Roberts, 2008; Westcott & Littleton, 2006). In adapting a similar child-centered approach, researchers studying youth's social media use have also embraced the use of qualitative methodologies, developing methods which span both online and offline to gain a fuller picture of youth's social media worlds (boyd, 2007, 2012, 2013, 2014; Itō et al., 2010; Livingstone, 2008; Ringrose & Barajas, 2011; Rueda, Lindsay, & Williams, 2014).

boyd (2007, 2012) and Itō and colleagues (2010) combined online observations of youth social media spaces with offline ethnographic observations and interviews to capture youth's social media use from multiple perspectives. Pushing the methodology forward, Ringrose and Barajas (2011) conducted initial interviews, gained access to participants' social media profiles, and then conducted a follow-up interview to discuss the social media content with participants. Livingstone (2008) went even further, as she looked at the participant's social media profiles with her participants during the interview (Livingstone, 2008). More recently, Rueda and colleagues (2014) videotaped teenage couples' conversations about their relationship-related social media performance. Moving these methodological approaches forward, this study uses a technologically enhanced semi-structured interview that involves a youth-centered and participant-driven social media tour.

As part of a larger qualitative study, 82 youth shared their lived experiences and perspectives about social media use. Data collection occurred during the summer of 2013 at four youth centers in a major metropolitan area in the Northeast region of the United States. The youth centers are located in an urban environment and offer after-school and summer programming for youth (6–17 years of age), ranging from athletic activities to college-preparatory programming. During their summer programming, all eligible youth (9–17 year olds) received an invitation to participate in this study.[3]

The data-collection process involved an initial interview (open to 9–17 year olds) and a follow-up interview about social media use (open to 10–17 year olds).[4] During the audio-recorded initial interview, youth discussed peer social life, including their lived experiences and perspectives about social media. During the social media follow-up interview, participants gave the interviewer a youth-centered and participant-driven tour of their social media accounts across social networking sites (here after social media tour). Thus, this analysis will focus on youth's discussion about social media during the initial interview and the social media tour.

Within the population at the youth centers, youth varied in their social media use and parental knowledge of their social media activities, which ranged from no use to hidden, semi-hidden, and open social media use. Thus, youth's navigation of adult authority affected their decision to participate in the social media tour. Each interview had its own parental informed consent and child assent forms; thus, participation in the social media tour required parental knowledge of youth's social media use. In considering participation, youth reflected on their willingness to share their social media with an adult interviewer, their concerns with issues of privacy, whether their parents knew about their social media use, and whether to tell their parents about their social media use in order to acquire parental informed consent. Of the 82 youth participating in the overall study, 10 nine year olds were automatically excluded due to the required technological literacy and 22 chose not to participate in the social media tour without disclosing the basis of their decision.

Subsequently, it is possible that youth's participation in the social media tour is limited to those who keep open lines of communication with their parents about their social media use. To combat this limitation, discussions about social media in the initial interview have been included in this analysis, adding youth who do not use social media and youth whose parents do not know about their social media use. Moreover, parental knowledge about youth's social media use varied across social networking sites, inadvertently leading to the inclusion of youth with semi-hidden social media use in the social media tour. Ultimately, participants varied in their social media use, ranging from no use to hidden, semi-hidden, and open social media use.

The interviews took place at the participants' respective youth centers, between 9am and 3pm for 9–12 year olds and between 9am and 7pm for 13–17 year olds. The data-collection team included a 32-year-old Latina woman, two 21-year-old white/Caucasian women, and an 18-year-old Latino man. We spent two weeks participating in programming activities, never exercising our adult authority and continuously aiming to build rapport with prospective participants. Most importantly, we prioritized the interviewer–participant match with the most rapport to facilitate the interview process. While navigating space availability at the youth centers, the initial interviews occurred in private and semi-private office spaces and at the edge of the out-door football field to secure audio privacy. Due to its technological demands, the social media tours occurred in private and semi-private rooms with Internet access at the youth centers, where only the youth participant and the adult interviewer had visual access to the participant's social media accounts.

Once in the interview space and with parental consent in hand, the interviewer reviewed the assent form with each participant, answered any questions, and continuously encouraged the participant to exercise his or her agency. Participation in the study was confidential, with participants' identities protected through self-selected pseudonyms. The only exception to confidentiality was clearly stated in the parental informed consent form and the youth informed assent form. "When the evidence of *significant harm* is present, the principal investigator (Campos-Holland) will act towards reducing or eliminating harm" (Field Notes, summer 2013).[5] The interviewer then initiated the data-collection process with the following script:

> In this interview, you are my teacher and I am your student. I am very interested in learning about your experience and opinion about social media ... Although I am an adult, I am not here to tell you what to do ... *It's okay if we run into something other adults consider inappropriate during the tour. I am here to learn from you about how youth experience social media, including what youth see.* (Field Notes, summer 2013)

The last two sentences were added after the first three social media tours were conducted to respond to participants who scrolled quickly past others' posts. The fast scrolling occurred over posts that participants considered *inappropriate* for adult view, primarily posts involving *bad words*. The edited introductory script did reduce the fast scrolling, but we cannot confirm that participants stopped self-editing. After all, it is a *participant-driven* social media tour.

During the social media tour, the interview setup included a computer with Internet connection, screen-video-recording software, and an audio recorder for backup, all of which were introduced to each participant. With accounts open and in control of the mouse, the participant engaged in the social media tour, taking the interviewer from one social networking site to another and reflecting as a content producer, audience member, and critic of social media use.

For the purpose of this paper, we analyzed the social media section within the initial interview and the overall social media tour. However, five participants chose to skip the social media section in the initial interview, but fully engaged during the social media tour. Of the 82 participants, 32 youth participated only in the initial interview, 45 youth participated in both the initial interview and the social media tour, and 5 youth participated only in the social media tour. The 82 participants are 9–17 years of age, 43 females, and 39 males, and of African American, Latino/a, Jamaican, mixed, or other descent (see Table 1 for details). Thus, this is an analysis of how minority youth perceive and experience adult authority within networked publics. Their social media experience included no use, Facebook, Twitter, Instagram, Ask.fm, Tumblr, Pinterest, Muzy, Vine, YouTube, ooVoo, kik, and/or other social mobile apps.

This technologically enhanced qualitative approach involves complex data management and analysis. The initial interviews (average of 57.73 minutes) and social media tours (average of 43.10 minutes) went through

Table 1. Sample Demographics.

Race/Ethnicity	Gender	Age									Total
		9	10	11	12	13	14	15	16	17	
African American	Girl	4	3	3	3	3	1	2	–	–	19
	Boy	3	6	3	3	3	1	2	1	1	23
Latino/a	Girl	1	–	1	2	3	2	2	1	–	12
	Boy	–	–	–	–	1	1	1	–	–	03
Jamaican	Girl	–	–	1	1	1	–	–	–	–	03
	Boy	–	–	1	–	–	–	1	1	1	04
Mixed[a]	Girl	1	1	1	1	–	1	–	–	2	07
	Boy	–	1	2	–	1	1	–	–	–	05
Other[b]	Girl	–	–	–	–	–	–	–	–	2	02
	Boy	1	1	1	–	1	–	–	–	–	04
Total		10	12	13	10	13	7	8	3	6	$N=82$

[a]The "mixed" category includes 4 white/Latino, 2 African American/Latino, 1 African American/Native American, 1 African American/white, 1 African American/Filipino, 1 African American/Jamaican, 1 African American/Nigerian, and 1white/African American/ Latino youth.

[b]The "other" category includes 2 white/Caucasian, 2 Saint Lucian, and 2 unidentified youth.

first transcription by person one and second transcription by person two to check for accuracy. In addition, the social media tour went through a third transcription process involving one of the transcribers weaving the participant–interviewer conversation with the social media content, including the social media item being discussed and responses to the item by the social media audience.[6]

In a collaborative process, we used Atlas.ti to carry out a thematic analysis. The first and second author established a coding approach that takes into account the different types of data: general discussion about social media, discussion of their and others' social media use, live social media use during the interview, and social media content never discussed. The first and second author then engaged in general thematic coding, using very broad and inclusive codes. Among other themes, youth's discussion about adults was prevalent, producing a strong "adults" code. For the purpose of this paper, the first and fourth author then sat side-by-side and engaged in a detailed and reflective coding process of the "adults" code. We developed a coding system – adding, defining, merging, and separating codes – eventually returning to recode data with codes that emerged in the later part of the analysis. Upon difference in perspective, we listened to each other and adjusted the coding system to best reflect youth's perspectives. To finalize the analysis, the first author engaged in a comparative analysis to explore youth's situated perspectives and experiences across gender and age. Throughout the analysis, we will specify when differences are present – middle childhood, pre-adolescent, versus adolescent peer groups,[7] or girls versus boys.[8] Overall, youth discussed adult authority in relation to their social media use, their perspective on adults' social media performance, and their interaction or lack of interaction with adult authority across social networking sites.

RESULTS

Defying common perception, not all participating youth were social media users or were interested in social media use. However, social media is part of contemporary social life and participants were not immune to this reality. Active social media users in this local youth culture shared their talents, created and participated in social movements, bought and sold products, and performed an array of social roles within their networked publics. However, youth's social media use was subject to varying and at times conflicting normative expectations emerging from family life, the educational context, and the local peer culture. Intensifying the complexity, youth continuously worked toward achieving privacy from adults with power over their everyday lives. In today's hybrid online-offline social world, youth navigate adult authority across normative schemas and networked publics. In this section, we explore the variation in social media use across age groups, adult authority and youth's social media use within the familial and educational contexts, the collision of peer, familial, and educational normative expectations, and youth's tactics in this complex dance for freedom.

Age Gaps in Social Media Use

In this local peer culture, adults dominated Facebook, creating a cyber-family room for youth, but were lagging behind in other social networking sites. Youth used this gap in adult presence to achieve privacy from adult authority across their networked publics.

Youth's Social Media Use

With a few exceptions, introduction to social media use was a gradual process in this local peer culture, with youth in age-segregated peer groups varying in their social media exposure and use. Most youth initiated social media use during middle childhood. If the youth was in middle-childhood during data collection, they were being introduced through instant messaging. Nine-year-old girl Lany began the social media section in the initial interview with, "I don't have none of those," while 11-year-old boy Kevin stated, "Facebook? I know about it, but I don't have one. Like at my school, people pass time on Facebook and they say that they talk to people on it." Instead, youth in middle-childhood peer groups were slowly being introduced to social media through instant communication apps, such as ooVoo, kik, Tango, TextPlus, and Skype. Initiation was local and peer-driven, with youth asking one another, "What is your kik?" As Sponge Bob, 9-year-old boy, illustrated:

> Teens have Facebook. I've never seen any little kid my age with Facebook. I have an ooVoo account [*video chat and instant messaging app*]. Well, [*my friend's*] dad has a tablet, but he doesn't know where it is. So, I'm guessing we can't really talk.

Similarly, 9-year-old girl Benia stated, "I don't know about Facebook. I just go on TextPlus to text my sisters, my friends, and my brother. The app is free." The youth in middle-childhood peer groups were focused on the instant communication apps provide.

As participants transitioned into the pre-adolescent peer group, social networking sites began to infiltrate their local peer culture. Tamrah, 13-year-old girl, illustrated:

> When I was 10, I had Oovoo. Facebook, I just started when I was 11. I was really excited. I was like, "Please mommy, make me an account, please, please!" So it was almost like near Christmas and she told me, "On Christmas, you can get it." I'm like, "No, I wanna get it now!" At first I didn't know how to work it. And then when I was 13, I had Instagram.

Similarly, Coffee, 12-year-old girl, experienced a family-initiation into social media use, but her use became peer-driven:

> My mom and I used to play the games on Facebook. There used to be this game with pets and they were adorable. I haven't been on it in years, [*points at the end of her Facebook timeline*], 2007. And then when I got older, 4th or 5th grade, I started using it more 'cause everybody [*peers*] started getting one.

Whether it was a family-initiation and/or peer-driven initiation, youth in the adolescent peer group began to engage in peer-driven use on social networking sites.

Youth in the adolescent peer group were moving or already engaging across social networking sites in search of one that meets their current social needs. Seventeen-year-old boy Superman got creative in narrating these movements, "The first thing was Myspace, and then it went to Facebook, Twitter, Instagram, Vine and back to Instagram. Vine and Instagram were in a car together, crashed, and Vine died!" Harvey, 15-year-old boy, discussed differences across social networking sites:

> I don't go on Facebook because it gets boring after a while. You're looking at the same thing every 24/7. *So, how is Instagram not boring?* 'Cause you're taking pictures of your shoes, getting a lot of likes and comments. It's another Facebook, but better.

Overall, at the time of data collection, the middle-childhood peer group was moving from no social media use to instant messaging, the pre-adolescent peer group was adapting to social networking sites, and the adolescent peer group was moving or engaging in different types of use across social networking sites.

Adult Social Media Presence

Youth in this local peer culture experienced a variety of adult social media presence, from no use to family-oriented use. Sixteen-year-old girl Blue explained, "Some adults I know have social media and they like using it. And some adults are like, 'Eh [*not interested*], I'm not really on that page.' It takes up time and sometimes just kids use it." Reflecting on adult social media use, 13-year-old boy Aries explained, "Adults use it to find friends and their family members, and to talk to them. That's how my mom uses it, to check messages." Adult differences in use influenced how youth experience social media.

From youth's perspectives, adult presence dominated Facebook, creating a cyber-family room where adults shared family life with friends and family members. Khalise, 10-year-old boy, illustrated:

> This is my dad's page. [*As he clicks through his father's Facebook photos*], that's when he was in the army, [*next*] from when he was a kid, [*next*] him and my baby brother, [*next*] my football team, [*next*] and that's after a football game [*pointing at his father in a restaurant*]. We went out eating to celebrate.

Pricella, 13-year-old girl, experienced her mother's social performance in a different way:

> [*Scrolling down her newsfeed on Facebook, Pricella stops and points at a series of posts*], this is all my mom. She be sharing things:
>
> Mom: cooking dinner for the fam. cant wait to go on vacation two more days ... yes relazing time, beach, pool, wine, beer, sand, and my hubby and kids – with [*Pricella*] [*received 14 likes*]

> Mom: spending time with my baby [*Pricella*] watching movies loving the time with my boo [*received 10 likes*]
>
> Mom: at the park with the family [*with Pricella and other family members*] [*received 5 likes*]
>
> I don't think that she should have [*a Facebook*]. I was like, "No, 'cause then you're gonna embarrass me."

Adult dominance made Facebook the cyber-family room, and youth responded with tactful engagement or annoyance.

While adults dominated Facebook, this family experience did not transfer across social networking sites. As 13-year-old girl Furtenfur explained, "Like for Facebook, some adults might see it as a way to reconnect with high school friends, but like Instagram might be stupid to them." The difference in adult and youth's perspectives on Facebook was part of family conversation. Harvey, 15-year-old boy, recalled, "I tell my dad, 'Facebook is getting boring.' He was like, 'No it's not. It's fun.' I said, 'Okay, you stay on boring life.'" In this local context, youth were simultaneously experiencing shifts in their social media use and a variation in adults' social media use. Of the adults in participants' lives, not all adults were using social media and those who did were taking a family-oriented approach, dominating Facebook and hesitating about using other social networking sites.

Adult-Authority Gap

This variation in adult social media use translated into the uneven distribution of adult authority across social networking sites. Thus, most youth in this local peer culture experienced dominant and at times invasive adult authority on Facebook, with no or weak adult authority on Instagram, Twitter, Ask.fm, and other social networking sites. As Noodles, 11-year-old girl, responded, "*Does your mom have your password?* Only Facebook. *Does she know about your Instagram?* She knows about Instagram, but she doesn't really care about it." The lack of adult presence created a gap in adult authority across social networking sites. Iman, 15-year-old girl, illustrated:

> I think teens use [*social media*] to communicate in ways that parents can't really see. 'Cause a lot of parents have Facebook, but parents don't really make Instagram, Twitter, Vine, Tumblr and all those things. So teens have like their space where they can express themselves: how they feel, however they want, and to anyone.

Thus, in their continuous negotiation for freedom, participating youth achieved privacy from adult authority within the adult-authority gap across social networking sites. However, adult authority was a strong force and maneuvering around it was not an easy task, especially when adult authorities in the educational and familial contexts were constantly imposing normative expectations onto youth's social media use.

Familial Authority and Youth's Social Media Use

In this local peer culture, adult authority was concentrated within the familial context, with parents (legal guardians) holding the primary authoritative role and other adult family members offering supplemental assistance. However, the way adult authority shaped youth's social media use varied in conjunction with adults' varying experiences with social media use. Interestingly, some parents initiated youth's social media use on Facebook, but in general parents prohibited, guided, and/or surveilled youth's social media use.

Facebook Initiation

With a strong adult presence on Facebook, youth's initiation into social media use was a family matter in this local peer culture. With limited access to a phone or computer, a portion of the youth was first exposed to social media on family members' accounts during middle childhood. At the time of data collection, Ella, 9-year-old girl, was experiencing Facebook through her father's social media use, "I see my dad on the computer, Facebook. It looks like people are posting pictures, writing about their families. I think it's cool. It's just that I really want my own, but I don't have an email." Beyond exposure, 10-year-old boy Khalise annoyed his way into getting a Facebook account, "I was going on my dad's Facebook, talking to my friends, and he just gave me one because he wanted me to stay off his." Family-initiation into social media thus transitioned into a more peer-driven use.

Once social media infiltrated peer culture, youth's negotiation with adult authority for access began. Furtenfur, 13-year-old girl, recalled, "Everyone [*peers*] was bugging me about it and then I asked my mom and my dad if I could have one. That was 2011, literally be-g-g-ged them for one. They said yes and I made one on Easter." With youth pleading their way through the negotiation, some parents eventually gave in. Sebble, 11-year-old boy, responded, "*Why do you have Instagram?* Well, 'cause everyone else had one. My mom wouldn't let me for a while, 'cause I wasn't mature enough. *What made her change her mind?* I asked her, 'Can I please get one?! Please!

Please!'" Family-driven initiations were beginning to compete with peer-driven initiations in this local peer culture.

For a significant portion of youths with parents who use social media, parents were involved in the enrollment process. Elaine, 11-year-old girl, had recently gotten her first Facebook account, "My mom signed me up last month. *What was it like to get it?* I had to lie about the year I was born. If I put 2001, it wouldn't let me sign up." Other parents held off until the required age. Maria, 17-year-old girl, recalls, "I got Facebook on my 13th birthday. It was like, 'The only way I'm gonna get a Facebook is if I could create one on my own.' The agreement: add my mom so she could see what I was posting." Just as Facebook was the cyber-family room, initiation into Facebook was a family matter in this local peer culture. This being the case, a portion of the youth requesting permission to access social media were met with prohibition.

Social Media Prohibition

Youth in this local peer culture came to understand prohibition as an extension of adults' concern for their age status, anticipated misconduct, and anticipated exposure to inappropriate social media content. Most social networking sites require the minimum of 13 years of age to open an account, including Tumblr, Snapchat, Instagram, Twitter, and Facebook. Thus, youth experienced adults' struggle with the "old-enough" dilemma. Noodles, 11-year-old girl, explained, "Adults think, 'Oh they need to be safe. I don't think they should be having it until they're a teenager.' I think 10 should be the limit." Teal, 11-year-old girl, described her experience with the "old-enough" dilemma:

> My mom said I'm not old enough. *So, when are you old enough?* I have no clue. I asked if I could have a Facebook. She said, "No." "Twitter?" "No." "Instagram?" "No." *Why does it interest you?* Because when I'm on there, I can talk to my family, my friends, put quotes up there, and she still said, "No."

The "old-enough" dilemma clashed with youth's strong desires for access to social media.

Beyond the "old-enough" dilemma, parents expressed to participants a concern for youth's potential misconduct online. Jadah, 10-year-old girl, explained, "Posting half naked pictures, that's not good. And adults are scared that their kids would do that. *What should adults do?* Talk to their kids." In trying to make sense of it, adolescents contrasted adults' fears with the everyday realities of youth's social media use. Astrici, 15-year-old girl, stated:

> Adults should be more open to it, but they're not. And I don't see why. They feel like kids are going to be doing stuff on here that they're not supposed to, but we do not. Like half the time we're on [*social media*] bored anyway. Like you be up at like 2:00am [*addressing peer networks*], "I'm up. I'm hungry. Somebody call me. Who is up?!" It's harmless, completely harmless, but they don't realize that.

Adult fear of youth misconduct lost power when it did not match youth's lived experiences with social media.

In addition to citing fears of potential misconduct, adults expressed to youth, especially girls, their desire to protect them from others' inappropriate social media performance. Aaliyah, 10-year-old girl, illustrated:

> Like my grandmother, if she sees something like a boy put one of those cute-type photos [*ab selfies*], she'll be like, "No, come off. I don't want you hanging around with all 'em boys, talking to all these boys." She says, "Whatever you said in the past can affect you in the present, and whatever you say now can affect you in the future." He-he-he! [*Sarcastic laugh*].

Beyond girls, adults expressed to youth concerns about anticipated victimization. Maria, 14-year-old girl, explained:

> Some adults view it as a terrible way for children to hang out. "That's awful. No child should hang out on social media." It's because of all the bad things that actually do go on, on social media. *What kind of bad things?* Um, like the drama, the picking on, and stuff they really don't wanna get their kid involved in. Like that's one of the reason my mom was very hesitant about it.

Whether it was the "old-enough" dilemma, anticipated youth misconduct, or anticipated victimization, a portion of adults in participants' lives moved towards the prohibition of social media use.

Adult Guidance on Social Media

Beyond family-initiation and prohibition, a portion of the adults in participants' lives became invested in guiding youth's social media use. This was a prevalent experience for participating youth with family-initiated accounts and/or those who engaged in open negotiations and eventually gained access, both usually on Facebook. Through guidance, adults in participants' lives encouraged social media use in moderation. Tiara, 13-year-old girl, explained, "My grandma [*legal guardian*] was like, 'Don't be on it too much, 'cause you're gonna make it your whole life. Be careful.' She wants me to post every two months, not every day." Guidance moved beyond adults expressing concern and onto a constant negotiation. Blue, a 16-year-old girl, illustrated:

> My mom thinks I'm on it a lot. She's like, "You need to do something; get off of that." I'm just like, "Eh" [*uninterested*]. She thinks that is all that teenagers do, and sometimes it's true. Like, I'll just spend time on Instagram, Facebook, and Vine, just in my room. She be like, "I wanna see your face!"

In addition to guiding youth's amount of use, adults in participants' lives encouraged girls to protect themselves from danger and adolescents to control their self-presentations.

Adults in girls' lives were especially expressive about issues of safety and encouraged careful information sharing and audience management. Thirteen-year-old girl Tiara recalled her family's advice, "My uncle and my aunt said, 'Be careful! Don't broadcast yourself too much. Like don't say, 'Okay, I'm at Pizza Hut with my sister.'" Participants embraced this advice, especially since sharing one's location could open the door to vulnerability. Girls were also advised to be careful about sharing identifiers on social media. Noodles, 11-year-old girl, responded:

> *How do people choose their name* [*on Twitter*]*?* Like I usually do it based on my name, but not really giving out my name. On Facebook, I spell my name wrong. Like my mom says, "Do something else in your name, because I don't want people tracking you down." I think she's right about that, because people will do the craziest things to children and teenagers.

Possibilities of danger were embedded in this guidance and girls carried this fear across social networking sites.

In addition to not sharing current locations and identifiers, girls were encouraged to carefully manage their audiences through selectivity. Eleven-year-old girl Dee responded, "*Have adults in your family given you advice?* Yeah. Like don't follow someone you don't know 'cause you don't know who that person is." To determine whether a pre-existing connection existed, 11-year-old girl Noodles carefully examined the requester's profile, "Like I gotta look at their pictures, how they look, what they talk about. My mom doesn't want me to request older men, only my friends." Although a few male children did receive similar advice, this safety-based advice was especially prevalent in girls' lived experiences.

Beyond moderate use and girls' safety, participants in their adolescence experienced direct adult guidance on self-presentation. Personal images can be powerful and adults encouraged adolescents to use favorable images in their profiles. ZStar, 13-year-old boy, shared his experience, "On Facebook, when my mom doesn't like my picture, if it's an embarrassing one of me, she wants me to erase it. If I like it, she still tells me not to post it. Once she said I looked immature." This advice expanded to images of everyday objects, even if it was an expression of an interesting experience. Tamrah, 13-year-old girl, illustrated:

> [*Scrolling through her Facebook images*], that is, Ew!! [*Participant is grossed out by a close-up picture she posted of an orange Gatorade cap with a large bug on it; the item received 0 likes and 0 comments*]. I put that picture up because I found my Gatorade top on the floor. I'm like, "How did that bug get in here?" So I took pictures. At least nobody commented on it. My mom was like, "You should delete that picture 'cause that's nasty."

Focused on the experience, adolescents at times shared content that adults in their lives considered inappropriate, especially prevalent for adolescents sharing popular media. Superman, 17-year-old boy, illustrated:

> [*Clicks onto the next image*], Ed, edd, and eddy. [*The image was of a Canadian cartoon about three adolescent boys - ed, edd and eddy – all in a police line-up with text,* "THUG LIFE nigga." *Someone else posted the image and the participant shared it with the caption,* "Love this cartoon. #awesome." *The shared item received 39 likes and 1 comment*]. I had that as my background, but I changed it 'cause my mom saw it. She was like, "Young man, what is this language?"

Attracted to the cartoon, Superman shared an image with inappropriate content and was guided to remove it. Similarly, 12-year-old girl Carmela was listening to *2 live crew* and wrote the title of the song as a status, "face down ass up." Her aunt saw it and, "She told me to delete it. So, I had to delete it." Perception is everything on social media, and adolescents experienced adults' direct guidance on self-presentation, primarily involving the deletion of unflattering images and references to popular culture.

Girls' Experience with Overt Surveillance

Although adult surveillance was generally a part of youth's social media use, overt adult surveillance of their social media performances and that of their networks was primarily the experience of girls in this local peer culture. For 10-year-old girl Khalise, surveillance was part of her negotiation for access, "My dad said, 'I'm gonna be looking at your Facebook.' 'Cause he don't want me doing nothing bad, like cyberbullying or anything. I don't post anything bad or anything." Beyond potential surveillance, girls' parents engaged in overt surveillance, as 12-year-old girl Sticks explained, "My mom hacks my page, trying to find out the password, 'Are you sure this is your password?' 'No!' And then she searches for me and it shows people with my name. So, she clicks on me to see what is up." In addition to overtly monitoring participants' social media performances, girls experienced adults surveilling their social media networks. As 12-year-old girl

Lala explained, "My [*older*] sister, she likes to monitor my Instagram. Like, she knows who I follow. She just checks it a lot. Oh like she doesn't want me to follow this person." Adults in girls' lives also surveilled audience's interaction with youth's profiles. Tamrah, 13-year-old girl, illustrated:

> Yeah, I don't know who tagged me in this, but I don't like it. [*Pointing to a meme image of an exaggeratedly old looking woman with white hair and lots of facial wrinkles with text,* "THIS PICTURE IS CURSED SHARE AND LIKE, IF YOU DON'T SHE'LL KILL YOU IN YOUR SLEEP TONITE"]. My mom like, "Who keeps tagging you in these nasty pictures?"

In girls' experiences, overt adult surveillance involved adults' control of youth's social media performances and protection from others' social media performances. Interestingly, youth's social media profiles were constantly changing, including participants' self-presentations, their networks, and others' self-presentations. With such dynamic networked publics, youth's experiences with overt adult surveillance were always varying in intensity.

Adult Authority is Fragile

Without youth's adherence, adult authority becomes a fragile phenomenon in the familial context. Participants, especially adolescents, began to refrain from requesting permission to access social media or were able to access it on their own. As Astrici, 15-year-old girl, explained, Facebook has lost its value for adolescents in this local peer culture, making prohibition irrelevant:

> I wanted Facebook when it first came out and everybody was on it. I was like, "Oh my God! Facebook this, Facebook that. I wanna get Facebook." My mom was like, "No, for what? You don't need one." "What?!!" So she lets me make one this year. I'm like, "I don't want this no more." That's why I'm like hardly ever on it. Everybody is like on Instagram.

As youth peer culture outpaced adult prohibition, adult authority lost strength.

In the complexities of prohibition, some youth came to perceive it as *unfair*. Courtney is a 13-year-old girl who lives in a blended household with her father, stepmother, two stepsiblings, and a baby sibling. Unlike her teen stepsister, who has a family-initiated Facebook account and is open about her other social media accounts with her mother, Courtney did not negotiate for access with her father. Instead, she created social media accounts without his consent. After being found out, her father permanently banned her from social media, "I was like, 'Ah, man! It's done for me!' I kinda think that's so unfair because they don't look through [*my sister's*]. She has a Facebook, Instagram, kik, Oovoo, like all these social media. And I can't have one without getting in trouble." In response, Courtney began to engage in hidden social media use.

In addition to prohibition, adult authority lost value when participants' perceived adult guidance as *unfair*. Harvey, 15-year-old boy, is a passionate Celtics fan, and his half-brother from his father's side enjoys teasing him:

See, my brother did that [*referring to a photo shared at the top of Harvey's Facebook timeline*]:

> Harvey's account shared ***Adding Games***' image [*4-picture collage with a picture of Jesus knocking on a door with imposed text "Like," a picture of Ariana Grande and Nicki Minaj taking a selfie together with imposed text "Like," a picture of four excited basketball players holding a trophy (Miami Heat's Dwyane Wade, Lebron James, Chris Bosh, and Norris Cole) with imposed text "Share," and a picture of a zombie face with decomposing flesh with text "Ignore"*]. With a caption from the original source [*not from participant's account*] that read, "Vote Fast!! Dad: Suck my dick. Daughter: Ewww, wtf!? Dad: Listen to me now!! Daughter: If Mom saw us? … See more."[9] [*received 3 likes*].

> I let my brother use my phone. He was trying to talk to my girlfriend, "Are you mean to my brother?!" And I guess he liked the wrong thing [*on Facebook*]. I got blasted at for that. My mom thought I liked all that stuff, but I didn't. I was like, "No, it wasn't me. It was my brother, 'cause he is using my phone right now, while I'm outside."

Taking into account that his brother posted the item and that the content was not created by his brother, Harvey perceived his mother's reaction as unfair. With this being such a recent event, he began to avoid Facebook. Whether parents created family-initiated accounts for youth, took the prohibitionist, guidance, or surveillance approach, adolescents came to question the relevance and fairness of familial-adult authority.

Educational Authority and Youth's Social Media Use

In addition to the familial context, adult authority is a prominent feature of the educational context. Thus, youth in this peer culture were experiencing an ongoing anti-cyberbullying campaign, an incapacitation approach to social media, and a few participants encountered hyper surveillance through social media by school personnel.

Deterring Cyberbullying

While all youth in this local peer culture were experiencing a heavy-handed anti-cyberbullying curriculum, their perception differed across age-segregated peer groups. The middle-childhood peer group, without access to social media, equated social media to cyberbullying. Refractor, 10-year-old boy, demonstrated this in his response, "*What is social media?* That is a thing called cyberbullying. That wrestler came and said, 'This year 4,000 to 5,000 kids will lose their lives to bullying.' I shouldn't bully. If I am bullied, tell a parent or a teacher." This anti-cyberbullying campaign instilled fear in the experiences of the middle-childhood peer group. As Jadah, 10-year-old girl, illustrated, "The only time they talk about social media is when they're talking about cyberbullying, 'Block the person or delete your account if you can't handle it.' I feel scared to have a Facebook." The pre-adolescent peer group however, was getting tired of the repetitive message. Lala, 12-year-old girl, mimicked the excitement, "'Speak up! Just ignore it … don't continue the argument.' I think they're right, 'cause a lot of people are killing themselves for that, and blah, blah, blah, blah." Unlike participants in middle childhood or pre-adolescence peer groups, adolescents in this local peer culture were veterans of the anti-cyberbullying campaign and were experienced social media users.

As such, youth in the adolescent peer group reflected on the anti-cyberbullying campaign's relevance to their everyday lives. Workshops on cyberbullying took a deterrence approach, warning youth about the dangers of social media with extreme examples. Superman, 17-year-old boy, explained:

> This guy came to our school who talked about Snapchat and how when you send a picture, they can send it to anyone they want. And then that person sends it to another person. It could be like a nude picture, and then they send it to the whole school. That's just what happens. *Have you seen cyberbullying happening to anyone you know?* No.

Although adolescents in this local peer culture can conceptualize the existence of these extreme victimizations, they had not experienced such events. Thirteen-year-old girl Tiara illustrated:

> It's like a warning. They're not trying to get us off the Internet. They're just trying to say, "Be careful with what you're doing 'cause there's a lot of people who are gonna say stuff about you. So, don't do too much while you're on the sites." *Has cyberbullying actually happened?* Yes there's—, not with me. Nobody I know has cyberbullied before, but like I know that people around there have probably cyberbullied. I'm not sure.

Even when participants discussed events within their local context that somehow were perceived to be related to the extreme examples, these events were too distant from the participants' social circles. Uncle Drew, 17-year-old boy, recalled:

> I remember, in freshmen year, there was a big situation about Facebook. I guess, they put this girl out, because somebody had hacked the page. So those people got suspended for like more than ten days.

The extreme examples in the anti-cyberbullying campaign were powerful, but too distant for the adolescents in this local peer culture. Thus, fear was definitely not the immediate response to social media for the youth in the adolescent peer group.

Incapacitation through a Social Media Ban

In addition to the anti-cyberbullying campaign, youth in this local peer culture, especially adolescents, were experiencing adult authority's struggles with the balance between prioritizing pedagogical goals and youth's social media use within the educational context. Blue, 16-year-old girl, explained, "At school, [*social networking sites*] are blocked from the computers. *Why?* Like if we're supposed to be doing work on the computers and we're on Facebook." However, educational policies continuously involved a cultural lag, as 17-year-old girl Kate stated, "In middle school, they were like, 'You're not supposed to go on Facebook or any social media on school computers.' But everyone pretty much has a laptop, and they do whatever they want." For youth in this local peer culture at the time of data collection, the dominant means of access were mobile apps on youth's cellphones. Vanessa, 14-year-old girl, excitedly explained, "School can't handle it! 'Cause it's all on our phone … once we're at lunch, we can take out our phones and use it then. They can't control what we do on our phones during lunchtime." A few schools however were creating more relevant policies, as 15-year-old girl Iman illustrated:

> At my school, they bag-and-tag you if they catch you on your phone. First time, they take your phone, put it on a little plastic bag, fill out a slip, and at the end of the day you go to the main office to get it. Second time, your parent has to come to get it. Third time, you get detention. *You tweet stuff at school though?* Yeah, sometimes, like if something funny happened, even though you're not supposed to.

Beyond new technology and relevant policies, a few participants had overcome the technological barriers. Seventeen-year-old girl Roxanne clarified, "It's blocked from the computers, but we found ways around it. On the http, you put an 's' on it." Based on participants' lived experiences, adult authority's incapacitation approach within the education context was ineffective.

Hyper Surveillance through Social Media
In addition to the anti-cyberbullying campaign and the social media ban, a few adolescents in the sample discussed adult authority's hyper surveillance in the educational context. In this process, youth experienced how adult authorities slowly stepped into conflicting normative expectations. Aries, 13-year-old boy, illustrated, "Sometimes the principal takes people's phones, goes on their Facebook to see the fight; who was videoing it; and who was next to the fight and did not stop it. That's why teens delete the videos on their phone." Thirteen-year-old girl Tamrah's principal applied hyper surveillance to the whole student body:

> My school is like checking my Facebook. They check the whole school, whoever has Facebook. Yeah, you have to add them as a friend. I be careful what I say on Facebook, too. I tell my friends, "Y'all have to be careful 'cause our principal was listening and looking at our Facebook." *So, earlier you said boys said nasty things on Facebook?* But some of them don't go to my school, but I keep telling them to stop because that's going on my page and then they go and think that I was the one that's writing. And the principal keeps telling me to delete it. I stop going on Facebook. *'Cause the boys or 'cause of your principal?* 'Cause the boys, but sometimes the principal. I know it's not good to keep looking at people's pages. I feel like, it's too much business going on. Like my friend has a picture of her family member who died, and then the principal is all up in people's business. I don't like it.

Regardless of the adult authority's motives, the few adolescents experiencing hyper surveillance began to reflect on whether this is an appropriate approach in the educational context.

Familial, Educational, and Peer Expectations Collide

Within this local peer culture, youth's social media use was governed by conflicting normative schemas that emerge from the familial, educational, and peer contexts. In today's hybrid online-offline social world, these normative schemas eventually collided. Participants experienced the struggle between familial and educational-adult authorities, used adult authority in the educational context to navigate peer conflict, and elicited the help of adult authority in the familial context to fight off peer threats.

Familial Authority Meets Educational Authority
Adult authorities in the familial and educational contexts did not necessarily take the same approach. Zeriah, 14-year-old girl, illustrated this

possibility, "*What does your school say about social media?* They say that as long as you're smart and safe with it: 'cause see, a lot of the kids, they don't have parents who are adults." Interestingly, the indirect or direct clashes between these separate authorities occurred between parents taking the guiding approach and school personnel engaging in hyper surveillance. Sebble, 11-year-old boy, experienced this very tension, "*Have adults at school said anything?* Yes! Right after I got in trouble the principal said, 'You're not even supposed to have Instagram 'til you're 13.' My mom said she didn't care. Well, she didn't say it to the school." With a guiding experience in the familial context and a principal engaging in hyper surveillance, 13-year-old girl Tamrah faced a much more invasive tension:

> Sometimes if we wear inappropriate clothes, you go to the principal's office. I had these shorts on, not little shorts. And she tells me I'm trying to show things. "No, I'm not." It made me mad. And then, in one picture on my Facebook, I had my bathing suit on. My auntie took that picture, and the principal was like, "You gotta cover that." I'm like, "Oh my god." I got mad at her. She like, "Y'all got…." 'Cause some girls like wearing belly shirts and taking pictures, and like I'm not them. I'm wearing perfect shorts, but she told me to stop wearing 'em. I told my mom. My mom is looking at my principal's page and she's telling me to unfriend her. But if I do, I'm the one that's gonna get in trouble. 'Cause my principal wants to continue looking at our pages. And then my cousin wrote something like cussing [*on my page*]. I was like, "Stop!" And then my principal saw it and I had to go to the principal's office. She like, "Tell me the word—." I'm like, "No. I'm not." She was really trying to force me. I'm like, "I'm not saying it!" She's like, "Oh, you staying here." I'm like, "No, I'm not. I'm-a say it and then I'm-a get in more trouble. Are you, the principal around here, trying to boss people to say a bad word?" And then she told me to get out of the principal's office 'cause she couldn't take it no more. *What status was she so upset about?* You know the song Party Rock? And people write that status a lot. I can't say that word. I'm-a just spell it out, L-M-F-A-O. So, she's like, "Tell me what the meaning is." I'm like, "My gosh." She's like, "You're banned from Facebook." I'm like, "Excuse me! My mom is not gonna let me be banned from Facebook." I asked the dean, "Can I call my mom?" I called immediately. The next day my mom came to school. "Why are you trying to make my daughter say the bad word? Are you a principal? She is younger than you. It's not your job to look up Tamrah's Facebook. It's my job."

In addition to the collision of two adult authorities with different normative expectations, Tamrah's social media performance was taken out of context and she was made accountable for others' use of popular media on her timeline. Overall, for a few participants, their adult authorities collided at the crossroads of parental guidance and school's hyper surveillance.

Educational Authority Meets Peer Culture

For this local peer culture, the educational-adult authority failed to separate bullying from mundane conflict in their anti-cyberbullying campaign. Without differentiating, youth in the middle childhood and pre-adolescent peer groups used adult authority in the educational context as a tool in peer conflict. Ten-year-old girl Analiyah illustrated:

> Rachel, sick and tired of her. Anyway. He-he-he [*fake laugh*]! So, she went to my dean, "Oh Analiyah should've never been talking about me on Facebook." I'm like, "I don't talk about you on Facebook." And she was like, "Yes you did. You put it all over your status." I was like, "Okay, dean, can we please go on my Facebook page and see what this girl is talking about?" So she went on Rachel's page and didn't see anything. Then she went on my page, went down all my statues, everything that I've been saying. "Do you see anything?" No. Rachel got in trouble for it, got suspended. And then she kept saying, "Well, it is on there and I'm not lying." And then the dean was like, "No excuses. You made a mistake."

Youth in the middle childhood and pre-adolescent peer groups used their opposition's social media content as evidence to make their case to educational-adult authority and eliminate their enemy during peer conflict. Seventeen-year-old girl Roxanne is the only adolescent who experienced this directly, but it involved adult authorities in different schools:

> [*My cheerleading team*] couldn't go to [*Greenville High*] for the game because some argument the girls had on Twitter [*with the opposing team's cheer leaders*]. We were posting harmless stuff, "Yeah, we're gonna beat them." Then they told their administrators that we were gonna fight them. So they said, "Your cheerleaders can't come." It was stupid.

The lack of differentiation between cyberbullying and mundane conflict made adult authority in the educational context the ideal tool to win peer conflict.

Parental Guidance Meets Peer Conflict

Girls experiencing guidance in the familial context enlisted adult authority in their struggle with peer conflict. Astrici, 15-year-old girl, listed Monica as an ex-friend during her initial interview, but ran into Monica's post during her social media tour, "Remember Monica? *You're still friends with her on Facebook?* Uh-huh [*yes*]. Mommy said to leave her in case she says anything. *Has she said anything about you?* Nm-mm [*no*]. Not that I've seen. She knows I'll get her." Parents taking the guiding approach walked their children through difficult moments, at times encouraging the end of the conflict. As 13-year-old girl Tamrah illustrated, "I like posting pictures to express myself. I put my Jordan's up and someone was like, 'That's ugly.' My mom said, 'Don't keep arguing with them. It's gonna make it bad.'" Furthermore, a few participants experienced the direct collision of two familial authorities. Kyre, 12-year-old girl, recalled her friend's conflict:

> This is about my friend CJ and this girl Lisa. CJ tweeted, "Lisa is a hoe." And Lisa seen it and was like, "How am I a hoe?" And CJ was like, "Because you messed around with boys." Lisa showed her mom and the argument got big on social media. And then her mom went to CJ's house. CJ's mom deactivated CJ's Twitter for a couple of months, took her phone away, and all that. I think Lisa felt different about the fight because everybody on CJ's Twitter seen it. She got retweets and favorites and all that. And so Lisa, she just got put on the spot.

Relative to the positive aspects of peer culture for youth, peer conflict was minor in this local peer culture. However, girls primarily called upon guiding parents to intervene during instances of peer conflict. In the process, guiding parents took a strong role in helping their children, especially girls, respond to peer conflict on and off social networking sites.

Youth Resisting Adult Authority

While navigating adult authority, youth became very aware of their status in contrast to adults, especially in regards to power. ZStar, 13-year-old boy, illustrated, "Like sometimes adults wanna find out what kids are saying and everything. Sometimes they might read their kid's Facebook and Instagram, like their private stuff, to see what they're doing. That's kind of messed up. They shouldn't do that." In time, youth began to develop techniques to navigate the cyber-family room, prohibition, guidance, and surveillance. Experienced pre-adolescents and adolescents actively avoided adult authority across social networking sites and a few questioned the authority of extended family members.

Navigating the Cyber-Family Room Through Performance

With strong social media experience, pre-adolescents and adolescents in this local peer culture learned to navigate their audience(s). Youth who experienced family-initiation and guidance in the cyber-family room limited their presence and maintained a family-oriented social media performance on Facebook. Bubbles, 11-year-old girl, illustrated, "Honestly on Facebook, I don't post statuses, like rarely. If I do, they're nice. Like my mom puts pictures and tags me on them. I mean, there's nothing for me to hide." In addition to inactivity, youth engaged in self-editing. Nicole, 13-year-old girl, avoided specific posts:

> My aunt and my mom's friend are following me. *Does it change the way you post?* No. Well some things, yeah, but some things, no. Like those reading things, I wouldn't post it if it has bad words because I don't cuss. *What would your aunt say?* "What are you talking about?" Or she'll just show my mom.

Across other social networking sites, youth in this local peer culture either maintained a consistent family-oriented performance or began to separate family and peer audiences.

Twelve-year-old boy Zack used a consistent family-oriented performance, "*Does your mom know you have a Twitter?* Mh-mm [*yes*]. She doesn't say anything 'cause she knows I don't do dumb stuff on it, like making fun of other people." Similarly, Zeriah, 14-year-old girl, highlighted the importance of maintaining a family-oriented performance across social networking sites:

> I could be tweeting right in front of them, but they don't look because I'm not trying to hide anything. If they ask me a question, I'll answer their question. I remember one time I was tweeting something and they're like, "What're you doing?" I said, "I'm tweeting." They were like, "Oh, you have a Twitter?" I was like, "Yeah." My dad was like, "Why?" I was like, "'Cause it's fun!" And they shook their heads. *If they saw your Twitter account, what would they say?* I don't know. I mean, it's nothin' to look at. It's not like I have inappropriate stuff on it. It's not like I'm doing pornography or anything. Everybody's on Facebook. It's a family thing. So, I'm not tryin' to hide anything from my family on Twitter, just in case it blows up.

A consistent family-oriented performance across social networking sites appeared to be the least stressful approach for youth in this local peer culture.

However, a family-oriented performance limited youth's engagement in peer culture. Thus, youth began to separate peer and family audiences across social networking sites. Superman, 17-year-old boy, maintained very clear boundaries, "My cousin followed me on Instagram and I blocked her. *Why'd you block her?* No family is allowed to follow me on Instagram." The separation of family and peer audiences required youth's continuous vigilance and boundary reinforcement across social networking sites.

Girls Respond to Prohibition

A significant portion of participants faced adult authority's prohibition of social media in the familial context. While not all youth were obedient, participants within this local peer culture expressed a sense of appreciation for the adult authority's perspectives. However, girls, unlike boys, had to face being an outsider within their local peer culture. Teal, 11-year-old girl, illustrated:

> Well Able and Jaime, they talk to each other on ooVoo. I do not know what that is. Um, sometimes they use Skype, Instagram, Facebook, and Twitter. I feel left out [*distraught sounding*]. I miss out on the good pictures, on the funny quotes. They be like, "Oh it was hilarious! When are you gonna get an Instagram?" I was like, "I'm not allowed to." And so it's like, "Oh."

To make sense of their situation, girls began to question the value peer culture places on social media. Courtney, 13-year-old girl, strongly desired access, but reflected:

> Man! It puts me down! [*Silence*], well, it really doesn't matter to have an Instagram, because it's just posting pictures and liking them. So, that's not important. Facebook is done! And then Twitter is the same thing as Instagram. So like, hm [*thinking*], it really doesn't matter anymore. No, so I'm like, whatever, I don't need any of these [*both laugh*].

Through minimizing the importance of social media within their everyday lives, some girls came to terms with their parents' prohibitive approach.

Moving forward, some girls came to embrace their parents' approach. Iman, 15-year-old girl, recalled:

> I never had a Facebook. I was never really interested in making a Facebook. In middle school, my mom didn't really want me to have one. I was like, "Oh, everybody has a Facebook except for me." But after like a week or two, I didn't really care anymore. It's like, I didn't see the purpose. Like, "You're 12, talking to people that you can just text on your phone or you can just talk to them there."

Wanting to not go against adult authority, girls attempted to take on the adult perspective and reduced the importance of social media in their young lives.

However, peer culture was a strong force and some girls carefully went around prohibition. Zeriah, 14-year-old girl, made her case:

> My parents said no Facebook, nothing about the other [*social networking sites*]. They don't know I have a Twitter, 'cause they would tell me to deactivate it. They think it's not safe, which, if you're smart, then it is. People that follow me, I make sure I know what they're tweeting about. I look through every single picture, read their bio, look at their profile, everything. *So here is the name you have on Twitter? Can they find you?* Mm [*thinking*]—, they'll type in my real name though. So, no! When I first got Twitter, I locked my tweets so that only the people that have to request a follow could see. And then once I got comfortable with Twitter and my current follower, I unlocked it.

In this local peer culture, most prohibition was directed towards Facebook. With lessons learned from Facebook and with access to Twitter, Instagram, and other social networking sites, experienced youth took different strategies to navigate adult authority.

Risk-Prevention Guidance in Action

Youth in this local peer culture continuously received direct and indirect guidance from adult authority in regards to social media. Whether they were engaging in open, semi-hidden, or hidden social media use, participants took various steps to protect their social media activities, including the use of false identifiers to avoid being found. Although the inaccurate birth date was a common experience for those creating social media accounts before 13 years of age, arrival to the required age did not translate to correcting age on social media profiles. These practices separated youth's social media performances from their legal and residential identity, inhibiting strangers familiar with youth's social media performance from coming into their everyday lives, and preventing those in youth's everyday lives from becoming familiar with their social media performance. With adult authority residing in participants' everyday lives, techniques used to fence off anticipated perpetrators were also shielding youth from adult authority. Although participants, especially adolescents, were becoming masters of audience management on social media, the local-driven nature of social networks and the desire for social media popularity threatened youth's achieved privacy from adult authority. Regardless, participants used these foundational risk-prevention tactics to carefully manage their social media audience(s) across social networking sites.

Girls Self-Edit to Combat Heightened Surveillance

A few girls experienced the combination of overt familial surveillance and educational hyper surveillance, and responded with intense self-editing. Pricella, 13-year-old girl, was not allowed to date:

> If you go to your profile, settings, you can put that you are in a relationship or complicated relationship. *When you were with Luis or Alfredo, did you do it?* No! *How come?* My mom! My mom! She would have known 'cause it puts on other people's statuses, "Pricella is single. Pricella is in a relationship. Pricella is that." And I was like, "Can't put that."

In addition to self-editing their social media performances, girls remained vigilant about how others' social media performances interacted with their profiles. Aaliyah, 10-year-old girl, was experiencing overt surveillance at home and hyper surveillance in school:

> Like my friend Kiara, she doesn't like my other friend Aaron. I was talking to him [*on my Facebook timeline*] and I said, "Kiara says hi." And then they started going at it O-N my profile! And I'm like, "Get off my profile! If someone sees this, I'm-a get in trouble!" And then they kept doing it. I deleted the messages, and then they stopped.

Family's overt surveillance and school's hyper surveillance encouraged girls to maintain a self-presentation that carefully satisfied adult authority.

Youth Avoid Adult Authority across Sites

Although youth in this local peer culture were not genuinely interested in adults' social media performances[10] and actively avoided adult friends and followers as risk prevention, their social media audiences did include a few adults. Thus, youth's focus was not the avoidance of adult audience members, but the avoidance of adults with authority in their everyday lives. Fifteen-year-old girl Alexis made it evident on Twitter, "*Does your family follow you?* My aunts, my cousins do. Not my mom, she hasn't followed me yet. *Do you want her to?* No!" Beyond not desiring the presence of adult authority, some participants engaged in active avoidance. Iman, 15-year-old girl and active Twitter user, illustrated:

> My father, he recently made a Twitter. I was like, "Oh my goodness! I can't, I don't want him." I just told him, "I don't use my Twitter that much anymore." And then I went on my Twitter and locked it. [*Also*], I always change my @name 'cause sometimes I get bored looking at it. So, he was like, "Did you have a Twitter? Your @name is?" And I was like, "Oh yeah! No. I don't remember." So they don't question me about it. *What don't you want him to see?* I just know he's gonna question me about everything, like the people that follow me, the people that talk to me, like mention me or say something. He'll be like, "Whose Twitter was that? What tweet said this? Why did you say this? What do you mean by this?" That would make me real nervous. I don't want to have to deal with that. Like just, "It's my space. You can have yours."

Iman's social media experience allowed her to avoid adult authority on social media without outright challenging her father's authority. In addition to carefully avoiding adult authority, some participants openly managed their social media networks to block adult authority.

In this dynamic, peers could be a threat because they can call upon adult authority to control youth's social media performances in the familial and educational contexts. Twelve-year-old boy Roger was perceived as such threat, "My brother said he blocked me. *Why?* He doesn't want me to tell on him." Blocking familial peers was preventive or a response to a situation. Coffee, 12-year-old girl, responded:

> *Does your mom know that you have Instagram?* Yeah. She didn't find out until 3 months after. I didn't want her to know. My sister knows, but she is a snitch. She tells my mom certain things that she wasn't supposed to tell her. My sister, I don't follow her for a reason. She is extremely nosy.

Furthermore, youth experiencing school-based adult surveillance began to reject school-based peers in order to block educational-adult authority. Kyre, 12-year-old girl, carefully separated peer groups:

> I don't follow anybody from my school. If I was going to tweet something, they would probably screenshot it and show it to the teacher. I've seen that happen to my friend Ella one time. She tweeted about this girl and the girl screenshot-ed it and showed the principal. And she got suspended. I don't want that to happen to me.

Youth blocked peers from the familial and educational contexts to avoid adult authorities.

In addition to covert avoidance and eliminating peers from their networks, older adolescents began to openly deny adult authority access to their social networks across social networking sites. Pricella, 13-year-old girl, responded, "*Does your mom know you have an Instagram?* Yeah, I just didn't accept her as a friend. I was like, 'I'm not accepting you as a friend everywhere.'" Superman, 17-year-old boy, was similarly direct about his boundaries regardless of networking site, "*Do you follow family?* Nm-mm [*no*]. Well, two of my cousins have Instagram. My mom has Facebook. She knows I have Facebook, but she knows I'm not gonna friend request her. She's not allowed to friend request me on Facebook." Although youth were not interested in adults' social media performance and continuously avoided adult audience members, the primary concern was adult authority. Thus, experienced adolescents openly rejected adult authority on social media.

Blocking Illegitimate Authority

Beyond managing adult authority across social networking sites, family dynamics affected youth's audience management. Although not a common experience in this local peer culture, a few participants experienced multiple and at times conflicting adult authorities within their families, creating an ideal context for youth to question the authority of extended family members. Cassidy, 15-year-old girl, responded to her complicated family life through social media:

> [*Cassidy is a daughter of a teen single mother, with adult family members usually intervening in her mother's parenting practices. With multiple adult authorities, Cassidy has been subjected to conflicting expectations. Recently, her mother managed to depart from the*

> *family. However, the departure was not amicable and the family conflict moved to social media. Cassidy is friends with her mother on Facebook, but most other adult family members*] are blocked. I don't want them knowing what's going on in my life at all. They were in my business. I told them off and they stopped talking to me. I know that some of them blocked me.

Although not as conflict-driven, Vanessa, 14-year-old girl, had to manage what she perceived as illegitimate authority:

> [*Vanessa lives with her grandmother, mother, older brother, and younger siblings. Her mother takes a guidance approach, following her daughter's social media performance across Facebook, Instagram, and Twitter, except* Ask.fm. *Within the family dynamic, her older brother and grandmother exercise overprotective adult authority*]. Like if my grandma calls me and I don't pick up 'cause I'm doing something, she will call over and over and over. And then when I call her back, it's like a whole issue 'cause she's mad. My mom, she just tells me to call her, when to call her, and just texts me, "What are you doing? Where are you at? Just be home by this certain time." So, only my mom and little brother are on my Facebook. *Did you defriend your older brother or did you just never add him?* [*Whispers*], I blocked him. *Does he know?* [*Shakes head to signify no*]. I had to. I just don't want him to see anything. I can show you. [*She clicks on account, settings, and privacy settings*]. I have a lot of my family blocked. [*Under Privacy Settings and Tools, she goes to the list of blocked people, which includes 2 brothers, brother's friend, grandmother, great aunt, 3 uncles, 2 aunts, and an ex-friend.*] They can never, ever, ever, ever, ever, ever find me. *Does your mom know that you blocked them?* No!

When adult family members attempted to exercise an authority that youth did not recognize, adolescents with social media experience actively blocked these family members. Thus, for a few participants, it is beyond managing adult presence and adult authority, but avoiding the surveillance of what they perceive to be illegitimate adult authority.

Summary of Findings

With a technologically enhanced methodology, we explored how youth perceive and experience familial and educational-adult authorities across networked publics in relation to the local peer culture. An age gap in social media use set the stage for a complicated story. At the time of data collection, the middle-childhood peer group was transitioning from no social media use to instant messaging apps, the pre-adolescent peer group was being introduced to social networking sites, and the adolescent peer group was engaging in diversified social media use across social networking sites. Simultaneously, adults in participants' lives were not using social media or were dominating Facebook, creating a cyber-family room for youth on

Facebook and hesitating to move across other social networking sites. Within this age gap, youth navigated adult authority across their networked publics.

Familial-adult authority was dominant in this local peer culture, with parents (legal guardians) holding the authoritative role and extended adult family members providing assistance. Within this context, a portion of participants were initiated into social media through the cyber-family room on Facebook, while others experienced prohibition. Youth came to understand prohibition as rooted in adult fear: the "old-enough" dilemma, controlling youth's anticipated misconduct, or protecting girls from victimization. Other youth experienced parental guidance, with adults encouraging youth to use social media in moderation, girls to engage in risk prevention, and adolescents to control their self-presentation. Lastly, girls experienced overt surveillance in the familial context, with adult authorities actively managing their social media use. In response, adolescents began to question adult authority, especially as the prohibition of Facebook lost value in the local peer culture and adult authority's decisions came to be perceived as *unfair*.

In the educational context, youth experienced adult authority's attempts to control their social media use. Subjected to the ongoing anti-cyberbullying campaign, the middle-childhood peer group was scared of social media use, the pre-adolescent peer group was getting tired of the repeated message, and the adolescent peer group reflected on the extreme examples that lacked relevance to their everyday lives. Furthermore, adolescents were experiencing adult authority's struggle with pedagogical objectives and youth's social media use, involving the ban of social media from school computers and the school policies' continuous cultural lag. Although not a common experience, a few adolescents were subjected to school's hyper surveillance through social media.

Youth live daily life in the familial and educational contexts and are subject to peer cultural expectations. Thus, conflicting normative expectations eventually collided, including the clash between parental guiding and educational hyper surveillance, children and pre-adolescent girls' use of school personnel to settle mundane conflict, and girls' use of parental guidance to combat peer conflict on social media.

With stronger agency, especially during adolescence, participants reacted to adult authority's attempts to control their social media use. Youth who experienced initiation and guidance on Facebook engaged in family-oriented performance or separated peer and familial audiences across social networking sites; adolescent girls responded to prohibition through minimizing the importance of social media, taking on the adult perspective, or finding a way around prohibition; experienced youth used the risk-prevention tactics to shield their social media use from adult authority; girls who experienced overt familial surveillance and educational hyper surveillance engaged in careful self-editing; experienced adolescents actively guarded their non-Facebook accounts; older adolescents began to openly deny access to adult authority; and a few adolescents blocked illegitimate adult authority.

DISCUSSION

To conclude, we will briefly discuss this study's relevance to the field, limitations and future research, and policy recommendation.

Contribution to the Literature

Moving the field forward, we argue that adult fear is powerful, but fragile to youth's interpretation; networked publics are now adult-regulated and youth's ability to navigate them is based on their social location; and youth's social media use must be contextualized to be holistically understood.

Adult Fear in Action

Fear is at the core of the social construction of childhood and adolescence, prompting adults to protect and control youth (Gannon, 2008; Valentine, 1997). This risk perspective dominated participants' experiences with adult authority in the familial and educational contexts, aiming to prevent youth's victimization on social media and to control youth's social media use. Whether adults took a warmth approach, in this case through guidance, or a control approach, in this case through prohibition, incapacitation, or surveillance (Livingstone, 2007), youth experienced all as the enactment of adult fear through adult authority.

Similar to boyd (2007, 2014), youth exercised their agency to achieve privacy from adult authority. More specifically, youth navigated and re-appropriated adult fear to meet their peer cultural needs. The familiarity with adult authority's power and an awareness of the gap between adult fear and the reality of everyday social media use fostered youth's resistance to adult authority. Moreover, the perception of the *unfair* enactment of adult authority further nurtured youth's resistance. Similar to Valentine (1997) however, youth did not disregard adult fear. Instead, they worked on their risk-management competence, which they carefully practiced while avoiding adult authority.

Thus, adult fear is a dominant characteristic of youth's experience with adult authority and social media use, but youth's interpretation of adult's fear shapes its power.

Adult-Regulated Networked Publics

The social construction of childhood and adolescence involves adults' negation of youth's agency (Coleman et al., 2004; Gannon, 2008; Valentine & Holloway, 2001). Thus, youth are continuously in search of unregulated public spaces to achieve a sense of personhood (Coleman et al., 2004; Corsaro, 2014; James, 2009; Valentine, 1997). At its onset, social networking sites were conceptualized as unregulated publics within adult-regulated socio-spatial environments (boyd, 2007). However, this study encourages the field to rethink this earlier conceptualization.

If adult authorities gain access, youth's social performance is more vulnerable to adult regulation in networked publics than in socio-spatial environments. Networked publics' asynchronous communication, *searchability*, replicability, and invisible audiences (boyd, 2007, 2014) allow adult authority to unexpectedly carry out an extensive evaluation of youth's social media content at any time and from anywhere. If adult authority does not presently have access, there is the possibility that adult authority will unexpectedly gain access in the near future. Thus, networked publics are the new diaries, where youth's self-expression is subjected to everlasting adult authority.

Although networked publics in this local peer culture are adult-regulated, adult authorities are not fluent in the nuances of youth's social media use. Thus, social media content can easily be taken out of context or misinterpreted, launching an *unfair* use of adult authority over youth's lives. For example, a shared item can be interpreted as the youth's content, what was acceptable social performance in one context can easily be decontextualized and subjected to the normative expectation in a different context, and other's social media performance on youth's profile can be interpreted as the youth's performance.

However, motivated youth can easily make networked publics work in their favor (boyd, 2014; Coleman et al., 2004). When youth understand the power of adult authority and the limits of adult social media use, youth can turn adult control mechanisms on their head and slide their way into an unregulated space within or across networked publics. In this local peer culture, youth in middle childhood and pre-adolescent peer groups had a limited understanding of adult authority and opened their social media performance to adult authority's evaluation. In contrast, experienced youth in the adolescent peer group understood the power adult authority could have on their daily lives and creatively avoided its control within and across social networking sites.

In essence, this analysis confirms boyd's (2007, 2014) argument that privacy is not in opposition to public, but achieving privacy from adult authority is an expression of youth's agency. However, the ability to exercise agency is not accessible to all youth. For instance, similar to previous findings (Davies, 2011; Valentine, 1997), a significant portion of the children and girls identified with the adult perspective, limiting the motivation to establish an unregulated space. It was the experienced and motivated adolescents who were able to achieve privacy from adult authority. Youth across social locations varied in their ability to resist adult authority on social media.

Thus, youth navigate adult authority across networked publics; however, youth's resistance needs to be analyzed in relation to their social locations.

Youth's Social Media Use in Context

Similar to the youth migration from Myspace to Facebook in the 2000s (boyd, 2013, 2014; Livingstone, 2008), youth's movements to and from instant messaging apps, Facebook, Twitter, Instagram, Vine, Ask.fm, and other social networking sites was peer-driven in this local peer culture. It was especially evident in participants' persistent requests for access to social networking sites with peers in mind and their continuous navigation of adult authority in relation to the local peer culture.

Thus, this paper reaffirms the need to analyze youth's social media use in relation to the local peer culture. For instance, adult presence on Twitter is a well-established phenomenon (Lasorsa, Lewis, & Holton, 2012; Rane, & Salem, 2012); most relevantly, parents have been known to use Twitter to discuss their children's well-being (Mullen et al., 2015). However, in this local context, youth perceived Twitter as an adult-free space, where peer culture and youth self-expression could flourish without the constraints of adult authority. It is this local contextualization that allowed for this analysis to capture the complex power dynamics that youth navigate and the creative re-appropriation process youth engaged to achieve privacy from adult authority.

Furthermore, this analysis highlights the importance of exploring social media and peer culture in relation to its institutional contexts (Corsaro, 2014; Pascoe, 2011). In this case, youth faced adult authority in the familial and educational contexts, continuously tackling barriers that blocked access to social media and attempted to tame their freedom of expression on social media. It is in relation to these institutional barriers that youth collectively developed and shared tools to navigate authority. More specifically, peer culture developed normative expectations in contrast to its institutional surroundings, such as adolescents' family-oriented performances on Facebook and the relative freedom of expression on Twitter for its peer audience. Thus, youth's social media use within a local peer culture is not independent of its institutional context; rather, it responds to and uses its institutional terrain to meet its own needs.

Thus, youth's social media use needs to be contextualized in relation to the local peer culture and its institutional setting.

Limitations and Future Research

Although this paper makes a methodological contribution to the study of youth's social media use, its research design is not all encompassing. Thus, there is a need for future research. First, this methodological approach only captures youth's perceptions of adult motivations. Youth's narratives cannot fully capture how parents and school personnel exert control. Instead, this approach captures youth's lived experiences and perceptions of power in these institutions. Thus, it is important for future research to explore *educational personnel and parents' lived experiences and perceptions* about issues of adult authority and youth's social media use. Second, although this paper addresses youth's lived experiences and perceptions about adult authority and social media use in the *educational context*, other methodological approaches might be able to focus on this context and capture it more holistically. Third, the sample was recruited from *youth centers*, which included parents' active involvement in youth's enrollment and possibly attracted youth from families with specific parental practices over others. Thus, future research exploring adult authority and youth's social media use should include the recruitment of youth from other institutional contexts. Fourth, our sample primarily includes *racial/ethnic-minority youth*. With minority youth facing the heightened possibility of being transformed from 'at risk' to 'the risk' (Rios, 2011), is it possible that youth's experiences with parental fear are shaped by the risk of discrimination? Is the rigorous parental involvement captured in this study a product of anticipated tensions with other adult authorities over youth's social media use? Since this paper leaves these questions unanswered, it is important to interview the parents of minority youth about youth's social media use, or interview non-minority youth to contrast with the findings in this study. Fifth, our sample is gendered balanced (42 girls and 39 boys), but *more girls participated in the social media tour* (30 girls and 20 boys). Thus, it is important for future research to explore the gendered difference found in this analysis with a more gendered balance sample that uses a social media tour. Lastly, the adult interviewers engaged in a very extensive rapport building and encouraged youth to exercise their agency during the interview process. However, this enactment of agency might have involved *self-editing*. Thus, future research designs involving the social media tour need to develop further techniques to overcome this possible barrier to access youth's experiences and perceptions about social media.

Policy Recommendations

Despite its limitations, this study is a strong empirical foundation for policy, especially relevant to families and schools. First, we strongly recommend that schools and families communicate with one another about their approaches to youth's social media. If possible, schools and communities should organize to develop contextualized expectations for youth's social media use. In doing so, it will be important to move beyond adult fear and engage youth's positive experiences with social media. Second, as adult authorities engage in interpreting youth's social media content, it becomes imperative to not decontextualize it nor make it subject to outside normative expectations. Thus, adult authorities need to become more familiar with the nuances of social media use to avoid engaging in *unfair* practices that prompt the loss of authority. Third, returning to the conceptualization of childhood and adolescence, it is important that adult authority moves beyond fear. This is not to say that fear needs to be reduced or eliminated. Rather, evidence over fear needs to be at the center of adult practices in regard to youth's social media use. Lastly, implementing the latter, schools and families need to clearly differentiate mundane conflict from bullying (Pyżalski, 2012), and engage youth to develop separate and relevant tactics. Overall, it is vital to acknowledge youth's agency and incorporate youth's perspectives in social media policies.

NOTES

1. To maintain participants' identities confidential, youths' legal names have been replaced with youth-selected pseudonyms.

2. Here after, we will use the term youth to refer to both children and adolescents, unless there is a clear distinction between children and adolescents.

3. Based on past data-collection experiences, Campos-Holland (principal investigator & first author) chose to exclude 6-8 year olds due to the research design's inability to fully capture their perspectives. The best methodological approach to use with this age group is participant observation (Corsaro, 2014).

4. Nine-year-old children were excluded due to the technological demands of the follow-up interview. Participants received a $10 gift certificate per interview to the store of their choice as an expression of our gratitude.

5. This was acted upon only once, in the case of a female adolescent who made a post on Ask.fm about her cutting experience and lack of desire to stop. The item was found during data management. Concerned about the "lack of desire to stop," Campos-Holland contacted the teen coordinator at the research site. During the initial conversation, Campos-Holland encouraged the youth coordinator to consider the possibility that social media performance does not necessarily reflect real lived experiences and that youth sometimes use each other's social media accounts. To address the issue, the teen coordinator enlisted the help of an in-house specialist who already had a connection with the participant's family. The in-house specialist then visited the family, and the parent discussed her previous awareness of the participant's mental health status and the family's multiple attempts to help the participant in her recovery. In response, the in-house specialist provided the family with community resources and offered continuous support. Today, the participant is getting help and continues her engagement with activities at the youth center.

6. The social media tour produced complex data with extensive identifiers; thus, the data is held in a secured institutional server. Several steps have been taken to secure the data, including the heightening of security, the use of additional passwords, and limited access.

7. This comparative analysis demonstrates how youth social media use, as other social phenomenon, is contextualized within age-segregated peer groups (Corsaro, 2014; Nenga & Apgar, 2011).

8. Initially, we attempted to conduct a comparative analysis across race/ethnic groups. However, the gender and age variables (see Table 1), and similarities in parental practices across racial/ethnic groups, broke down the sample into subsamples that were too small for comparison.

9. During data management, we visited the ***Adding Games***' Facebook account and discovered that all images on this profile have the same caption.

10. The exception was celebrities' social media performances.

ACKNOWLEDGMENTS

We would like to extend our deepest gratitude to the youth for granting us access to their private digital worlds, the youth centers for granting us access to this local peer culture, the student researchers and IT personnel who contributed to the data collection and data management processes and the reviewers who helped us strengthen this paper. This research was funded by Connecticut College's R. F. Johnson Faculty Development Fund, Research Matters Faculty Grant, Susan Eckert Lynch '62 Faculty Research Fund, Center for the Comparative Study of Race and Ethnicity's Faculty Research Grant, the Margaret Sheridan '67 Research Initiative Career Enhancing Life Skills Program, and the Connecticut College Social Sciences, Humanities and Arts Research Program. Gracias.

REFERENCES

Adler, P., & Adler, P. (1998). *Peer power: Preadolescent culture and identity*. New Brunswick, NJ: Rutgers University Press.

Ahn, J. (2011). The effect of social network sites on adolescents' social and academic development: Current theories and controversies. *Journal of the American Society for Information Science and Technology*, *62*(8), 1435–1445.

Allen, K. P. (2012). Off the radar and ubiquitous: Text messaging and its relationship to 'drama' and cyberbullying in an affluent, academically rigorous US high school. *Journal of Youth Studies*, *15*(1), 99–117. doi:10.1080/13676261.2011.630994

Álvarez, M., Torres, A., Rodríguez, E., Padilla, S., & Rodrigo, M. J. (2013). Attitudes and parenting dimensions in parents' regulation of Internet use by primary and secondary school children. *Computers & Education*, *67*, 69–78.

Bhat, C. S. (2008). Cyber bullying: Overview and strategies for school counsellors, guidance officers, and all school personnel. *Australian Journal of Guidance and Counselling*, *18*(1), 53–66.

boyd, D. (2007). Why youth (heart) social network sites: The role of networked publics in teenage social life. In *MacArthur foundation series on digital learning: Youth, identity, and digital media volume* (pp. 119–142). Cambridge, MA: MIT Press.

boyd, D. (2013). White flight in networked publics?: How race and class shaped American teen engagement with Myspace and Facebook. In L. Nakamura & P. A. Chow-White (Eds.), *Race after the internet* (pp. 203–222). New York, NY: Routledge.

boyd, D. (2014). Making sense of teen life: Strategies for capturing ethnographic data in a networked era. In *Digital research confidential: The secrets of studying behavior online*. Cambridge, MA: The MIT Press.

Brown, J. D., & Bobkowski, P. S. (2011). Older and newer media: Patterns of use and effects on adolescents' health and well-being. *Journal of Research on Adolescence*, *21*(1), 95–113.

Brown, J. D., Keller, S., & Stern, S. (2009). Sex, sexuality, sexting, and sexEd: Adolescents and the media. *Prevention Researcher*, *16*(4), 12–16.

Cobb-Moore, C. (2012). 'Pretend I was mummy': Children's production of authority and subordinance in their pretend play interaction during disputes. *Sociological Studies of Children and Youth*, *15*, 85–118.

Coleman, J., Catan, L., & Dennison, C. (2004). You're the last person I'd talk to. In J. Roche, S. Tucker, R. Flynn, & R. Thomson (Eds.), *Youth in society* (pp. 227–234). Thousand Oaks, CA: Sage.

Corsaro, W. A. (2014). *The sociology of childhood.* Thousand Oaks, CA: Sage.

Corsaro, W. A., & Molinari, L. (2008). Entering and observing in children's worlds: A reflection on longitudinal ethnography of early education in Italy. In P. Christensen & A. James (Eds.), *Research with children: Perspectives and practices* (pp. 179–200). New York, NY: Routledge.

Davidson, E. (2011). *The burdens of aspiration: Schools, youth, and success in the divided social worlds of Silicon Valley.* New York, NY: NYU Press.

Davidson, J., & Martellozzo, E. (2013). Exploring young people's use of social networking sites and digital media in the Internet safety context. *Information, Communication & Society, 16*(9), 1456–1476. doi:10.1080/1369118X.2012.701655

Davies, C. (2011). Digitally strategic: How young people respond to parental views about the use of technology for learning in the home. *Journal of Computer Assisted Learning, 27*(4), 324–335.

De Ridder, S., & Van Bauwel, S. (2013). Commenting on pictures: Teens negotiating gender and sexualities on social networking sites. *Sexualities, 16*(5–6), 565–586. doi:10.1177/1363460713487369

Eastin, M. S., Greenberg, B. S., & Hofschire, L. (2006). Parenting the internet. *Journal of Communication, 56*(3), 486–504.

Eder, D. (1995). *School talk: Gender and adolescent culture.* New Brunswick, NJ: Rutgers University Press.

Engel, S. (2005). Narrative analysis of children's experience. In S. Greene & D. Hogan (Eds.), *Researching children's experience: Approaches and methods* (pp. 199–216). London: Sage.

Erdur-Baker, Ö. (2010). Cyberbullying and its correlation to traditional bullying, gender and frequent and risky usage of Internet-mediated communication tools. *New Media & Society, 12*(1), 109–125. doi:10.1177/1461444809341260

Gannon, S. (2008). 'Twenty-four seven on the computers': Girls, ICTs and risk. *Gender & Education, 20*(4), 361–373. doi:10.1080/09540250802190198

Gao, G. (2015, April). On social media, mom and dad are watching. In *Pew internet & American life project.* Washington, DC: Pew Research Center. Retrieved from http://www.pewresearch.org/fact-tank/2015/04/10/on-social%20media-mom-and-dad-are-watching/

Gleason, B. (2013). #Occupy wall street: Exploring informal learning about a social movement on Twitter. *American Behavioral Scientist, 57*(7), 966–982. doi:10.1177/0002764213479372

Greenhow, C. (2011). Online social networks and learning. *On the Horizon, 19*(1), 4–12.

Greig, A. D., Taylor, M. J., & MacKay, T. (2007). *Doing research with children.* Thousand Oaks, CA: Sage.

Hinduja, S., & Patchin, J. W. (2008). Cyberbullying: An exploratory analysis of factors related to offending and victimization. *Deviant Behavior, 29*(2), 129–156. doi:10.1080/01639620701457816

Irwin, K., Davidson, J., & Hall-Sanchez, A. (2013). The race to punish in American schools: Class and race predictors of punitive school-crime control. *Critical Criminology, 21*(1), 47–71.

Itō, M. (2006). Japanese media mixes and amateur cultural exchange. In *Digital generations: Children, young people and new media. Online Information Review, 31*(5), 706–707.

Itō, M., Baumer, S., Bittanti, M., Cody, R., Herr-Stephenson, B., Horst, H. A., & Tripp, L. (2010). *Hanging out, messing around, and geeking out*. Cambridge, MA: MIT Press.

James, A. (2009). Agency. In J. Qvortrup, W. A. Corsaro, & M. S. Honig (Eds.), *The Palgrave handbook of childhood studies* (pp. 34–45). New York, NY: Palgrave Macmillan. doi:10.1111/j.1099-0860.2011.00420.x

Katz, V. S. (2014). *Kids in the middle: How children of immigrants negotiate community interactions for their families*. New York, NY: Rutgers University Press.

Kent, A. (2012). Responding to directives: What can children do when a parent tells them what to do? *Sociological Studies of Children and Youth*, *15*, 57–84.

Kowalski, R. M., & Limber, S. P. (2007). Electronic bullying among middle school students. *Journal of Adolescent Health*, *41*(6), S22–S30.

Lange, A., & Mirendoff, J. (2009). Method and methodology in childhood research. In J. Qvortrup, W. A. Corsaro, & M. S. Honig (Eds.), *The Palgrave handbook of childhood studies* (pp. 78–96). New York, NY: Palgrave Macmillan. doi:10.1111/j.1099-0860.2011.00420.x

Lasorsa, D. L., Lewis, S. C., & Holton, A. E. (2012). Normalizing Twitter: Journalism practice in an emerging communication space. *Journalism Studies*, *13*(1), 19–36.

Lee, S. (2013). Parental restrictive mediation of children's Internet use: Effective for what and for whom? *New Media & Society*, *15*(4), 466–481. doi:10.1177/1461444812452412

Lenhart, A., Duggan, M., Perrin, A., Stepler, R., Rainie, L., & Parker, K. (2015, April). Teens, social media, and technology: Smartphones facilitate shifts in communication landscapes for teens. In *Pew internet & American life project*. Washington, DC: Pew Research Center. Retrieved from http://www.pewinternet.org/files/2015/04/PI_TeensandTech_Update2015_0409151.pdf

Lingel, J. (2013). The digital remains: Social media and practices of online grief. *Information Society*, *29*(3), 190–195. doi:10.1080/01972243.2013.777311

Livingstone, S. (2007). Strategies of parental regulation in the media-rich home. *Computers in Human Behavior*, *23*(2), 920–941.

Livingstone, S. (2008). Taking risky opportunities in youthful content creation: Teenagers' use of social networking sites for intimacy, privacy and self-expression. *New Media & Society*, *10*(3), 393–411.

Livingstone, S., & Bober, M. (2013). Regulating the Internet at home: Contrasting the perspectives of children and parents. In D. Buckingham & R. Willett (Eds.), *Digital generations: Children, young people, and the new media* (pp. 93–113). New York, NY: Routledge.

Livingstone, S., & Helpsper, E. (2007). Gradations in digital inclusion: Children, young people and the digital divide. *New Media & Society*, *9*(4), 671–696. doi:10.1177/1461444807080335

Madden, M., Lenhart, A., Cortesi, S., Gasser, U., Duggan, M., Smith, A., & Beaton, M. (2013, November). Parents, teens, and online privacy. In *Pew internet & American life project*. Washington, DC: Pew Research Center. Retrieved from http://www.crossing-guardconsulting.com/wp-content/uploads/2013/01/PIP

Marwick, A. E. (2012). The public domain: Social surveillance in everyday life. *Surveillance & Society*, *9*(4), 378–393.

Matthews-Juarez, P., Juarez, P. D., & Faulkner, R. T. (2013). Social media and military families: A perspective. *Journal of Human Behavior in the Social Environment*, *23*(6), 769–776. doi:10.1080/10911359.2013.795073

Mayall, B. (2008). Conversations with children: Working with generational issues. In P. Christensen & A. James (Eds.), *Research with children: Perspectives and practices* (pp. 120–135). New York, NY: Routledge.

Mazzarella, S. R. (Ed.). (2005). *Girl wide web: Girls, the Internet, and the negotiation of identity*. New York, NY: Peter Lang Publishing.

Meeder, R. (2005). Access denied: Internet filtering software in K-12 classrooms. *TechTrends*, *49*(6), 56–58.

Mesch, G. S. (2009). Parental mediation, online activities, and cyberbullying. *Cyber Psychology & Behavior*, *12*(4), 387–393.

Mitchell, K. J., Finkelhor, D., & Wolak, J. (2001). Risk factors for and impact of online sexual solicitation of youth. *JAMA*, *285*(23), 3011–3014.

Mitchell, K. J., Finkelhor, D., & Wolak, J. (2005). Protecting youth online: Family use of filtering and blocking software. *Child Abuse & Neglect*, *29*(7), 753–765. doi:10.1016/j.chiabu.2004.05.008

Mitchell, K. J., Finkelhor, D., & Wolak, J. (2007). Youth Internet users at risk for the most serious online sexual solicitations. *American Journal of Preventive Medicine*, *32*(6), 532–537.

Mullen, S. A., Gerbert, B., Heaivilin, N., Lin, B., Page, J., & Tsoh, J. Y. (2015). What do parents say about their children's oral health on Twitter? *Journal of Pediatric Dentistry*, *3*(1), 17–23.

Navarro, J. N., & Jasinski, J. L. (2012). Going cyber: Using routine activities theory to predict cyberbullying experiences. *Sociological Spectrum*, *32*(1), 81–94. doi:10.1080/02732173.2012.628560

Nenga, S. K., & Apgar, L. A. (2011). The age of love: Dating and the developmental discourse in a middle school summer camp. *Sociological Studies of Children and Youth*, *14*, 109–131.

Notley, T. (2008). Online network use in schools: Social and educational opportunities. *Youth Studies Australia*, *27*(3), 20–29.

Padilla-Walker, L. M., & Coyne, S. M. (2011). "Turn that thing off!" Parent and adolescent predictors of proactive media monitoring. *Journal of Adolescence*, *34*(4), 705–715.

Padilla-Walker, L. M., Coyne, S. M., Fraser, A. M., Dyer, W. J., & Yorgason, J. B. (2012). Parents and adolescents growing up in the digital age: Latent growth curve analysis of proactive media monitoring. *Journal of Adolescence*, *35*(5), 1153–1165.

Pascoe, C. J. (2011). *Dude, you're a fag: Masculinity and sexuality in high school*. Berkeley, CA: University of California Press.

Peter, J., Valkenburg, P. M., & Schouten, A. P. (2006). Characteristics and motives of adolescents talking with strangers on the internet. *Cyber Psychology & Behavior*, *9*(5), 526–530.

Pini, M. (2004). Technologies of the self. In J. Roche, S. Tucker, R. Flynn, & R. Thomson (Eds.), *Youth in society* (pp. 160–167). Thousand Oaks, CA: Sage.

Poyntz, S. R., & Hoechsmann, M. (2011). Children's media culture in a digital age. *Sociology Compass*, *5*(7), 488–498 .doi:10.1111/j.1751-9020.2011.00393.x

Pyżalski, J. (2012). From cyberbullying to electronic aggression: Typology of the phenomenon. *Emotional and Behavioral Difficulties*, *17*(3–4), 305–317.

Qvortrup, J. (2009). Childhood as a structural form. In J. Qvortrup, W. A. Corsaro, & M. S. Honig (Eds.), *The Palgrave handbook of childhood studies* (pp. 21–33). New York, NY: Palgrave Macmillan. doi:10.1111/j.1099-0860.2011.00420.x.

Rane, H., & Salem, S. (2012). Social media, social movements and the diffusion of ideas in the Arab uprisings. *Journal of International Communication, 18*(1), 97–111.

Reich, S. M. (2010). Adolescents' sense of community on Myspace and Facebook: A mixed-methods approach. *Journal if Community Psychology, 38*(6), 688–705. doi:10.1002/jcop.20389

Ringrose, J., & Barajas, K. E. (2011). Gendered risks and opportunities? Exploring teen girls' digitized sexual identities in postfeminist media contexts. *International Journal of Media & Cultural Politics, 7*(2), 121–138. doi:10.1386/macp.7.2.121_1

Rios, V. M. (2011). *Punished: Policing the lives of Black and Latino boys*. New York, NY: New York University Press.

Roberts, H. (2008). Listening to children and hearing them. In P. Christensen & A. James (Eds.), *Research with children: Perspectives and practices* (pp. 225–240). New York, NY: Routledge.

Rueda, H. A., Lindsay, M., & Williams, L. R. (2014). "She posted it on Facebook": Mexican American adolescents' experiences with technology and romantic relationship conflict. *Journal of Adolescent Research, 30*(4), 1–27. doi:0743558414565236

Shin, W., Huh, J., & Faber, R. J. (2012). 'Tweens' online privacy risks and the role of parental mediation. *Journal of Broadcasting & Electronic Media, 56*(4), 632–649.

Staksrud, E., Ólafsson, K., & Livingstone, S. (2013). Does the use of social networking sites increase children's risk of harm? *Computers in Human Behavior, 29*(1), 40–50.

Sumiala, J., & Tikka, M. (2011). Reality on circulation: School shootings, ritualized communication, and the dark side of the sacred. *ESSACHESS: Journal for Communication Studies, 4*(8), 145–159.

Thorne, B. (2015). Boys and girls together ... But mostly apart. In R. Arum, I. R. Beattie, & K. Ford (Eds.), *The structure of schooling: Readings in the sociology of education* (pp. 462–469). Thousand Oaks, CA: Sage.

Valcke, M., Bonte, S., De Wever, B., & Rots, I. (2010). Internet parenting styles and the impact on internet use of primary school children. *Computers & Education, 55*(2), 454–464.

Valentine, G. (1997). "Oh yes I can." "Oh no you can't": Children and parents' understandings of kids' competence to negotiate public space safely. *Antipode, 29*(1), 65–89.

Valentine, G., & Holloway, S. (2001). On-line dangers?: Geographies of parents' fears for children's safety in cyberspace. *The Professional Geographer, 53*(1), 71–83.

Van Dijck, J. (2013). 'You have one identity': Performing the self on Facebook and LinkedIn. *Media, Culture & Society, 35*(2), 199–215. doi:10.1177/0163443712468605

Vandebosch, H., & Van Cleemput, K. (2009). Cyberbullying among youngsters: Profiles of bullies and victims. *New Media & Society, 11*(8), 1349–1371. doi:10.1177/1461444809341263

Vandoninck, S., d'Haenens, L., De Cock, R., & Donoso, V. (2012). Social networking sites and contact risks among Flemish youth. *Childhood, 19*(1), 69–85. doi:10.1177/0907568211406456

Werner, N. E., Bumpus, M. F., & Rock, D. (2010). Involvement in Internet aggression during early adolescence. *Journal of Youth & Adolescence, 39*(6), 607–619. doi:10.1007/s10964-009-9419-7

Westcott, H. L., & Littleton, K. S. (2006). Exploring meaning in interviews with children. In S. Greene & D. Hogan(Eds.), *Researching children's experience: Approaches and methods* (pp. 147–167). Thousand Oaks, CA: Sage.

Williams, A. L., & Merten, M. J. (2011). iFamily: Internet and social media technology in the family context. *Family & Consumer Sciences Research Journal*, *40*(2), 150–170. doi:10.1111/j.1552-3934.2011.02101.x

Wohn, D. Y., Ellison, N. B., Khan, M. L., Fewins-Bliss, R., & Gray, R. (2013). The role of social media in shaping first-generation high school students' college aspirations: A social capital lens. *Computers & Education*, *63*, 424–436.

Wolak, J., Finkelhor, D., & Mitchell, K. (2004). Internet-initiated sex crimes against minors: Implications for prevention based on findings from a national study. *Journal of Adolescent Health*, *35*(5), 424.e11–424e20. doi:10.1016/j.jadohealth.2004.05.006

Wolak, J., Mitchell, K. J., & Finkelhor, D. (2007). Does online harassment constitute bullying? An exploration of online harassment by known peers and online-only contacts. *Journal of Adolescent Health*, *41*(6), S51–S58.

Wridt, P. J. (2004). An historical analysis of young people's use of public space, parks and playgrounds in New York City. *Children Youth and Environments*, *14*(1), 86–106.

Ybarra, M. L., & Mitchell, K. J. (2008). How risky are social networking sites? A comparison of places online where youth sexual solicitation and harassment occurs. *Pediatrics*, *121*(2), e350–e357.

ALL THE WEB'S A STAGE: THE EFFECTS OF DESIGN AND MODALITY ON YOUTH PERFORMANCES OF IDENTITY

Harry T. Dyer

ABSTRACT

Purpose – *Online Social Networking Sites (SNSs) such as Facebook and Twitter have become increasingly popular in the last decade. Each SNS varies somewhat, with different forms of expression, communication and customization. Different sites may have different priorities, methods of interacting, social features and definitions of what it means to be 'social' on their sites.*

Methodology/approach – *This paper reports on 2 months of exploratory observations and interviews with participants using two of the most popular SNSs; Facebook.com and Twitter.com. Paying attention to the modal nuances of the sites and their effect on social interaction and identity portrayal, the focus of analysis is upon how these two sites are interacted with as 'stages' for identity performances, and how the varying aspects of design and modality on these interactive sites can result in different multimodal identity performances and social interactions.*

Technology and Youth: Growing Up in a Digital World
Sociological Studies of Children and Youth, Volume 19, 213–242

ISSN: 1537-4661/doi:10.1108/S1537-466120150000019007

Findings – *Data revealed that youth are adeptly able to negotiate the different modal options presented to them online, yet the temporal aspects presented by the design of the site, the differing definitions and priorities in the framing of identity presented by the SNSs, and the modal choices present across the two sites resulted in markedly differing presentations of identity to markedly differing audiences.*

Originality/value – *This research demonstrates the impact of modality and design on how we act and interact, and highlights that as Digital Sociologists and Researchers, we should be careful not to treat all Online SNSs the same, but pay attention to the plethora of nuances these sites offer as stages for identity performances.*

Keywords: Multimodality; Identity; Facebook; Twitter; Digital Sociology

'What media are needs to be interrogated, not presumed'.
– Larkin (2008).

Much has been made and written in the last decade of youth engagement with social networking sites, and rightly so. They are increasingly a part of daily social life; they are accessed through a growing range of technologies; they are accessible at home, on the bus, at work, and at school; and they not only reflecting current social practices, but are creating and encouraging new and novel social methods of action and interaction (Miller, 2011; Wilson, Gosling, & Graham, 2012). Facebook claims as of 2014 to have 1.35 billion active users, 864 million of whom log on daily, a 19% year over year rise (Facebook.com Newsroom, 2014). Similarly, Twitter claims as of 2014 that 500 million tweets are sent daily, with 284 million active users on the site (Twitter.com About, 2014). The increasing popularity, ubiquity, and variety of online Social Networking Sites (SNSs) means that an understanding of their uses and functions has become increasingly vital for many diverse fields of research, and that an understanding of the importance of design and modal choices upon social interaction is crucial.

This paper adds to the growing body of knowledge surrounding youth engagement with SNSs to consider the importance of the many modal choices offered across the increasingly broad array of SNSs for youth to express and portray identity. Whilst research is progressively understanding these sites and spaces as increasingly important locations of social interaction for youth (Bennett & Robards, 2014; boyd, 2014, 2007), Wilson, Gosling and Graham in their review of current Facebook research in the Social Sciences (Wilson et al., 2012, p. 204) point out that '... there are dangers in treating (SNSs) as a single general category without drawing distinctions among them'. Despite the increasing variety and diversity of SNSs, little consideration is currently being given to the important role that the arrangement of these social arenas plays in shaping how youth interact, and how they construct and portray identity. As such, this research pays close attention to the design aspects and modal nuances of two prominent SNSs to understand how the layout, design and modal affordances can affect participation, identity construction, social interaction and social navigation online.

This paper begins by discussing the importance of SNSs in daily social life for youth, both online and offline, before moving on to defining identity performance and the role of multimodality online. The findings are then presented, with the research finding that the differences in design across the two sites affected the roles the participants found themselves fulfilling, the audience they felt they were performing for, and the way and modes through which the participants acted, interacted, and described themselves.

WHY YOUTH SOCIAL NETWORKING SITES

Much of the research conducted on SNSs has focused on the actions and interactions of youth. There are several reasons for this, the most often cited being that they make up a large portion of the active users of SNS. The proliferation of youth in SNSs has been noted by a number of researchers, for example Lenhart et al. (2011) found that Facebook, a site founded specifically for college students, was being utilised as a site for social networking by 93% of SNS users aged between 12 and 17. Livingstone, Mascheroni, and Murru (2011) further showed that this trend was not specific to Facebook, recording the large proportion of youth across a wide range of SNSs, confirming the findings of Baker and White (2010, p. 2), who proclaimed that 'adolescents represent the largest and fastest growing demographic sector using the Internet'.

However, it is not just the popularity of these sites that is important to the youth population, but rather the increasingly important role that this technology plays in modern social life. Researchers such as boyd and Ellison (2007), Lenhart and Madden (2007), Baker and White (2010), and Livingstone, Mascheroni, et al. (2011) have all noted that SNSs are increasingly important locations of social interactions for large numbers of youth, and that these sites play an integral role in how young people socialise, interact and form identities. Quinn and Oldmeadow (2013) attempted to confirm the increasingly important role of technologies and SNSs in young people's social lives by conducting a statistical analysis into how socially connected young people felt, noticing a strong positive correlation between mobile phone and SNS use, and the sense of social connectedness young people felt. Again, this is not specific to one site, and instead is a trend that has be reported across a number of sites. For example, in their study of gender differences on Twitter, Pujazon-Zazik and Park (2010, p. 78) noted that 'an important developmental task of adolescence is identity formation', and that this development was increasingly happening online. boyd (2014, p. 26) sums up the important and ubiquitous nature of SNSs for youth, stating that 'along with planes, running water, electricity, and motorized transportation, the internet is now a fundamental fact of modern life'. For the current youth generation, SNSs are increasingly important and useful tools for navigating the social landscape. So much so that in their report on a questionnaire of nearly 100 students, Mazzoni and Iannone (2014, p. 303) define SNSs as 'part of the functional organs that support emerging adults in their ability to connect and to be connected to a social network and to develop and maintain it over time'. SNSs have become another critical tool for peer-to-peer social interaction, and are increasingly playing a larger role as technology develops and becomes more common place in the lives of many young people.

The importance of SNSs for youth can in part be shown in the amount of time they spend on these sites. Agosto and Abbas (2013, p. 137) note that it is important 'to study youth as a separate sub-group, as there appear to be differences in the ways young people and adults use SNSs'. They noted that while there was little difference in the general subject matters discussed by this generation as compared to previous generations, the amount and the frequency of these interaction and activities differs greatly, and is likely to continue to grow given the popularity of these sites, and the growing ubiquity of technologies. This matches findings by other researchers who noted the increased amount of online engagement by younger generations (Christofides, Muise, & Desmarais, 2012b; Livingstone, Kirwil, Ponte, & Staksrud, 2014; O'Keeffe, Clarke-Pearson, & Council on Communications and Media., 2011; Palfrey & Gasser, 2013; Ralph,

Berglas, Schwartz, & Brindis, 2011; Rideout, Foehr, & Roberts, 2010). Indeed, the amount of time the youth spend engaging with this technology lead Clark (2005, p. 203) to define the emerging youth generation as the 'constant contact generation'.

An important aspect of SNSs for youth social identity development is the potential they offer for adolescents to create a space of their own. boyd (2007) suggests that the availability of youth-controlled spaces to explore aspects of social identity away from adults is of essential importance to today's youth, and is a main aspect of why adolescents use SNSs. She points out that 'what is unique about the Internet is that it allows teens to participate in unregulated publics while located in adult-regulated physical spaces such as homes and schools' (boyd, 2007, p. 21). Though admittedly much has changed in the SNS landscape since 2007, including a noted growth in adult users (Madden, 2010), Agosto and Abbas (2013, p. 117) make similar observations, stating that 'SNS provide youth-shared spaces for interaction and communication'. The implication here is that SNSs create a space for youths free from adults in which to explore and socially interact, allowing them to establish their own social rules and ideals. This suggestion however is called into question given the recent and increasing influx of adults onto popular sites such as Facebook and Twitter (IStrategyLabs, 2014; Christofides, Muise, & Desmarais, 2012a; Madden, 2010). Some researchers have suggested that this has led to an exodus of teenagers from Facebook as they seek other site to explore and claim as their own (Miller, 2013). This trend has been spotted as early as the beginning of this decade (Van Grove, 2010). While youth are still numerically high on these sites, the increasing adult presence may have some effect on how SNSs are used as sites of identity development and exploration for youth in the future, a topic worthy of further research and exploration.

It is worth noting that the use of SNSs for social interactions does not happen independently of existing social ideals, actions and interactions. In recent research there has been a shift towards conceptualising and grounding internet interactions by establishing their increasingly important role in offline social life, and by challenging the idea of an offline-online dichotomy, and 'digital dualism' (Bulfin & Koutsogiannis, 2012; Jurgenson, 2012; Fields & Kafai, 2009; Miller, 2011; Wilson et al., 2012; Young, 2011). This serves as an important move away from early research which optimistically suggested the potential freedoms of the new social environment offered online, highlighting the potentially freedoms and dangers a new 'blank' medium offered away from established social constructs, procedures and rubrics (Bruckman, 1996; Danet, 1998; Harasim, 1993; Rheingold, 1993;

Turkle, 1995). This ontological shift towards an understanding of the interwoven nature of online and offline lives is much needed and a fascinating area of research whose discussion is active and still ongoing, and suffice it to say is worthy of much greater consideration and lengthier treatment than can be provided in this paper.

For the purposes of this paper however, it is worth noting that Facebook has been seen as strengthening ties to pre-established offline contacts in a manner that presents a marked move away from previous earlier SNSs such as MySpace and Habbo, where potentially an online presence could be maintained with some degree of separation from offline, and where a certain degree of anonymity was present and expected (Lampe, Ellison, & Steinfield, 2006, Ross et al., 2009, Wilson et al., 2012). PEW research confirmed this emphasis on pre-established offline contacts, noting that 95% of American teens use Facebook to connect to known friends (Lenhart & Madden, 2007). This connection to known offline contacts can be seen as grounding interactions, and importantly, identities within pre-established social parameters (Young, 2009b, 2011; Haythornthwaite, 2005), meaning the purpose of interacting in online SNSs in often no longer grounded in anonymity but in extending and fostering existing social connections, and extending established offline identities (Baker & White, 2010; Quinn & Oldmeadow, 2013).

What is clear is that SNSs have quickly become key locations of social interaction for today's youth, who are spending increasing amounts of time using the tools provided to them by the SNSs to interact and shape social identities. Given the increased importance and the growing role of this technology for the development of social identities, and the increased reliance upon the tools presented by SNSs for identity development and portrayal, further research should be conducted into how the specifics arrangement and choices of social tools affect the manner in which social identity is shaped and formed.

PERFORMING IDENTITIES ONLINE: WHY THE INTERNET IS A STAGE FOR IDENTITY PERFORMANCES AND HOW THIS STAGE CAN AFFECT OUR PERFORMANCE

This paper's understanding of identity is informed by Goffman's work on performativity. Goffman's (1959) dramaturgical perspective is a widely influential method that considers identity to be a malleable, non-constant performance, and suggests that identity can be seen as the sum of the social actions and interactions undertaken by a person within a Social Space. It was Goffman's assertion that identity is formed in and through social interactions; rather than considering a fixed and innate core 'self', we should instead see identity as a malleable and plural socially constructed performance crafted for a specific social situation (Branaman & Lemert, 1997; Elliott, 2014). Informed by Goffman's (1959) influential work, this paper will employ a dramaturgical perspective to frame the social actions and interactions observed online as a performance of social identity.

This performative notion of identity has successfully been explored in an online context. Uski and Lampinen (2014) for example studied what they defined as 'profile work'; strategic self-presentation in SNSs. Observing Facebook and the music-sharing platform Last.fm, they focused upon how users maintained a feeling of authenticity in their presentations of identity. They noted that despite the users presenting differing identities across the two sites, the presentations were still seen as an 'authentic' presentations of self.

Goffman's notion of performativity is particularly useful when we carry on this metaphor of acting, and consider how the props and staging available to the user can shape interactions. Goffman noted that social situation will shape and affect many aspects of how we act and interact; we may use different language, mannerisms, and even accents when we find ourselves at the doctor compared to when we are down the pub. He suggested that identity can be viewed as 'a dramatic effect arising diffusely from a scene that is presented' (Goffman, 1959, p. 252). Goffman's attention to the importance of environment, and his understanding of how differing social spaces can result in differing social actions and interactions (see Van den Berg, 2008) has led to him being called 'the quintessential sociologist of the "situation"' (Jensen, 2006, p. 151).

This concept holds particular interest when we consider that each online site can contain potentially different methods and modes of interaction, and different stages and props for our identity performances, which may produce different performances, and potentially different identities. When we add this to the suggestion that online social identity performances are not separate entities, but are affected and shaped by social norms and cultural conventions present offline (Uski & Lampinen, 2014; Reid & Boyer, 2013), then the identity performances taking place online can be seen as extensions of offline social constructions, moved to and performed upon a new stage, with new props and new potential methods of performance.

Much like offline performances, identity presentation and performance online can be seen as a malleable concept, performed with and through a varying array of semiotic modes, whilst remaining grounded in offline social norms and expectations.

Van den Berg (2008) noted the importance of 'staging' online in regards to identity performance. Focusing upon what she defined as online 'scripts' – contextual clues that help performers understand the appropriate identity performances based upon the specific staging available to the performer – she found that alterations in situation and 'scripts' leads to changes in how the users understand their situation and their expected role within it. For Van den Berg (2008), identity becomes a concept that is dependent upon audience and situation, with any changes in situation leading to a change in the performances observed.

Following Goffman (1959) and Van den Berg (2008), identity can be understood as a malleable concept that is the result of social interactions with others within a specific environment. It is this fluid and flexible understanding of identity that underpins this paper, as we consider what role the varying environment of online SNSs play in the formation and presentation of an identity.

MULTIMODALITY – UNDERSTAND THE IMPORTANCE OF 'THINGS'

For Goffman (1959, p. 15) an identity performance was 'all the activity of a given participant on a given occasion which serves to influence in any way any of the other participants'. In order to understand all the activity of a given participant online, a myriad of potential choices and activities must be considered. SNSs offer a wide variety of activities and customisable options- from photos, videos, and textual interactions, to font size, shape, and colour. Given the focus of this paper on the impact of the specific online social environment on the performance and enactment of social identities and actions, a theory and methodology is needed to unpack these dense environments and to take into account the myriad choices and specificities made available to users online.

The different range of actions available to online SNSs users, and the arrangement of the actions, can be explored using the concept of multimodality (Kress, 2004). Multimodality concerns itself with the understanding how a myriad of 'modes' affect our actions and interactions. Kress (2009, p. 79) defines a mode as 'a socially and culturally shaped resource for meaning making'. Through this definition we not only see meaning conveyed and actualised through text, but also through mediums such as images and the arrangement of content (Luff, Heath, Kuzuoka, Yamazaki, & Yamashita, 2006). As such, a specific arrangement of modes of communication can create different meanings, and lead to different manors of social engagement, different social interaction, and different identity performances. (Kress, 2004; Kress & Van Leeuwen, 2001; Norris, 2004; Van den Berg, 2008).

This understanding of the importance of design, combined with a consideration of the available modes of communication, is particularly pertinent online where a wide mix of modes of communication are present, as Young (2009a) noted of the SNS Myspace, and boyd and Heer (2006) and boyd (2004) found on Friendster. When considering multimodality online, Young (2009a) highlighted that the choices made by the website designers may affect how we are able to and choose to act and interact online. This has been the focus of a growing number of studies for a number of disciplines, who are increasingly recognising the effect and the importance of the multimodal nature of online interactions. For example, Karimov, Brengman, and Van Hove (2011) looked from a business studies perspective at how design affected aspects of consumer trust. Similarly, Lafkioui (2013) found that the resources available on SNSs affected how identity was constructed on multilingual French-based sites, and Skog (2005) found that the features present on the site 'LunarStorm' had a large influence upon social actions and behaviour, and upon aspects of identity performance, such as how much data the users chose to share on the site.

The effects of design has been found to be relevant even within websites that share similar modes, as the specific arrangements and prioritising of modes of communication can shape our interactions. Van Dijck's (2013) study highlights this, using design to focus upon to the concept of identity presentation in two online SNSs, Facebook and LinkedIn. By comparing certain features of what she defines as a 'general' site in Facebook, and a 'specific' business-based SNS in LinkedIn, Van Dijck (2013) theorises how different social aims of these two sites may alter how we can present identity using the same modes. This study also highlights the importance of not only considering the modal options available to a user, but also the consideration of these modal options *in situ*. Van Dijck (2013) found the combination, arrangement, and accessibility of the modes to be an important factor in how they were utilised. As such, this research seeks to show that, despite their many similarities, we should not assume all SNSs function in a similar way, but that a detailed investigation of each site is useful, and in many cases, necessary.

STYLE AND DESIGN – TAKING CHOICE AND SELECTION INTO ACCOUNT

Van Leeuwen (2005, p. 281) defined multimodality as the 'combination of different semiotic modes – for example, Language and music – in a communicative artefact or event'. This highlights an important point, namely that multimodal analysis should not only focus upon the products available to the user, but also upon the process of producing and interacting with multimodal events. In order to fully explore the participants' interactions with the available multimodal options, Meier and Pentzold (2011) suggest using the concept of 'style' as proposed by Van Leeuwen (2005), who defined style as the manner in which 'a semiotic event is performed' (2005, p. 287). Meier and Pentzold (2011) suggest that we should be interested in not only the resulting communicative artefacts, but the multimodal event itself, as this combination and choice of modes plays an integral part in the creation and performance of an individual communicative artefact. In particular, they were interested in exploring the process of production, suggesting that researchers need 'to engage with the manifold practical procedures of enacting and accomplishing such communicative products' (Meier & Pentzold, 2011, p. 1).

Meier and Pentzold also noted that 'style also marks individual and social identity' (2011, p. 6). In other words, our modal choices and specific multimodal combinations are an integral marker of our social identity performances. Seen this way, the notion of style is useful for this study as it provides a combination of a focus upon identity, the various modes of interaction and communication available online, and the multimodal nature of identity performance in order to offer real insights into how we perform identity within the complex multimodal landscape of SNSs. By utilising style, we remove the assumption that these multimodal landscape will produce certain forms of interaction, and instead take a less determinist stance to consider how the user interact with the available modes to shape and produce their ever-changing social landscape.

However, there are forces other than personal agency that may affect how the users select and combine different modes in a stylised manner to form and perform social identity performance. We should also take the time to acknowledge the choices that are made for the user, and that a user is not presented with a blank space to produce a social identity, but a space in which many choices have already been made for them. By acknowledging the importance of the staging of social identify performances, and the choices made in the design of this staging, we can acknowledge identity performances online not only as the result of stylised combinations and choices of many multimodal artefacts, but also as a result of the availability of these modes and choices made *for* the performers. These choices may affect how the users are able to make messages and social actions communicable, and may shape and encourage certain combinations of modes. Hutchby (2001, p. 9) touched upon this point, noting that 'assemblages afford, that is, they enable and constraint certain style-full ways of doing and saying'. Put another way, users of SNSs are not presented with a limitless multimodal landscape from which to present their identity, but are bound and affected by the design and layout of the sites, which can affect how their identities are enacted. This is acknowledged by Meier and Pentzold (2011, p. 10), who point out 'signs are realised in and through media, that is, modes of communication come into being with regard to the affordances of particular media'.

Given this understanding the role of stylised uses of multimodal options, this paper will report on two months of exploratory observations and interviews focusing upon the stylised modal expressions of identity across two SNSs, Facebook and Twitter.

METHODOLOGY

Interviews and observations with the participants were carried out over a two month period starting May 2012, with nine participants aged between sixteen and twenty-three. The participants were made up of 6 females and 3 males and were selected via the 'snowballing' method (Goodman, 1961), with the initial participants, selected from known contacts, suggesting other appropriate participants. This method of sampling allowed for observations not only of public social actions, but also of online interactions between the participants. Participants knew at least one of the other participants, and this allowed for the observation and consideration not only of the participants' public posts, but of the any public participant-to-participant interactions.

Each participant, by their own account, accessed the sites for at least 75 minutes daily, often more than 5 times a day. For ethical reasons, no private messages were observed during the research period, and no public interactions with non-participants were recorded. The selected participants were 'followed' on each site for the duration of the study, with their public actions and inter-participant interactions observed. All data has been anonymised, and no images or traceable information will be presented in this paper to protect the participants' identities.

I observed public social actions made by the participants across the two sites for a two month period, as well as public interactions between the participants on the two sites. This observation was conducted once a day for the duration of the two month period during which notes were taken on the types of interaction and the content of the interactions. Data collection took two forms. Where appropriate, screen-captured images were collected and anonymised to record the actions and interactions directly. These screen-captures are not included in this paper in order to preserve anonymity, and to avoid copyright issues. Daily reflective field notes were also collected in order to track my observations. Notes were collected about the mode and content of any public actions and interactions made that day.

Two interviews were conducted with each participant during the two month period, during which we discussed some of the actions and interactions I had observed, how they felt they were constructing their identities, and what they felt were the important aspects of their performances. We also discussed my early analysis of the social actions and interactions, allowing them to explain their reasons for taking the observed and recorded actions. This two-way interaction helped triangulate my observations and allowed the participants to explain their actions and interactions in their own words, helping to reduce possible ethnocentrism (Babbie, 1986; Gold, 1958). Interviews were conducted in a semi-structured manner, during which we discussed and reviewed the data I had captured. The data was coded and analysed qualitatively via the thematic content of the posts, and the modes through which the participants interacted. The main topics that arose from these observations and presented and discussed below.

Though there may be questions about the authenticity of the identities observed and about the accuracy of the observed identities, this research holds that the performances, or more specifically in this case, the behaviour and social acts, are the unit of analysis, and not the individual. The identity observed, as mentioned in the literature review, is identity at it broadest point; a malleable and changing construct that it constantly considered and worked on by the performer. Rather than attempting to observe a fixed construct of a continuing 'whole' of identity, this research is interested in the moment-to-moment presentations of aspects of identity, and how the modes and content of these performances are stylistically selected by the participant and guided by the sites and the audience. As such, it does not hinder the research that only public interactions were observed, as these were the unit of analysis used to perform identity on the SNSs, and are viewed as public expressions of identity within this environment. 'Front stage' (Goffman, 1959) publicly-presented performances and interactions were observed, with justification and discussion of 'back stage' considerations of the performers (1959) unpacked during the interview process.

RESULTS: PERFORMING IDENTITY IN SNS

I will begin by presenting a brief overview of the data, foregrounding the key aspects that arose, before discussing their implications with regards to stylised multimodal identity performances. The names of the participants have been anonymised, and all images and screen-captures are described in order to avoid anonymity and copyright issues.

Facebook's Use of 'Friendship' and of a Shared Social Past Grounded in Offline Locations and Interactions

The interviews and observations suggested that Facebook's design is framed around the concept of engaging and maintaining social contact with pre-established contacts from the participants' pasts and presents. Facebook frames itself as a place 'to keep up with friends and family' (facebook.com, 01/2012). This message is further strengthened through the use of the 'friendship' motif throughout the design of the site. Users with whom the participants interact with are referred to as 'friends', and across locations on the site users are encouraged to 'find', 'search for', or 'invite' friends. Facebook also suggests list of 'People you may know'. This finding of friends on the site is tied to the offline world further by Facebook encouraging users to pick friends based upon their 'hometown, school, employer, and more'.

In their interviews, the adolescent participants commented upon these ideas, suggesting that Facebook was used to maintain relationships with known contacts. When asked to describe the audience on Facebook they described them as 'school friends' or 'old mates'. Even when in some cases their privacy filters meant the information was visible to anyone, they still suggested that 'just because anyone can see it doesn't mean everyone will, right? It's only for people who know me'. Oftentimes they noted that Facebook served as a useful tool when they otherwise couldn't interact with these contacts offline. They also commented that 'it would be weird, you know? To Facebook people that didn't know you. That's just … creepy'. This notion of pre-established contacts was a commonly discussed theme of the use of Facebook. Though I attempted to discourage direct comparison of Facebook and Twitter, the topic often fell back to how the sites were fundamentally different. One participant suggested 'friends on Facebook are probably more people you actually … know … or like went to school with, but on Twitter they tend to be more random people, you know, kinda more open, so you're sharing interests'. Others discussed how they felt the audience of Facebook was 'friends … and unfortunately, family', and knew that on these sites they could discuss certain events and topics that that audience would relate to.

This focus upon a 'friendship' motif was highlighted in the design of the page by the two separate locations on every user's page showing the number of friends they had, as well as a space located on the static bar at the top of the page enabling users are able to quickly 'Find friends'.

In the users 'about' sections (which lists information about them), there were further links to family, with a section listing the family members of the user who are on Facebook. This page also facilitated the continuation of the ties to real-world social actions and existing social connections, with list provided of work and educational background, and even locations in which the user has lived.

By making the participants define themselves by a list of real-world locations and achievements, and encouraging participants to add 'People you may know' based upon 'hometown, school, employer, and more', Facebook defines and shapes what they believe to be the ideal social interaction on their site, and establishes what they believe to be the core facts and important aspects of a social connection and a social identity. Facebook's definition of the social is one built around interactions with known offline contacts, as confirmed by the participants in their interviews. Facebook's 'social' is one that builds upon already established social interactions and friendships. Facebook highlights the importance of family and pre-established friends, and encourages social interactions based upon established real-world connections. One participant noted that it was 'hard to interact with and find new people on Facebook. There's no point really'.

Facebook's 'definition of social' has an interesting temporal element too. It is a definition that focuses on the present and the past; on established relationships, locations, and events, not on fostering new connections or creating future social relationships. Indeed, the participants frequently called their pages 'timelines', viewing them as diaries of their online social lives. This idea can further be shown to be present in the site design, with the inclusion of a digital time-line which allows a user to move to a specific moment in their Facebook history and observe the interactions at that period of time. This sets the site up as an ongoing historical biography. One participant remarked that Facebook was 'a diary of their life'. Though Facebook does not actively discourage the finding of new social connections, the participants noted that they found the idea of connecting with unknown users of Facebook strange.

Facebook's design shaped the acceptable boundaries of expected social interaction on these sites; with the amplification of the ease of certain types of social interaction (the interaction with known contacts based upon historical pre-established relationships) and the reduction of ease for other types of social actions and interactions (the fostering of new relationships). Through the use of a diary-like biography based upon the users' histories, the use of the time-line, and the establishment of the notion of friendship as being built from pre-established relationships, Facebook hindered the finding of unknown people, and encouraged and defined social interaction as being between known contacts.

Customisability of Design on Facebook

Users were given few options to customise the design of Facebook, though they were presented with one space for a potentially more open-ended description of themselves, with the option to fill out a section titled 'about you'. In this section, the users were free to write whatever information they want about themselves, rather than selecting from a list of predefined options, as was prevalent for many other aspects of the site. However, tellingly, only 3 of the participants made use of this section of their profile, with the majority choosing to leave this section empty.

The participants claimed this was due to the prevalence of pre-established social relationships taking place on the site. As one participant, named Chris, put it, his friends knew 'more about me than a bunch of choices on Facebook'. The established social connections present on Facebook led to Chris potentially flouting expectations and toying with many of the descriptions in his profile. He provided false information to many of the predefined options, with his languages and religious views listed incorrectly in what he described as a 'joke'. Chris confirmed that these languages are religions were chosen because he thought they were the 'craziest option I could find!' He also tellingly wrote a message in his 'about you' section confirming that he only expected to communicate with users who already knew him. Writing that 'if ur lookin at this, the chances are u already know me. So no point in telling u is there'.

When we discussed this during one of his interviews, Chris was rather hesitant to over-explain the joke, but did say that he just did this 'as a joke, kinda random I guess … I really didn't think about it much, I just thought it was funny. Everyone who knows me knows it's a joke. All my mates on Facebook know it's me from my photo, not from this stuff'.

Chris' playfully stylised use of this customisable section again serves to highlight the notion that Facebook's definition of 'Social' is one that is built upon pre-existing relationship, and the familiarity of the relationships allows for these spaces to be less 'truthful' than they were shown to be in similar modal affordances on Twitter, as discussed below.

The Discourse of 'Followers' and the Use of Topics on Twitter

Rather than defining other users as 'friends', as is seen on Facebook, Twitter used the term 'followers' and 'following' as a signifier for the relationship between the users. This term is potentially more open-ended than 'friend', as it implies that there can be many reasons that people choose to socially interact with you and to view you updates. Indeed, When asked why they felt other users followed them, the participants suggested that not only were their established offline friends following their updates, but that other users were following them due to the content of their updates, or 'tweets'. The participants suggested this was due to the type of friendships Twitter revolved around. A number of participants mentioned how Twitter interactions and followers were chosen due to the context. One participant claimed 'Twitter is all about the stuff I like, the TV shows and celebrities I like'. Another summed this up, saying 'Twitters more about interests, so yer, they're different (from Facebook)'.

Twitter's design facilitated this feeling, enhancing and encouraging the ability of the user to talk about topics of interest, and to find other users who were talking about these topics through the use of hashtags (#). Hashtags provide users with a way to trace who is talking about a topic at any given time. When a hashtag is placed before a phrase, it becomes searchable to other users who use the same phrase. Nearly all of the participants made use of the hashtag facility, 4 participants told me they check regularly to see what is trending, and often took part in producing tweets about highly trending topics.

The use of traceable topics presents another interesting aspect of Twitter's design, namely the temporal aspects of social interaction on the site. The trending topics of discussion are constantly changing as users discuss new topics. Participants claimed they felt Twitter was 'faster moving', as it encourages discussion and awareness of events that are happening and topics that are currently being discussed.

On Twitter, in order to interact with other users and respond to posts, users had to create a new post every time. This was markedly different from Facebook, which allowed users to comment upon other peoples' updates in situ, rather than creating a new post in response every time. Given that the participants all followed at least 100 other users, with one following 347 other users, this created a large amount of updates for the user to process, and resulted in a fast moving page that discussed topics in real time. Twitter also created a temporal feel by only allowing interactions in short 'tweets'; textual interactions that are restricted to 140 characters. The participants claimed this often made them consider the content of their messages carefully in order to convey the information they wanted to in the short amount of space afforded to them, and oftentimes lead to several tweets being sent to convey longer pieces of information, creating and adding to the fast temporal feel of the present on Twitter.

Twitter presented an alternate definition of Social interaction, one that was built around topics of interest rather than pre-established offline social connections. This can be shown by revisiting Chris, whose rather playful Facebook 'about you' section was discussed previously. On Twitter, Chris was noticeably more careful in his descriptions of himself, choosing to describe himself as a 'Big cricket and football fan and Player;' and later in the research period as a 'Part Time LAD, Full time Lover, Love a bit of Banter, HUGE CHELSEA FC FAN, BLUE IS THE COLOUR!!'

Twitter presented the user with only one area in which to describe themselves to other users. They were not presented with a list of option to choose from, but instead were given a section in which they could describe themselves in up to 160 characters. Many of the participants defined themselves in these sections via topics and areas of interest to them, and carefully maintained this section of their profile during the research period.

Here we see Chris defining himself in a more 'truthful' manner, based upon topics of interest to him. There is less of an assumption that he is interacting with a pre-established audience, and we see Chris presenting a more considered description of himself. He claimed in the interview that 'I guess I care a bit more on Twitter about what I say about myself, not everyone ... knows me'. He later expanded that 'I don't want wierdos talking to me on Twitter, you gotta be careful what you say cos anyone can read it'.

Twitter's motif of 'followers', and the deliberate choice to facilitate the discussion of topics that are happening in real time created for the participants a prioritising of interactions that are based upon interests, not necessarily upon pre-established offline relationships. The description sections seen in the examples above are more truthful, and present a list of topics that Chris feels best describes him. When this was discussed during the interview, Chris claimed that people wouldn't necessarily know who he was on Twitter, so he chooses to define himself via 'stuff that I like'. Much like Facebook, Twitter's design encourages and aids certain ways of socially interacting and potentially discourages others. In this case, due to the potential interactions with unknown users based upon mutual topics of interest, we see the encouragement of a truthful, succinct definition of identity based upon topics of interest, and a reduction in the participant's reliance upon already established relationships.

Changing Identities on Twitter and Static Identities on Facebook

When asked about the selection criteria for relevant information on their Twitter self-descriptions, one participant answered that it was an 'ongoing process, the information I want on Twitter changes with my interests'. This highlights another aspect of identities displayed on Twitter, namely, the care taken to maintain an up-to-date description.

The information provided on Facebook's about page were not changed during the research period by any participants, and revolved around the concept of past achievements, of an established social history, with users listing places they had worked and studied, places they had lived, their family, etc. On Facebook, these identities was presented as a constant; a fixed entity that would be added to over time. This suggests that the identity portrayed on Facebook had some aspect of permanence to them, or, alternatively, as two participants claimed, that information 'just isn't important on Facebook.'

On Twitter however, the identities were works in progress, evolving and changing, with a temporal pace matching the overall design of the website. Users took great care over how they described themselves, and all mentioned that they considered the context on their descriptions carefully and regularly.

The Use of Pictures

Throughout the observation period the participants all used a number of modes to communicate and interact. One of the most utilised modes was the use of photography and imagery. The participants tended to use Facebook to share their photos, as its design facilitated the easy sharing of images. One participant commented that they tended to share pictures on Facebook as they could 'tag' their friends, which would link their photos to their friends' pages, and in turn encourage social interaction and discussion of the image. The option to comment on an image in situ was not present on Twitter, and the participants also noted the temporal pace of Twitter meant photos could often get lost in the plethora of updates produced on the site.

Facebook's facilitation of photos, and the ability to comment and interact through them, lead to some interesting stylised modal choice made by the participants,. One participant in particular highlighted this during the research period. The participant, named Hannah, made two posts within an hour, one on Twitter and one on Facebook. Neither of the post can be shown to protect anonymity, therefore they will be described below. The post on twitter cannot be shown due to Twitter's searchable index of all public tweets made, but the twitter post was textual, and insinuated that there was so little food present in her house that there were mice in her parents' cupboards. Hannah's Facebook post, discussed a similar subject matters, with a picture of a near-empty fridge, with only milk, butter, eggs, cream, and Coca-Cola inside, and the words 'what the hell, parents?!!' above the image. Hannah chose to convey a similar message on both sites, a few minutes apart, but chose to highlight and express the point differently, using different modes on the two different sites.

When asked about this choice, Hannah said 'my family and everybody know what it's like at my house.' She explained that she was aware that her family members, and people who knew her and her family offline, used Facebook, and that she felt that the use of the image conveyed the message without having to explain or contextualise what was happening.

On Twitter, she felt she could be potentially more critical of her parents, and added textual details to explain the situation. The stylised choices made on Facebook suggest an awareness of appropriate content for the audience, and the ability to not have to explain or expand on the image provided. Hannah could use the modes of image and text to quickly relay the appropriate message to her perceived audience. On Twitter, her modal choices were potentially limited, and so we see the use of text to convey a message relating to her parents and food. These two different choices show an understanding of stylised modal selection to socially interact, as well as highlighting how these modal choices are framed by the site design.

Style – Timbre and Content of Performances

The previous example of modal differences highlights not only the stylised use of modal choices shown by the participants, as they select and use appropriate modes for each environment to convey their messages, but also the active consideration of appropriate content and appropriate identity for each site and each perceived audience.

The participants often had very different styles when approaching the two sites, and presented different aspects of themselves on each site. One participant, named Sue, claimed that she used Twitter as an 'escape from my boring real life'. This was reflected in her use of the two sites. On Facebook, she discussed health and family issues, and talked about issues in her daily life, often in a serious manner that elicited support and help from her friends. On Twitter however, her identity was optimistic, and often child-like. Sue chose a picture of a teddy-bear as her profile image on Twitter, and had images of the children's book 'the very hungry Caterpillar' on her page. She claimed that she felt Twitter was liberating for her, and that she could 'take a break from my real life to talk about fun stuff'.

Another participant, Mary, presented very different personalities on Facebook compare to Twitter. Whilst on Facebook she discussed family issues, and interacted with her established offline friends, on Twitter she discussed her interests, such as the works of Chaucer. She felt that her Twitter profile was more 'whimsical', and allowed her to maintain and attend to a part of her personality that she otherwise wouldn't have be able to adequately explore and express. She claimed that Twitter presented her with a receptive audience for her interest, which may have been lost on her current friends. As such, the identities presented on the two sites were, she felt, vastly different. An active separation of different identities and the thoughtful choice of content was reflected by many of the participants. For example, one participant noted that he separated events in his life into 'things just for Facebook, and things for Twitter'. This active consideration shows a careful selection and an awareness of the identities being performed on each site.

DISCUSSION

The findings show two complex sites of multimodal activity where users make various kinds of messages and meanings in and due to various situational contexts, design contexts, and contexts of culturally meaningful activity. The actions and interactions observed were influenced by a number of factors including: the allowances and modal choices available on the sites; the layout and design aspects of the sites; the discursive choices of the SNSs; and the communities and audiences the participants were performing to and for.

It is clear that the varying self-presentation strategies and social interactions were affected by several factors. The design aspects of the spaces of social interaction reflected and highlighted the use of specific Discourses of 'social' by each site, with the sites prioritising and encouraging certain aspects of social interaction and downplaying others. Facebook defined social interaction in terms of an extension of offline established contacts and events, an extension of already occurring past relationships. Twitter, on the other hand, based their understanding of social around the 'now', focusing on current events and shared topics of interest. This shaped the way the sites were approached and utilised by the participants, and affected the goals and aims of the participants on each site, the ways they used the sites, and the information they put on these sites. It also shaped the way the participants felt about and understood the maintenance of their identity. On Facebook, less care and attention was given to the identities as they were presented as static objects, based upon established past facts. On Twitter, identities were seen as ongoing objects, changeable and of increased importance in establishing and building new social relationship, rather than maintaining existing relationships as seen on Facebook.

The participants showed an awareness of these aims and displayed an adept ability to unpack these goals and 'play the game', navigating the social landscape effectively. They were aware of successful and unsuccessful strategies for dealing with each site, and showed an awareness of appropriate material and modal choices based on the audience they were interacting with and for, the platform and staging on which they were performing, and the modes available to them. The participants showed that the expectations and preferred uses of the sites could also be flouted, and that they had the ability not only to understand the implicit modal and social rules, but also, that they could unpack and test the limits of these rules, using them to their advantage. The participants displayed stylised interactions on the sites, producing individual identities that were shaped in part by the design and modal options of the sites, in part by their own agency and social aims, and in part by their social environment and offline social expectations.

Facebook presented an environment based upon family and friendship, and encouraged interactions based upon shared events and real-world issues. This echoes similar findings from other Facebook based research (Miller, 2011; Young, 2011). We can see that through the design of the site, Facebook presents a definition of 'Social Networking' that revolves around shared experiences, a temporal history, and known offline contacts and social relationships. Twitter on the other hand presented and framed 'Social Networking' as the exchange of information, social events and the discussion of current events. Given this, and the stylised manner in which each participant interacted with the sites, we should once again be careful to generalise all social networking online as similar. Though there are similarities on these sites, we should strive to look past these to fully recognise how users are approaching and using these sites, and how the performances are shaped by the SNSs and their understanding of 'Social Networking'. The modal nuances and the various definitions of 'social' present two sites that produced noticeably different performances, with each participant interacting through and with the available modes to produce a stylised performance of identity.

The intended uses, the design, and the modal choices affected many aspects of the participants' choices, actions, and interactions, as well as shaping their perceived audiences and expected social aims. Users omitted topics, tapered interactions, maintained some aspects of their identity, alter other aspects and expressed different identities based on their understanding of the sites. Topics were also presented differently on each site based upon the audiences, the participants' understanding of the offline context and the modal availabilities. This highlights that not only does the offline/online dichotomy break down on SNSs, but that it is a malleable state that the user can use to emphasise or support other aspects of identity performance. This is highlighted through the complex, site and user specific presentation of the same topic, as highlighted in Hannah's decision to use images of Facebook and text on Twitter to discuss the same topic.

The differences in the design and audience fostered by the two sites also affected the way the participants defined and described themselves. On both sites users were afforded several indices of information for performance of identity such as personal information, historical and location based information and preference information. These were utilised differently based upon the perceived audiences and the messages the participants were trying to impart. Participants suggested that on Facebook not all this information was pertinent to the environment, the types of interaction taking place or to the performance of identity to an audience who already possessed much of this information. However, on Twitter the participants filled and maintained this information carefully. The participants suggested that the main reason for this disparity was the types of audiences that the staging and modal choice that the two sites fostered. Though both sites provided a space in which to 'write' identity, the space on Twitter was more important to the participants and was considered more integral to social interaction. Despite the similarities in certain aspects of design, when considered within the context of the sites of social interaction, we can see that the modes were used, engaged, and interacted with differently.

The participants' social interactions were also affected by the modes available on each site. Facebook allowed and favoured the posting of images, videos and longer sections of text that could be re-visited and continually added too. This variety of modes meant the participants had more flexibility and options, allowing for stylised manner to emerge when displaying messages and perform identity. The participants suggested they preferred using Facebook to share photos as they felt they were relevant to the audience they were interacting with, and liked the interactive elements present on Facebook that allowed continued engagement and commentary with the photos. The temporal pace of Facebook also aided the use of a variety of modes as the participants felt that the slower pace made the pictures and videos more permanent and easily located in comparison to Twitter.

The participants felt they all managed to successfully perform their identities and successfully interact despite the modal, temporal, design and audience differences between the sites. However, it was clear that the differences between the sites affected the manner in which they performed, and the choices they made when considering modal choices and appropriate content for the site and audience. Despite the character restrictions on Twitter, and the heavy favouring of text-based interactions, the participants still managed to effectively navigate the site and perform stylised identities, using, adapting and understanding the potential restrictions of the sites.

CONCLUSION

Goffman (1959) suggested that social interaction involved the continuous strategic expression of identity and self. When the self is presented, certain aspects are emphasised and others are concealed for the audiences and contexts we find ourselves in. The same social norms can be seen extending online (Uski & Lampinen, 2014), manifesting themselves in this new modal environment. This conscious expression of self can be more considered online, given the delayed temporal nature of the internet (boyd & Ellison, 2007; Uski & Lampinen, 2014). The data presented in this paper suggests that the context and audience affect what is emphasised and what is concealed. This is guided by the modes available and the social norms implied by the site design and ethos. Stylised performances take place online, tapered not only by the users themselves, but by who they are interacting with and the modes and environments they are interacting through. The design of SNSs did not hamper the ability of the user to successfully enact and perform identity, but it did change how they approached the performance, which subjects they presented, how they understood and prioritised the tools at their disposal, how they used the tools at their disposal, and how they understood their role as a performer.

The findings here match those of previous research in noting that design and situation effect social action (De Saint-Georges, 2004; Karimov et al., 2011; Lafkioui, 2013). Of particular note is Van Dijck's (2013) work, which noted that though SNSs can share similar modes, the arrangement and presentation of these modes will affect how they are utilised and how identity is presented. Van den Berg (2008) provides perhaps the most pertinent explanation as to why this happens through the notion of online 'scripts' – the contextual clues that help users understand their role within the social landscape. In the case of this data, Facebook and Twitter both presented differing scripts and differing definitions of 'social', meaning that despite some modal similarities, the user were presented with different roles for them to fulfil, and interacted with these modes accordingly, producing differing performances that were sensitive to the perceived audience and situation.

The findings presented here suggests that Goffman's (1959) performative theories still hold relevance online, and are strengthen by the inclusion of a focus upon modality. The adolescent participants showed an ability to navigate the social landscape and produce flexible identity performances tapered to meet the requirements of the sites, the expected social interactions upon the sites and the audience to whom they were performing. The participants showed an awareness of the flexibility of social identities and an ability to unpack to 'scripts' (Van den Berg, 2008) of a social landscape and perform their identities accordingly. Rather than displaying one unified self, the participants navigated the social landscape, ably and competently picking and choosing aspects of identity to highlight, and modes through which to best express themselves. This data highlights that identity increasingly must be seen as a performance realised through many modes, shaped by a number of aspects including the availability and prevalence of certain modes, the audience, and the social aims and ideals presented through site design.

The use of 'style' (Meier & Pentzold, 2011) as a concept proved particularly useful when considering multimodal activity. By not only considering and observing the multimodal outputs produced by the users but also by discussing the way the participants considered and combined the available modes, an understanding can be reached of the how the multimodal landscape is approached and navigated. The users selected and combined the available modes based on their own needs and upon the suggested social expectations of the sites. The choice of modes both made for and by users are important and relevant to understanding a social situation (Kress, 2004; Young, 2011). The research highlights that choice of modes can be a manner of self-expression, with different modes used to send different messages. In the potentially diverse modal landscapes online, users can express themselves through the embodiment and use of a number of modes. Previous studies have noted users showing inventive uses of sites and modes, and the formation of un-anticipated modes of self-expression (Miller, 2011; Uski & Lampinen, 2014; Young, 2011). Similarly, this research observes the flouting of expected uses of the available modes, and presentation of stylised manners in which to use the available modes and perform identities. As websites continue to grow and change, the manner in which we use and interact with them will continue to adapt, pushing and creating novel, innovative identity performances.

Though there may be commonalities in these sites, the interactions vastly differed. Even when there were modal similarities the temporal choices, the arrangement of these modes, the favouring and discouraging of different social activities, and the goals of social networking on these sites differed greatly. This echoes previous findings such as Van Dijck's (2013) observations of modal similarities between Facebook and LinkedIn, two sites displaying vastly differing aims and attitudes towards social interaction. As such, the nuances and arrangements of websites should always be considered as they can play an important role in shaping the data collected from interactions on the sites. Social Networking can act as a useful general term of study, yet as shown in this research, the meaning and aims of Social Networking can be interpreted differently from one site to the next, which in turn affects the users and uses of the site.

As our everyday social interactions become more multimodal (Jewitt, 2008; Kress, 2004, 2009), so much our understanding of the social. A consideration of modality, and how it is utilised, can wield useful and in-depth findings, helping to understand how and why people are socialising in a number of diverse ways across a number of modes. Similarly, a consideration of style in modality can help unpack how similar modes can be enacted in different manners, across different stages of performance. Though the focus presented here is narrow, the application of multimodality and of stylised identity portrayal could be very broad. Other affecting factors should also be studied to add to an in-depth understanding of multimodal identity performance, such as the routes of entry to these sites – as many users are accessing these sites through a number of devices which may present differing functions, and may affect how they are accessing and using SNSs.

This paper highlights the importance of design and the effects it can have upon the interactions and identity performances taking place on these sites. As this technology becomes increasingly ubiquitous globally, the impact of design upon our social interactions becomes more and more pertinent. The data collected displays two sites that differ vastly in their design, their use and in how they 'script' and present the concept of social interaction. This not only affects how the user interacts, and the topics and modes the participants discuss on the sites, but also how the users frame and shape their identities, and how they understand their social roles and the needs of the audience. The arrangement and availability of these modes, and the design ethos of the SNSs, can be seen as affecting many aspects of how the users view and interact with and on the site, which in turn affects the tone, type and form of the messages the users are sending. Or, to butcher the words of the bard; all the web's a stage, and all the men and women merely players. How we play and perform, and how we understand our roles is, importantly, shaped by this staging, as we perform on, in and through it.

REFERENCES

Agosto, D. E., & Abbas, J. (2013). Youth and online social networking: What do we know so far? In J. Beheshti & A. Large (Eds.), *The information behavior of a new generation: Children and teens in the 21st century* (pp. 117–143): Lanham, MD: Rowman & Littlefield.

Babbie, E. R. (1986). *The practice of social research*. Belmont, CA: Wadsworth Publishing.

Baker, R. K., & White, K. M. (2010). Predicting adolescents' use of social networking sites from an extended theory of planned behaviour perspective. *Computers in Human Behavior*, *26*(6), 1591–1597.

Bennett, A., & Robards, B. (2014). *Mediated youth cultures: The internet, belonging and new cultural configurations*. Hampshire, UK: Palgrave Macmillan.

boyd, D. M. (2004). Friendster and publicly articulated social networks. Presented at the conference on human factors and computing systems (CHI 2004), ACM, Vienna, April 24–29.

boyd, D. M. (2007). Why youth (heart) social network sites. The role of networked publics in teenage social life. In D. Buckingham (Ed.), *Youth, identity, and media volume*. MacArthur Foundation Series on Digital Learning. Cambridge, MA: MIT Press.

boyd, D. M., & Ellison, N. B. (2007). Social network sites: Definition, history, and scholarship. *Journal of Computer-Mediated Communication*, *13*(1), 210–230. doi:10.1111/j.1083-6101.2007.00393.x

boyd, D. M., & Heer, J. (2006). Profiles as conversation: Networked identity performance on Friendster. *System Sciences, 2006. HICSS'06. Proceedings of the 39th annual Hawaii international conference IEEE*.

Branaman, A., & Lemert, C. C. (1997). *The Goffman Reader*. Cambridge, MA: Blackwell.

Bruckman, A. (1996). Gender swapping on the internet. In P. Ludlow (Ed.), *High noon on the electronic frontier: Conceptual issues in cyberspace* (pp. 317–326). Cambridge, MA: MIT Press.

Bulfin, S., & Koutsogiannis, D. (2012). New literacies as multiply placed practices: Expanding perspectives on young people's literacies across home and school. *Language and Education*, *26*(4), 331–346.

Christofides, E., Muise, A., & Desmarais, S. (2012a). Hey mom, what's on your facebook? Comparing facebook disclosure and privacy in adolescents and adults. *Social Psychological and Personality Science*, *3*(1), 48–54.

Christofides, E., Muise, A., & Desmarais, S. (2012b). Risky disclosures on facebook the effect of having a bad experience on online behavior. *Journal of Adolescent Research*, *27*(6), 714–731.

Clark, L. S. (2005). The constant contact generation – Exploring teen friendship networks online. In S. R. Mazzarella (Ed.), *Girl wide web: Girls, the internet, and the negotiation of identity* (pp. 203–223). New York, NY: Peter Lang.

Danet, B. (1998). Text as mask: Gender, play, and performance on the internet. *Cybersociety*, *2*, 129–158.

De Saint-Georges, I. (2004). Materiality in discourse: The influence of space and layout in making meaning. *Discourse and Technology: Multimodal Discourse Analysis*, 71–87.

Elliott, A. (2014). *Concepts of the self*. Cambridge: Polity.

Facebook.com Newsroom. (2014, November 7). Retrieved from http://newsroom.fb.com/company-info/. Accessed on November 7, 2014.

Fields, D. A., & Kafai, Y. B. (2009). A connective ethnography of peer knowledge sharing and diffusion in a tween virtual world. *International Journal of Computer-Supported Collaborative Learning*, *4*(1), 47–68.

Goffman, E. (1959). *The presentation of self in everyday life*. New York, NY: Doubleday Dell Publishing.

Gold, R. L. (1958). Roles in sociological field observations. *Social Forces*, *36*, 217–223.

Goodman, L. A. (1961). Snowball sampling. *The Annals of Mathematical Statistics*, 148–170.

Harasim, L. M. (1993). *Networlds: Networks as social space. Global networks: Computers and international communication*. Cambridge, MA: MIT Press.

Haythornthwaite, C. (2005). Social networks and Internet connectivity effects. *Information, Community & Society*, *8*(2), 125–147.

Hutchby, I. (2001). Technologies, texts and affordances. *Sociology*, *35*(2), 441–456.

IStrategyLabs. (2014). *IStrategyLabs: 3 Million teens leave Facebook in 3 Years: The 2014 Facebook demographic report*. Retrieved from http://istrategylabs.com/2014/01/3-million-teens-leave-facebook-in-3-years-the-2014-facebook-demographic-report/

Jensen, O. B. (2006). "Facework", flow and the city: Simmel, Goffman, and mobility in the contemporary city. *Mobilities*, *1*(2), 143–165.

Jewitt, C. (2008). Multimodality and literacy in school classrooms. *Review of Research in Education*, *32*(1), 241–267.

Jurgenson, N. (2012). The IRL fetish. *The New Inquiry*.

Karimov, F. P., Brengman, M., & Van Hove, L. (2011). The effect of website design dimensions on initial trust: A synthesis of the empirical literature. *Journal of Electronic Commerce Research*, *12*(4), 272–301.

Kress, G. (2004). Reading images: Multimodality, representation and new media. *Information Design Journal*, *12*(2), 110–119.

Kress, G. (2009). *Multimodality: A social semiotic approach to contemporary communication*. London: Routledge.

Kress, G., & Van Leeuwen, T. (2001). *Multimodal discourse. The modes and media of contemporary communication*. London: Edward Arnold.

Lafkioui, M. (2013). Multilingualism, multimodality and identity construction on French-Based Amazigh (Berber) websites. *Revue Française de Linguistique Appliquée*, *18*(2), 135–151.

Lampe, C., Ellison, N., & Steinfield, C. (2006). A Face (book) in the crowd: Social searching vs. social browsing. *Proceedings of the 2006 20th anniversary conference on computer supported cooperative work* (pp. 167–170). ACM.

Larkin, B. (2008). *Signal and noise: Media, infrastructure, and urban culture in Nigeria*. Durham, NC: Duke University Press.

Lenhart, A., & Madden, M. (2007). *Teens, privacy & online social networks: How teens manage their online identities and personal information in the age of MySpace*. Pew Internet & American Life Project. Retrieved from http://www.pewtrusts.org/en/research-and-analysis/reports/2007/04/18/teens-privacy-and-online-social-networks-how-teens-manage-their-online-identities-and-personal-information-in-the-age-of-myspace. Accessed on March 2015.

Lenhart, A., Madden, M., Smith, A., Purcell, K., Zickuhr, K., & Rainie, L. (2011). *Teens, kindness and cruelty on social network sites*. Pew Internet and American Life Project. Retrieved from http://www.pewinternet.org/2011/11/09/teens-kindness-and-cruelty-on-social-network-sites/. Accessed on March 2015.

Livingstone, S., Kirwil, L., Ponte, C., & Staksrud, E. (2014). In their own words: What bothers children online? *European Journal of Communication*, *29*(3), 271–288.

Livingstone, S., Mascheroni, G., & Murru, M. F. (2011). Social networking among European children: New findings on privacy, identity and connection. *Hermès*, *59*, 89–98.

Luff, P., Heath, C., Kuzuoka, H., Yamazaki, K., & Yamashita, J. (2006). Handling documents and discriminating objects in hybrid spaces. *Proceedings of the SIGCHI conference on human factors in computing systems* (pp. 561–570). ACM.

Madden, M. (2010). *Older adults and social media*. Pew Internet & American Life Project.

Mazzoni, E., & Iannone, M. (2014). From high school to university: Impact of social networking sites on social capital in the transitions of emerging adults. *British Journal of Educational Technology*, *45*(2), 303–315.

Meier, S., & Pentzold, C. (2011). *Multimodal online discourse: A sociosemiotic clarification*. Retrieved from http://www.medkom.tu-chemnitz.de/mk/online-diskurse/pdf/

Miller, D. (2011). *Tales from facebook*. Cambridge: Polity.

Miller, D. (2013). *Facebook's so uncool, but it's morphing into a different beast*. Retrieved from http://theconversation.com/facebooks-so-uncool-but-its-morphing-into-a-different-beast-21548. Accessed on November 7, 2014.

Norris, S. (2004). *Analyzing multimodal interaction: A methodological framework*. London: Routledge.

O'Keeffe, G. S., Clarke-Pearson, K., & Council on Communications and Media (2011). The impact of social media on children, adolescents, and families. *Pediatrics*, *127*(4), 800–804.

Palfrey, J., & Gasser, U. (2013). *Born digital: Understanding the first generation of digital natives*. New York, NY: Basic Books.

Pujazon-Zazik, M., & Park, M. J. (2010). To tweet, or not to tweet: Gender differences and potential positive and negative health outcomes of adolescents' social internet use. *American Journal of Men's Health*, *4*(1), 77–85.

Quinn, S., & Oldmeadow, J. (2013). The Martini effect and social networking sites: Early adolescents, mobile social networking and connectedness to friends. *Mobile Media & Communication*, *1*(2), 237–247.

Ralph, L. J., Berglas, N. F., Schwartz, S. L., & Brindis, C. D. (2011). Finding teens in TheirSpace: Using social networking sites to connect youth to sexual health services. *Sexuality Research and Social Policy*, *8*(1), 38–49.

Reid, G. G., & Boyer, W. (2013). Social network sites and young adolescent identity development. *Childhood Education*, *89*(4), 243–253.

Rheingold, H. (1993). *The virtual community: Homesteading on the electronic frontier*. Cambridge, MA: MIT press.

Rideout, V. J., Foehr, U. G., & Roberts, D. F. (2010). *Generation M^2: Media in the lives of 8-to 18-year-olds*. Menlo Park, California: Henry J. Kaiser Family Foundation.

Ross, C., Orr, E. S., Sisic, M., Arseneault, J. M., Simmering, M. G., & Orr, R. R. (2009). Personality and motivations associated with Facebook use. *Computers in Human Behavior*, *25*(2), 578–586.

Skog, D. (2005). Social interaction in virtual communities: The significance of technology. *International Journal of Web Based Communities*, *1*(4), 464–474.

Turkle, S. (1995). Ghosts in the machine. *The Sciences*, *35*(6), 36–39.

Twitter.com About. (2014). Retrieved from https://about.twitter.com/company. Accessed on November 7, 2014.

Uski, S., & Lampinen, A. (2014). Social norms and self-presentation on social network sites: Profile work in action. *New Media & Society*, 146–164.

Van Den Berg, B. (2008). *Should read: Self, script, and situation: Identity in a world of ICTs. In the future of identity in the information society* (Vol. 262, pp. 63–76). FIP – The International Federation for Information Processing.

Van Dijck, J. (2013). "You have one identity": Performing the self on Facebook and LinkedIn. *Media, Culture & Society*, *35*(2), 199–215.

Van Grove, J. (2010). *Teens experiencing facebook fatigue*. Mashable. Retrieved from http://mashable.com/2010/06/30/teens-social-networks-study/. Accessed in March 2015.
Van Leeuwen, T. (2005). *Introducing social semiotics*. New York, NY: Psychology Press.
Wilson, R. E., Gosling, S. D., & Graham, L. T. (2012). A review of Facebook research in the social sciences. *Perspectives on Psychological Science*, *7*(3), 203–220.
Young, K. (2009a). Applying multimodal analysis to popular websites to develop students' digital literacy. *Proceedings of the 17th international conference on computers in education*, Hong Kong.
Young. (2009b). Online social networking: An Australian perspective. *International Journal of Emerging Technologies and Society*, *7*(1), 39–57.
Young, K. (2011). Social ties, social networks and the Facebook experience. *International Journal of Emerging Technologies and Society*, *9*(1), 20–34.

n00bs, TROLLS, AND IDOLS: BOUNDARY-MAKING AMONG DIGITAL YOUTH

Matthew H. Rafalow

ABSTRACT

Purpose – *This study illustrates how youth and young adults use boundary-making processes to create a regulated community online.*

Methodology/approach – *Ethnographic methods are used to compare deviance models of internet participation with work on digital youth culture.*

Findings – *This paper finds that digital youth draw boundaries around three categories of participation (n00bs, trolls, and idols) to identify new people who need help, ward off bullies, and uphold community ideals.*

Originality/value – *Contrary to deviance perspectives, this study finds that digital youth use boundary-making processes to cultivate a civil online community.*

Keywords: Digital youth; culture; symbolic boundaries; deviance; internet; qualitative methods

Technology and Youth: Growing Up in a Digital World
Sociological Studies of Children and Youth, Volume 19, 243–266

ISSN: 1537-4661/doi:10.1108/S1537-466120150000019009

More often than not, youth participation online is constructed through a lens of risk and deviance. Countless news stories and scholarly publications focus attention to widespread accounts of youth victimization online by perpetrators such as adult predators as well as peers of similar ages (ABC News, 2007; Olweus, 2012; Smith, 2012, 2014). Although literature from the sociology of childhood provide many examples of how young people are strategic actors in their social worlds, such as through peer boundary-making processes and developing status orders, rarely is this agent-centered framework applied to understand empirical examples of young people inhabiting online environments (Corsaro, 1997; Crosnoe & Johnson, 2011; Thorne, 1993).

Contemporary work on digital youth suggests that although young people are indeed capable of transgressive behavior online (Bardone-Cone & Cass, 2007; Biddle, Donovan, Hawton, Kapur, & Gunnell, 2008; Bond, 2012; Mitchell & Ybarra, 2007), such as through cyberbullying or interacting with dangerous strangers, youth often use the internet to simply pursue their interests and find like-minded peers, like through video game communities (Ito et al., 2010). As a result, these social environments may be less hostile than many sociological studies of youth cultures in school environments where participation can be mandatory. In this paper, I examine such a possibility. Through in-depth fieldwork in a youth-created, multigenerational gaming community online, I find that participants engage in boundary-making processes with regulatory effects. As a result, this peer culture establishes an interaction order that provides support for new community members (n00bs), wards off bullies (trolls), and cultivates community ideals (idols). This study thus attempts to broaden our understanding of how youth and young adults collaborate with one another online and in ways that drive social order in digital contexts.

DEVIANCE APPROACHES TO YOUTH PARTICIPATION ONLINE

A growing literature focuses attention on young people's use of the internet, a fascination due, in part, to the fact that youth are among the fastest adopters of digital technologies in the last decade (Madden, Lenhart, Duggan, Cortseni, & Gasser, 2013). Popularly termed "digital natives," or children who have grown up with technologies that permit communications online, this generation is among the most internet-connected population in the United States (Prensky, 2001; Zickuhr, 2011). National surveys indicate that 95% of teens and 97% of young adults are online, a figure that has been stable since 2006 (Madden et al., 2013). Additionally, teens and young adults access the internet using a wide variety of methods, including desktop computers, mobile phone technology, and internet-connected video game systems (The Nielsen Company, 2009).

Scholarly work on the subject of digital youth often examines the potential for harm online, particularly when in the presence of adult strangers. For example, studies of risky behavior among youth online have examined its capacity for promoting self-harm (Mitchell & Ybarra, 2007), anorexia (Bardone-Cone & Cass, 2007; Bond, 2012), and suicide (Biddle et al., 2008). Research also explores how children's access to the internet may be related to cyberbullying (Olweus, 2012; Smith, 2012, 2014) and exposure to pornography (Peter & Valkenburg, 2007). In particular, the topic of cyberbullying is at the forefront of the public agenda after a series of stories emerged in the media (ABC News, 2007). Related studies largely position youth as victims, even when they are the perpetrators of such offenses, as cyberbullies are often targets of such aggression, as well (Tokunaga, 2010). Although research in the interest of promoting safety for children is incredibly important, this work rarely takes a more agent-centered approach like that articulated by the sociology of childhood. As a result, scholars miss opportunities to identify not only risk factors of internet use but also potential avenues for youth to engage online with each other and with older adults in productive ways.

YOUTH CULTURE AND SOCIAL BOUNDARIES

In contrast to the aforementioned literature on youth risk online, the sociology of childhood and youth argues that youth are actors of their own social worlds and conceives of childhood as a socially constructed, variable, and contested category (Corsaro, 1997; Fine, 2004; Prout & James, 1997). The "youth at risk" narrative, for example, is often bound up in a developmental construction of childhood that positions youth as victims rather than active social participants (Corsaro, 1997; Prout & James, 1997; Lesko, 2001). Rather than seeing childhood only through the lens of adult-driven socialization, Corsaro (1997) suggests an interpretive reproduction model that frames peer interaction as a site where youth actively negotiate, share, and create culture among one another. Thus, in the present study, we may expect that young people are meaningful participants of their online habitats rather than simply at risk.

Studies of youth and youth culture in offline contexts document how youth construct and negotiate complex status orders, and engage in social stratification and socialization (Adler & Adler, 1998; Crosnoe, 2011; Pascoe, 2007; Wilkins, 2008). Of particular note, Thorne (1993) finds that elementary school children engage in boundary-making processes to identify status differences and delineate groups based on their cultural standards. Through various types of activities, such as contests, chasing, and other rituals, children affirm gendered boundaries and promote heteronormativity within their own peer cultures. Indeed, these boundary-making processes have been widely documented in other cases, such as among teens of mixed social classes (Eckert, 1989; Willis, 1977), gender performances and sexual orientations (Pascoe, 2007), and across racial and religious statuses (Wilkins, 2008). The form that boundary-making may take can vary, too. For example, Milner (2004) finds that youth employ labels such as "freak," "geek," and "cool kids" to delineate the status order at school. Although this scholarship by and large focuses on the negative consequences of boundary work, we may nonetheless expect that young people engage in similar boundary-making processes when they participate online with their peers.

DIGITAL YOUTH ONLINE

Contemporary work on digital youth suggests that although young people are frequently connected to the internet they participate online in ways that are typically an extension of youth cultural participation. Ito et al. (2010) observe that youth hang out with friends online casually not unlike their face-to-face interactions with peers. Additionally, they find that young people often use the internet to pursue their interests and find communities of other peers who share their interest when peers in their local environment are not supportive of such pursuits. Thus, young people can use the internet to connect with like-minded others in ways not limited by geographic distance. Moreover, youth opt-in to some of these peer cultural environments based on their interest rather than what has been documented in existing work that focuses on peer dynamics in mandated school settings.

Another avenue in which research on digital youth distinguishes itself from more traditional studies of youth culture is by arguing that digital youth participation online presents opportunities for intergenerational interaction. Thus, while youth can and do pursue friendships online with same-aged peers that they know from both local contexts as well as other online publics, young people have more opportunities to connect with young adults and older individuals as they pursue their interests of choice (Ito et al., 2010, 2013). In keeping with tenets of the sociology of childhood that position youth as active agents, these contemporary studies of digital youth position them as meaningful participants even in online communities of mixed age groups (Martin, 2014; Pfister, 2014).

At present, video gaming is among the most populated genres of interest pursued by youth and young adults. Recent reports suggest that 97% of youth in the United States ages 12–17 play computer, console, portable, or cell phone games, and more than half of teens play for at least an hour daily (Lenhart, 2008). Additionally, 81% of young adults ages 19–29 play games (Lenhart, Jones, & MacGill, 2008). Research on gaming shows that online gaming communities are worthy sites for empirical investigation of how young people and young adults interact with one another. For example, education researchers document how when young people play massively multiplayer online games, like *Lineage* or *World of Warcraft*, these games become platforms upon which participants socialize, collaborate, and even develop multimodal skills and other valued literacies (Steinkuehler, 2005, 2008, 2010). Thus, we may expect that online gaming communities where youth and young adults participate may be worthwhile sites of study for social phenomenon like peer socialization and boundary-making practices (Boellstorff, 2008; Kow, Young, & Tekinbas, 2014; Nardi & Harris, 2006; Rafalow & Tekinbas, 2014).

CASE AND METHOD

This study focuses on interactions between members of a youth-created community devoted to an interest in LittleBigPlanet 2,[1] a video game for the Playstation 3, with the intent to explore whether and how youth and young adults engage in boundary-making processes among their peers online and for what purposes. In what follows, I describe the video game, its companion online community, and methods for data collection and analysis.

LittleBigPlanet 2: The Game

LittleBigPlanet 2 is a side-scrolling puzzle platform game created by Media Molecule. It was released for Playstation 3 in 2011 and it sold over 350,000 games in the first week of availability (Leack, 2011). The game follows a character named Sackboy as he tries to save his home world, Craftworld, from an evil inter-dimensional vacuum cleaner called the Negativitron. The video game was marketed and designed for a wide audience across age and gender.

Among the game's key features, players have opportunities to craft levels and share them within the galaxy of Craftworld. Players can create new levels and share them in the game and in forums online, and players earn publicly visible badges for creating popular levels. Of note is that the video game was produced with mind to what Ito et al. (2013) term "openly networked design." In other words, the game provides a means players to share their pursuits within the game – such as game levels they create – with their peers in other online environments. Thus, player creations and activity are relayed between the game and online forums like *Sackboy Planet*.

Sackboy Planet is a website centered on an interest in the Playstation 3 game LittleBigPlanet 2. This website was selected because it is similar to other online gaming communities that have been examined in other studies in that it has a shared purpose (interest in the video game) and is a youth-inclusive population (Ito et al., 2010, 2013; Martin, 2014; Rafalow & Tekinbas, 2014). *Sackboy Planet* is an active web forum devoted to the game, and has over 1,000 active users at any time and a membership list of nearly 23,000 users in the last six years. Although it was not possible to obtain representative samples of the *Sackboy Planet* participant population, the forum leader and all interviewees suggest that it serves mostly teenagers and young adults (ages 13–29) and skews male. While this case does not exclusively focus on youth, it offers a lens into how digital youth, with their young adult peers, are key contributors to a regulated online environment. This case thus stands in contrast to some deviance perspectives that suggest that young people are at risk when in the presence of adult strangers online.

Method

To evaluate the culture of the online community, including any boundary-making processes employed by its participants, data was collected through two means: observation in the forums, and interviews with community members. Fieldwork took place during a nine-month period (September 2011–May 2012) during which the websites were visited several times a week to observe public interactions in the forums and chatrooms. The study involved 60 hours of observation by its conclusion. Although the observed activities on *Sackboy Planet* are public for anyone with internet access to view, I made considerable efforts to notify the forum leaders and community members by several different means. First, I contacted the creator and administrator of *Sackboy Planet* and shared information about the study, including the research questions and methods of data collection, for their permission to participate. Second, in consultation with this administrator, I posted a public message describing the study purpose and period of observation to the community. It should be noted that this is not a perfect means by which to reach all participants of a community, as it would be incredibly difficult to individually gain consent for passive observation on a platform where thousands of people interact. However, this strategy maximized community awareness of the study through the aid of the community administrator to develop a public notice of research activity and direct any participants with questions to myself and/or my institutional review board.

Observation in the forums was coupled with interviews of community members throughout the period. Because I was interested in status and boundary-making dynamics in the forums, I began by interviewing the creator of *Sackboy Planet* (a 17-year old male) as well the moderators of the community. With the permission of the creator, I posted an open call for interviewees, noting that I was looking to speak with players both new and more experienced. Interview participants were provided a link to a study information sheet, including details about the research design and how the interview will be used, and I reviewed this sheet with them and asked for verbal consent (either by voice or chat text). In consultation with my institutional review board, participants were not asked for written consent. No personal identifiers were collected about interviewees, and signed consent forms from subjects would require them to write down their names, breaching anonymity. For participants between ages 13 and 18, I followed a modified consent protocol. For these youth, I reviewed the study information sheet and attained verbal assent from both the interviewee and their parent or guardian. Risks for this population were minimized by not collecting subject identifiers, or linking interview data to their online identities (e.g., aliases), as well as by attaining verbal and not written assent from participants.

I conducted interviews through Skype audio calls or by Skype text chat (depending on the participant preference), and used a semi-structured interview protocol to gain an understanding of how participants conceived of their experience in the community and evaluated other participants. The interview protocol followed a narrative of participation in the forums, first asking how they found the community and what their daily habits were as participants. Later, I asked how they determined how they sorted among forum content in order to determine who to follow or what to play. As participants answered these questions, I probed for examples or stories to illustrate related boundary-making processes. Interviewees were also asked structured questions about their own status characteristics, including age, race-ethnicity, gender, class, education level, and geographic region. Ultimately, I report on interviews with 19 youth and young adults, as well as 5 additional respondents who are either 30 years or older or preferred to not disclose their age (see Table 1 for more detailed sample characteristics).

A limitation of this study is that it is not possible to know the exact demographics of participants on *Sackboy Planet*, including age, gender, and race-ethnicity, so that I can compare how the characteristics of this interview sample may align with statistics of overall participation. While the game is marketed toward both youth and adults, it is possible – if not likely – that the community is heterogeneous in age. Thus, this paper ultimately does not report on an essential "youth culture" in the sense that I report on data collected only from youth. For example, observed individuals in the forums could include both youth and adults. However, contemporary work on youth participation in online publics can rarely claim that they are only studying observed patterns of youth as most online

Table 1. Interview Respondent Characteristics.

Age	
13–19	6
20–29	13
30+	2
Unknown	3
Gender	
Female	3
Male	21
Race-ethnicity	
Asian	1
Black	1
Latino/a	3
White	16
Unknown	3
Nationality	
United States	21
Other	3
Total *N*	24

publics are not youth-only, unlike peer locations in face-to-face environments like schools (Ito et al., 2010). Rather, this study shows that regulatory boundary-making practices exist in an online community that youth, among others, inhabit, and through interviews illustrates how youth and young adults *both* exert agency in this boundary-making process. In this way, I find that existing deviance-oriented perceptions of digital youth are complicated by the fact that young people can hang out online with older peers and meaningfully participate in social regulation online.

I imported anonymized data from field notes and interviews into Dedoose, a mixed-methods data analysis software. Field notes and interview data are arranged into a database within the software program, and the data analysis tool allowed me to create codes and apply them to particular sections of each note or interview document. I used "open" coding to identify broad categories that emerged (i.e., "mention of boundaries"), followed by "focused" coding to unearth more detailed categorizations of the data (i.e., "n00b," "troll," and "idol") (Charmaz, 2014). This coding process allowed me to identify the most common emergent themes across the sources of data, and then categorize coded excerpts into more specific typologies that are discussed in the results.

WELCOME TO SACKBOY PLANET

In the sections that follow, I illustrate the results from a grounded approach to documenting and analyzing boundary-making processes that exist within this community online. This exploratory research illuminates my process of discovery as I interfaced findings with existing research.

The Role of Community Policing

What informs relations on *Sackboy Planet* and among its membership, and how is interaction regulated? In order to fully participate, visitors must register and create an account on the website. Registration only requires that participants create an alias, or username, and password, but it also importantly provides the first glimpse into regulatory expectations to every community participant. The leadership of the website present a number of rules upon registration with the community. After registering an account, users are re-directed to a website e-mail dashboard that includes a welcome message automatically generated by the forum system. The message informs users that they need to read the "Rules & Regulations" post on the website before participating in the community. Upon clicking the attached link, users find that the post elaborates on two forms of stated regulation: human moderators, and non-human, systematic regulation via the forum design. This Rules and Regulations post states that moderators, as officials of the forums, police interaction and watch out for cheating the reputation system. The second section of the rules and regulations describes the formal reputation and infraction system in more depth:

> Any time you break a rule, you will get a private message (called an "infraction") ... Once you reach a predefined number of points, the system (read: not a human being) will automatically ban you for a predefined number of days ... Overall, this system will be much more standard and automated, so you can keep track of how close you are to a ban.

Immediately upon registration with the community participants are notified of policies aimed to ward off bad behavior, which stands in contrast to existing research that positions youth engagement online as constantly fraught with risk. This rules and regulation page indicates that the primary forms of social control are through designated authorities (moderators) and more systematic reputation and infraction systems established through the forum design code.

Subsequent interviews with moderators of the forum, however, revealed that the sources and standards of the community etiquette are much more complex. Moderators do assign infractions when they witness bad behavior, but they are not the first line of defense. One moderator, ModBob (18 years old), explained: "If [community members] are doing something wrong and someone takes offense to it generally the person will let us know. Otherwise we really can't tell unless someone reports it." When describing their role, moderators regularly discussed how their policing is usually secondary to other forms of community regulation. Another moderator, Josephina (19 years old), elaborated: "Moderators just keep order and assist the new people as they come in and make them feel welcome ... everyone pretty much polices themselves, helps each other out." Moderators on *Sackboy Planet* viewed their role as important to the overall function of the community by assigning infractions and serving as greeters to new members, but because there are many users and so much activity in the forums they actively rely on members to report bad behavior and maintain community standards. As other work has suggested, the environment for interaction online is, in fact, different than many face-to-face environments that youth inhabit because online forums can be host to potentially thousands of other anonymous or pseudononymous others. Yet, these moderators describe community regulation as one means to address this structural difference.

When I asked non-moderator community members about community regulation, they confirmed the prevalence and importance of community policing. I asked if non-moderators ever step in and do something when they see bad behavior, and Turtle (14 years old) wholeheartedly agreed: "Plenty of times. I've done it once or twice too. I also report a lot of spam and advertising. So yea all of *Sackboy Planet* try to make it a nice place to be." Another member, Ponytale, responded to the same question by explaining his own self-professed role on the site as an unofficial community regulator: "I've never been a moderator or administrator or anything and I don't really care to. But I'm kind of like a non-moderator moderator. I tell people what I think or diffuse situations. Or if I don't do it publicly I'll write them a private message." For these participants, community policing – not moderator-driven regulation – was an important component of maintaining collective standards and is a shared responsibility. Contrary to the framing of literature on digital youth and risk, young people can be active agents in managing behavior online and building a community regulated around a shared purpose of kindness.

How Community Policing Occurs in Practice

How does social regulation occur on *Sackboy Planet*, and what does community policing look like in practice? As noted in the "Rules and Regulations" post described earlier, *Sackboy Planet*'s most prominent features include an algorithmically developed community management system. This system includes reputation markers such as various experience bars, stars, and badges on their profiles obtained through the accumulation of different social metrics, such as how many times they log in or how many other players like or thank their posts. In an interview with the forum leader (17 years old) and designer of the social metrics, he explains the role of these various status indicators in shaping forum interactions:

> In terms of rewards, we have the experience system … if they use the site and interact with people and help people they will earn experience for it. By posting to the site and posting blogs, by replying and making new threads users will gain experience … I guess we're printing our own kind of money.

The community leader's design of the website created a type of market based on interaction in the forums. While the leader himself does not see it as a "real money" currency, he portrayed it as a kind of market that relies primarily on social capital: being an active member of the community social network, creating discussions with other members, and replying to others' posts.

After observing in the forums for several months, I myself began to accrue trophies, or awards, that signaled my participation. One award was for being registered with the community for three months. I received a message notification alerting me to my award, and the forum system automatically inserted a small picture of a trophy with the number 3 on my profile page. Other, more senior members, had much more diverse symbols of achievement on their profiles. Some of them include titles such as "Veteran" to refer to the user's extensive participation over a long period of time. Others accumulate experience across one of several bars that increase or decrease depending on the behavior of the member in the forum. Lucidio, for example, had a full orange bar to reflect that many people "liked" or thanked his comments on the blog. He had a low green bar, however, which indicated that he was a relatively infrequent participant on the forums. The different types of trophies, or badges, and reputation bars players accumulate might function as a type of status currency within the forums.

In order to interrogate the collective meaning behind these awards, I asked every interviewee to describe the various reputation bars and points that appear on forum members' publicly visible account profiles. Some said that the game metrics matter: "They do care … if your profile has 2,000 hearts then everybody knows you." Respondents described others as completely unimportant: "I personally never even look at that stuff … I don't really notice it, basically." When they expressed such indifference to the reputation system, I followed up by asking what then actually determines who is considered good or bad. Community members like Poseidon (14 years old) revealed that other status indicators inform reputation:

> High reputation bars just mean that they participate a LOT in the forums. I suppose if someone is a higher level, they'd be less likely to have their posts skimmed. But if it appears to be a n00b then their post will most likely be skimmed/skipped. After a while, you really learn who the big names are on the site. :P

While it is clear that participants in this forum are active agents in creating and reproducing forms of social control on *Sackboy Planet*, how it is that they do so is rather complex. Although the stated rules and regulations clearly delineated certain expectations of the forum, moderators rely on the community to maintain its standards. Moreover, these participants see that enforcing community etiquette is a responsibility and an important way to maintain the collective good. Further interrogation of the community management system revealed that while these formal metrics do in some ways matter in shaping who is good or bad, such as when a user's level has earned thousands of hearts or plays from other players, others do not. This last quote suggests that other status indicators, such as n00b, may more meaningfully inform interaction in the community.

n00bs: New and Inexperienced Participants

Observation in the forums indicated that socialization into the appropriate standards of behavior is key not only to avoid having their behavior corrected by other participants. The following is an excerpt from a forum post:

> Bluebell: "Hey guys im new to all of this, but if you could just spend 30 seconds of your time to click this link and subscribe to me, it means a lot."
>
> Rangerguy: "Hi and welcome to *Sackboy Planet*! It's great to have you here, but here is a suggestion. Before asking people to subscribe to your youtube channel and play your level, maybe tell a tiny bit about yourself. What brought you to the site, your experience, etc. Most people here are nice and more than happy to help out. Look forward to seeing you around!"
>
> Flowerpower: "Welcome to *Sackboy Planet*. Lots of good information above. I agree with Rangerguy, let people get to know you first."

In the above example, requesting that others' view your content before following other rules, such as sharing more about yourself and letting others get to know you first instead of making anonymous requests, results in getting ignored. Knowledge of the social rules or community etiquette represents a type of cultural knowledge that must be attained and demonstrated to attain important resources from other members, such as guidance or attention to one's shared content in the forums.

A very frequent subject of discussion by community members from both forum activity as well as interviews was the "n00b." Upon further questioning about the n00b category, I learned from community members that n00bs are identified through discourse in the forums, and come to represent people and behaviors that signal being new to the community and unsocialized into community etiquette. Turtle (14 years old) explained:

> Interviewer: "How do you know someone is a n00b?"
>
> Turtle: "If they're like ZOMGZ PLAY MY LEVELZ PLZ! YOU WILL
>
> LOVE IT! OMG! :D"
>
> Interviewer: "hahahaha. So all caps probably!"
>
> Turtle: "Or all lower case :D"
>
> Interviewer: "ah, like me :)"
>
> Turtle: "run on sentences :P"
>
> Turtle: ":D"

On *Sackboy Planet*, n00bs are the least reputable members of the community and embody numerous statuses at once. As in the above interview excerpt, n00bs use standards of grammar that are not proper English as understood by the community. n00bs use all capital lettering, run on sentences, or, as I mistakenly did, all lower case lettering. Posts from community members suggest that the n00b label also referred to a relative lack of experience. One member created a post titled, "Alert: n00b!" She then noted that she was "fairly new to the game and just wanted to say hello," noting that she "would love to receive advice on how to improve at the game." In another post, a user expressed that he is a n00b and others responded by offering guidance and support:

> Banjo Bill: "Hi everyone! I'm a n00b!"
>
> Sassypants: "Hi! Welcome n00b! You could actually probably still say I'm a n00b myself, but I'm pretty active here."
>
> Basketball99: "Welcome to the community, n00b. ;) We know you are new to this place, so if you want to know more about the rules there's a section you can check out on it. If you have any other questions, feel free to let me know!"

Community members frequently described n00bs as inexperienced members, and in public posts they typically offer to help professed n00bs noting that they themselves were once n00bs or inexperienced members when they first joined. In this way, these participants engage in boundary-making processes as they wield the n00b label – sometimes in a jocular tone – to separate people who seem inexperienced and are in need of help to be socialized into community standards.

Trolls: Intentionally Harmful Users

Another cultural category that surfaced during observation and interviews was the "troll." On *Sackboy Planet*, trolls represent people who *intentionally* breach community etiquette and are perceived as harming other users. In one thread, members talked about a user who was trying take away the attention of people from *Sackboy Planet* and direct them to another website:

> Frog51: "The troll who claimed to be a community member is some idiot who is swearing over and over and telling people to go to a different forum."
>
> Goalie1991: "He's now posting this comment in every part of the forum. Someone please stop him."
>
> BigReader: "He's trying to steal people here to go to other places, and his actions are very irresponsible and he is spamming everyone. This is unforgivable behavior!"

In this example, swearing and drawing attention away from the forums was identified as a flagrant violation of community standards. In another post, a member complained about unkind reviews he had received on a game he created and shared with the community:

> An evil troll gave my shared game a bad review. I worked really hard on it, and I got really offensive feedback. My friends remind me that not everyone will appreciate stuff you share with the community, but this troll said mean things on purpose.

In contrast to the n00b category, which described users who may be inexperienced or immature, trolls exhibited behaviors that break the rules of the community intentionally. While giving someone a bad review in another community may be seen as acceptable behavior, giving a bad review in this way on *Sackboy Planet* is seen as an intentional breach of etiquette and justified, for this user, calling them a troll. Behaviors that are identified as troll-like have consequences for their engagement on *Sackboy Planet*. In another post, community members discussed what to do when one encounters a troll:

> Bluebird: "Has anyone ever had to deal with a troll? I'm having trouble because this dude is trolling me … what should I do?"
>
> Smackattack: "Ignore them."
>
> Starfighter: "Report them to the moderators."
>
> Rainbowbrite: "Don't feed the trolls. They don't deserve the attention."

Members of *Sackboy Planet* report that attention from other members is a scarce resource. In the above example, the most common response to others' troll displays is simply not responding at all. In other cases, it involves reporting them to moderators who can take more decisive action, such giving them a violation or removing them from the community. In an interview, a community member named Ninjadude (16 years old) described to me an experience when others labeled him a troll:

> I called [other forum members] n00bs and all these people were sort of like in response they were making jokes and making fun of me … I edited my post to get rid of it. Basically what I'm saying is people won't be mean but they'll be sarcastic or make fun of it and it makes you stop … like this is basically what happens if you're like a troll or being mean on the forums.

For this member, both the n00b and troll status categories are enacted through discourse in the community: the community demarcates particular interactions as associated with one of the cultural categories. He described his behavior, which was interpreted as violating community standards of decency and subsequently heavily policed, as troll-like behavior. Being labeled a troll results in community retaliation in ways that correct or modify the behavior of the identified user. Like the n00b category, this user never fully identified himself as a troll. This participant's experience reflects the moderating power that discursive boundary-making has on users' future behaviors. Such status categories in this community are transient, and are primarily used through discourse to police behavior.

While trolls and n00bs are similar categories in that they both can be used to moderate behavior online, n00bs represented players who are new or in the process of being socialized, whereas trolls are assumed to have intentionally violated the cultural etiquette. Participants use these labels to engage in boundary-making among participants in their online community, delineating types of behavior that do not meet the cultural etiquette or represent the cultural competencies valued by the community. On *Sackboy Planet*, categories such as n00b and troll are labels that are used strategically to uphold community standards. Cultural etiquette on the forums informs user reputation and whose content is viewed and evaluated in an attention-scarce environment.

Idols: Revered Participants

While n00bs and trolls are two emic identity categories used to identify bad behavior and socialize members into expected cultural competencies, how does the community identify ideal cultural competencies or those who display such competencies? By interrogating the last category employed in the forums – idols – I show how some users are identified and celebrated.

On *Sackboy Planet*, idols are celebrities in the community that often are depicted as role models. In a post titled, "Who is your idol?" community members shared people they looked up to, and why:

> MisterSack: "If I had to say anyone was my idol it would be the creators of this game. They are very creative. And also the leader of this community!"
>
> Lipglossgurl: "My idol makes the most amazing game levels and has such a distinct style, always innovative!"

In another post, Raindrop1985 explained why she joined the site to begin with: "Hi everyone! I couldn't resist joining this site anymore because of Johnjohnn. He's my idol and makes the best game levels!" Another user, TennisFan, described his community idol: "She's my idol because she's a super nice person and because she makes wonderful levels. She's my favorite creator on *Sackboy Planet*. The designs are really nice." Community members describe idols as people connected to the *Sackboy Planet* community who demonstrate styles or competencies valued by the community. One important competency with the game is producing and sharing game content that the community believes reflects great skill.

In both the forums and interviews community members would regularly refer to Johnjohnn (25 years old), a user who they held in high esteem. In an interview with Johnjohnn, he explained that some time ago he had an idol in the community before he achieved celebrity himself:

> JohnJohnn: "DarkMatter9 was my idol."
>
> Interviewer: "Why was DarkMatter9 your idol?"
>
> JohnJohnn: "The levels he made were so awesome I wanted to know his secret. What was he able to do that I wasn't?"
>
> Interviewer: "Was he a celebrity? Was he hard to pin down?"
>
> JohnJohnn: "Yes. He was kind of a celebrity … I met him through a project and we became friends."

Johnjohnn used the idol category to describe someone who was inspirational because of his skill level with game design: he had a "secret," that is, a set of cultural competencies or skills, that Johnjohnn did not know but wanted to learn. Elsewhere, Johnjohnn described DarkMatter as "inspirational," "a huge motivator" as he sought out to improve himself as a designer. Although idols can take multiple forms, such as those who find undiscovered levels and give recommendations to the community, members regularly described idols as people who became famous because of their created levels and who have the highest level of skill with game design. Much like the literature on peer cultures, these participants also engage in boundary-making practices to identify and celebrate high status members that represent their cultural ideals.

The idol status grants benefits to those who inhabit it, including access to other high status community members. DarkMatter9 (21 years old) explains:

> DarkMatter9: "Celebrity status helped me reach more people and I could find those people I work well with."
>
> Interviewer: "So you could be choosier."
>
> DarkMatter9: "Yes … and the people I work with tend to be celebrities themselves … I've never had someone successfully solicit [level design] help from me through the fan sites. I get a lot of messages from the game networks like "Come help me fix this"… it became tiresome to be constantly helping."

In the above example, this idol describes how he only works with or helps people who are also idols. When asked why he does not regularly help the community anymore, he expressed that he had already created extensive YouTube tutorials that can provide direction to regular members. Thus, idol status provides rewards in the form of access to other community elites, but in this community it is achieved through meeting the community expectation of helping and producing valued resources, like tutorials, for others.

Members of *Sackboy Planet* provide many opportunities for participants to work toward becoming an idol. Contests are very exciting and emotional ritual in the community. For example, a community leader created a challenge contest to design "the most beautiful game level that takes place entirely underwater." In his announcement post, he explained: "Hi everyone! This time around the great Ocean gave us ideas for the contest. Are you ready to take the dive?" In the subsequent paragraphs, he explained that the challenge was to design "the most creative and fun fast-paced underwater game." In addition, like other contests, he stated how it would be evaluated: through quality of "Gameplay, Originality, Aesthetic, and Sound." Community members were allotted a set amount of time to create and submit levels, and eventually the contest garnered scores of submissions that were shared publicly. All players could submit their own designed levels to be considered for awards and commendation. Contests were run by idols, and users post levels in a thread designated for the contest. These threads are highly active, and once contest participants submit their level many other players will also play them and leave comments that are a mix of support and critique. In the case of the water level, some players left comments saying "Great job!" or "I can't wait to play these levels!" Others were more critical: "Submissions this time around weren't that great. :-\" or "I didn't really like X's level." Participation in community contests is an opportunity for players to display their skills publicly and receive feedback, good or bad, from other membership. This feedback is a form of participant-driven boundary-making in action for idol status.

Contest rituals provide an example of how the idol category is a boundary-making process organized by the members of this community. Participants assess worthy and unworthy performances of cultural competencies, and reward those who display such competencies by spotlighting that member to the community. Although the idol label occupies less of a staple in daily forum interactions between members of *Sackboy Planet*, it certainly is prevalent during interviews and associated contests draw the highest views from participants than any other discussion topic. Through these practices, community leaders are able to separate the community members into low and high status groups, distinguish the idols through peer and judge evaluation, and determine whose shared content receives the most and the least attention.

CONCLUSION

Through a discussion of "n00b," "troll," and "idol," emic cultural categories to an online marketplace and community of gamers, I show youth participation online can be more complicated than existing literature suggests. Previous work that employs a deviance or risk-centered orientation to the online activities young people pursue online largely ignores that youth, with their young adult peers, are active agents in their own pursuits on the internet. Similar to existing studies from the sociology of childhood on peer cultures, the youth and young adults in this online community engage in boundary-making processes that organize people based on their behavior in the interest of maintaining a civil community. This study illustrates how youth and young adults use labels discursively to help people in need, ward off bullies, and celebrate their ideals.

The cultural labels that youth and young adults employ have meaningful effects on the behavior of the online community toward its participants. As shown in Table 2, "n00b" and "troll" are emic labels used through

Table 2. Perceived Characteristics of Categories and Effect on Community Behavior.

	Perceived Characteristics of Cultural Category	Effect of Category on Community Behavior
n00b	Inexperienced Poor grammar New	Mocked for poor grammar, offered help to be socialized into community etiquette
Troll	"Idiot" Swears/uses profanity Irresponsible Intentionally causes harm	Participant comments are skipped/ skimmed, user is sanctioned and/or evicted
Idol	Leader Creative Role model Innovative Celebrity	Participant is celebrated and receives high attention

discourse to identify inexperienced players and players who exhibit bad behavior. n00bs are described by community members as new and inexperienced, and have poor grammar. Often, identified n00bs are provided with guidance as to how to better interact in their future communications on the forums and with others on *Sackboy Planet*. Community members who exhibit n00b-like behavior are rarely given the attention needed to their products, and thus their formal metrics (including ratings and plays/purchases) are relatively absent in any data.

"Trolls" are described by community members as "idiots," people who swear or use profanity, irresponsible, and people who intentionally cause harm to others. Content shared in the marketplace by both n00bs and trolls are typically ignored, though the reasons for ignoring are different. n00bs are believed to be new and inexperienced players who often need guidance to be socialized into the expectations of the community. Trolls, however, are thought to have *intentionally* violated the norms of the community. Trolls are generally handled with much more strict sanctions, including widespread attacks by other members or being kicked out of the community.

"Idol" is a cultural category used to identify players who are celebrated members and skilled level designers. They are described by the community as leaders, creative types, role models, and innovative. They are widely assumed to be celebrities on *Sackboy Planet*. Their creations or products are often held in high regard, and receive much attention from members in an otherwise attention-scarce environment. Formal metrics for these members are at the highest levels of any user. They are regularly at the top of all the highly rated and purchased charts and their work is widely known in the community. Although the idol label is not used as regularly during forum interactions as is n00b or troll, it is a significant to the interviewed participants. Moreover, its use in the forums during related rituals, like contests, attain the highest views of any discussion topic in the forums. Idols come to represent the ideals of those who inhabit *Sackboy Planet*.

This study has a number of limitations that merit further investigation in future work. I was not able to obtain demographic data for the *Sackboy Planet* community as a whole, which makes it difficult to know the ages of those interacting in the public forums. It is also not possible to determine if the interview sample represents a demographic portrait of the community as a whole. However, despite these limitations, interviews with the forum leader, as well as other participants, suggest that the community *includes* youth and young adults and is multigenerational. Thus, the findings illustrated in this paper contest deviance-oriented perspectives on youth participation online by showing that they are meaningful participants in online social regulation – even in the presence of older peers. Yet, a study of an online community where demographic data is known would provide a clearer portrait of the experience, and it would provide more substantial evidence that youth are powerful forces as compared with other membership in the community.

This work suggests that research on young people's participation online need be cognizant of the capacity for youth to engage in not only risky activities but also be active agents in the development of a space for civil and safe interactions with their peers. As compared with other deviance-oriented studies that paint youth participation online as associated with cyberbullying, anorexia, bulimia, self-harm, and pornography (Bardone-Cone & Cass, 2007; Biddle et al., 2008; Bond, 2012; Mitchell & Ybarra, 2007), I find that young people and their older peers collaboratively produce a regulated environment. Indeed, the challenges that youth face to manage their community are sometimes different online than has been identified by literature on peer cultures in schools (Adler & Adler, 1998; Crosnoe, 2011; Eckert, 1989; Pascoe, 2007; Thorne, 1993; Wilkins, 2008; Willis, 1977). As described by the youth moderators of the forums, there are sometimes thousands of people interacting with each other in their community at any given time. Thus, appointed moderators are not able to observe every interaction. The youth in this community employ crowd management boundary-making strategies that rely on the broad populace to participate and help make it, in the words of one youth, "a nice place to be."

NOTE

1. Although people may use online platforms for many purposes, such as to extend communications with offline contacts, those interviewed all reported that they found *Sackboy Planet* and joined the community for the sole purpose of pursuing and sharing their interest in the LittleBigPlanet 2 game with others.

ACKNOWLEDGMENTS

Data collection for this study was made possible by grants from the John D. and Catherine T. MacArthur Foundation in connection with a grant making initiative on Digital Media and Learning. I thank Mizuko Ito, Amanda Wortman, Ksenia Korobkova, Crystle Martin, participants at the UCI Statistics Reading Group, participants at the ASA 2013 annual meeting, and the editors and anonymous reviewers of *Sociological Studies of Childhood and Youth* for helpful feedback.

REFERENCES

ABC News. (2007). *Parents: Cyber bullying led to teen's suicide*. Retrieved from http://abcnews.go.com/GMA/story?id=3882520&page=1. Accessed on January 20, 2010.

Adler, P. A., & Adler, P. (1998). *Peer power*. New Brunswick: Rutgers University Press.

Bardone-Cone, A. M., & Cass, K. M. (2007). What does viewing a pro-anorexia website do? An experimental examination of website exposure and moderating effects. *International Journal of Eating Disorders*, *40*(6), 537–548.

Biddle, L., Donovan, J., Hawton, K., Kapur, N., & Gunnell, D. (2008). Suicide and the internet. *British Medical Journal*, *336*, 800–802.

Boellstorff, T. (2008). *Coming of age in second Life: An anthropologist explores the* virtually human. Princeton, NJ: Princeton University Press.

Bond, E. (2012). *Virtually anorexic – Where's the harm? A research study on the risks of pro-anorexia websites*. Retrieved from http://www.ucs.ac.uk/SchoolsAndNetwork/UCSSchools/SchoolofAppliedSocialSciences/Virtually%20%20Anorexic.pdf. Accessed on December 19, 2013.

Charmaz, K. (2014). *Constructing grounded theory*. London: Sage.

Corsaro, W. A. (1997). *The sociology of childhood*. Thousand Oaks, CA: Sage.

Crosnoe, R. (2011). *Fitting in, standing out: Navigating the social challenges of high school to get an education.* Cambridge: Cambridge University Press.

Crosnoe, R., & Johnson, M. K. (2011). Research on adolescence in the 21st century. *Annual Review of Sociology, 37*, 439–460.

Eckert, P. (1989). *Jocks and burnouts: Social categories and identity in the high school.* New York, NY: Teachers College Press.

Fine, G. A. (2004). Adolescence as cultural toolkit: High school debate and the repertoires of childhood and adulthood. *The Sociological Quarterly, 45*(1), 1–20.

Ito, M., Baumer, S., Bittanti, M., boyd, d., Cody, R., Herr-Stephenson, B., ... Trip, L. (2010). *Hanging out, messing around, and geeking out: Kids living and learning with new media.* Cambridge, MA: MIT Press.

Ito, M., Gutiérrez, K., Livingstone, S., Penuel, B., Rhodes, J., Salen, K., ... Watkins, S. C. (2013). *Connected learning: An agenda for research and design.* Irvine, CA: Digital Media and Learning Research Hub.

Kow, Y. M., Young, T., & Tekinbas, K. S. (2014). *Crafting the metagame: Connected learning in the StarCraft II community.* Irvine, CA: Digital Media and Learning Research Hub.

Leack, J. (2011, January 18). *January 2011 NPD results show some software surprises.* PlayStation LifeStyle. Retrieved from http://playstationlifestyle.net/2011/02/18/january-2011-npd-results-show-some-software-surprises/. Accessed on January 29, 2013.

Lenhart, A. (2008). *Teens, video games and civics.* Washington, DC: Pew Internet & American Life Project.

Lenhart, A., Jones, S., & MacGill, A. (2008). *Adults and video games.* Washington, DC: Pew Internet & American Life Project.

Lesko, N. (2001). *Act your age!: A cultural construction of adolescence.* New York, NY: RoutledgeFalmer.

Madden, M., Lenhart, A., Duggan, M., Cortseni, S., & Gasser, U. (2013). *Teens and technology 2013.* Washington, DC: Pew Internet & American Life Project.

Martin, C. (2014). *Learning the ropes: Connected learning in a WWE fan community.* Irvine, CA: Digital Media and Learning Research Hub.

Milner, M. (2004). *Freaks, geeks, and cool kids: American teenagers, schools, and the culture of consumption.* New York, NY: Routledge.

Mitchell, K. J., & Ybarra, M. L. (2007). Online behavior of youth who engage in self-harm provides clues for preventive intervention. *Preventive Medicine, 45*, 392–396.

Nardi, B., & Harris, J. (2006). Strangers and friends: Collaborative play in world of warcraft. *Proceedings of Computer-Supported Cooperative Work* (pp. 149–158).

Olweus, D. (2012). Cyberbullying: An overrated phenomenon? *European Journal of Developmental Psychology, 9*, 520–538.

Pascoe, C. J. (2007). *Dude, you're a fag: Masculinity and sexuality in high school.* Oakland, CA: University of California Press.

Peter, J., & Valkenburg, P. M. (2007). Adolescents' exposure to a sexualized media environment and their notions of women as sex objects. *Sex Roles, 56*(5–6), 381–395.

Pfister, R. C. (2014). *Hats for house elves: Connected learning and civic engagement in Hogwarts at Ravelry.* Irvine, CA: Digital Media and Learning Research Hub.

Prensky, M. (2001). Digital natives, digital immigrants. *On the Horizon, 9*, 1–2.

Prout, A., & James, A. (1997). *Constructing and reconstructing childhood: Contemporary issues in the sociological study of childhood.* Philadelphia, PA: RoutledgeFarmer.

Rafalow, M. H., & Tekinbas, K. S. (2014). *Welcome to sackboy planet: Connected learning among littlebigplanet 2*. Irvine, CA: Digital Media and Learning Research Hub.

Smith, P. K. (2012). Cyberbullying and cyber aggression. In S. R. Jimerson, A. B. Nickerson, M. J. Mayer, & M. J. Furlong (Eds.), *Handbook of school violence and school safety: International research and practice* (2nd ed., pp. 93–103). New York, NY: Routledge.

Smith, P. K. (2014). *Understanding school bullying: Its nature and prevention strategies*. London: Sage.

Steinkuehler, C. A. (2005). The new third place: Massively multiplayer online gaming in American youth culture. *Tidskrift Journal of Research in Teacher Education*, *3*(3), 17–32.

Steinkuehler, C. A. (2008). Massively multiplayer online games as an educational technology: An outline for research. *Educational Technology*, *48*(1), 10–21.

Steinkuehler, C. A. (2010). Video games and digital literacies. *Journal of Adolescent & Adult Literacy*, *54*(1), 61–63.

The Nielsen Company. (2009). *Youth and media ... Television and beyond*. New York, NY: The Nielsen Company.

Thorne, B. (1993). *Gender play: Girls and boys in school*. New Brunswick, NJ: Rutgers University Press.

Tokunaga, R. S. (2010). Following you home from school: A critical review and synthesis of research on cyberbullying victimization. *Computers in Human Behavior*, *26*, 277–287.

Wilkins, A. C. (2008). *Wannabes, goths, and Christians: The boundaries of sex, style, and status*. Chicago, IL: University of California Press.

Willis, P. E. (1977). *Learning to labor: How working class kids get working class jobs*. New York, NY: Columbia University Press.

Zickuhr, K. (2011). *Generations and their gadgets*. Washington, DC: Pew Internet & American Life Project.

THE IMPACT OF INFORMATION AND COMMUNICATION TECHNOLOGY (ICT) USAGE ON PSYCHOLOGICAL WELL-BEING AMONG URBAN YOUTH

LaToya O'Neal Coleman, Timothy M. Hale, Shelia R. Cotten and Philip Gibson

ABSTRACT

Purpose – *Information and communication technology (ICT) usage is pervasive among present day youth, with about 95% of youth ages 12–17 years reporting use of the Internet. Due to the proliferation of ICT use among this generation, it is important to understand the impacts of ICT usage on well-being. The goal of this study was to determine the impact of ICT usage on psychological well-being among a sample of urban, predominately African American youth.*

Methodology/approach – *Paper and pencil surveys were administered to fourth and fifth grade students enrolled in 27 elementary schools in the southeastern United States. Relationships between hours using various*

Technology and Youth: Growing Up in a Digital World
Sociological Studies of Children and Youth, Volume 19, 267–291

ISSN: 1537-4661/doi:10.1108/S1537-466120150000019008

types of ICTs and the frequency of Internet activities on depression, hopelessness, self-esteem, and belonging were examined using Ordinary Least Squares (OLS) regression.

Findings – *Results indicate that ICT usage has both positive and negative implications for psychological well-being, depending upon the type of ICT use and outcome being examined.*

Social Implications – *The proliferation of ICT usage among present day youth may actually lessen its impact on psychological well-being. Since the amount of ICT usage does not seem to influence psychological well-being, future research should examine the impact of ICT content on psychological well-being.*

Keywords: Information and communication technology (ICT); psychological well-being; urban youth; minority youth

INTRODUCTION

The Internet has become a "central and indispensable element in the lives of American teens and young adults" (Lenhart, Purcell, Smith, & Zickuhr, 2010, p. 5). Today's youth have been described as "digital natives" – born and raised in a world where the use of Information and Communication Technology (ICT) is nearly ubiquitous (Palfrey & Gasser, 2008). ICTs such as mobile phones, television, and the Internet, among others, are an integral part of the daily lives of many youth. The typical youth spends more time using ICTs than the typical adult spends at work (Rideout, Foehr, & Roberts, 2010), making knowledge regarding the impact of ICT usage on the development and well-being of this group increasingly necessary. While there are positive social and educational benefits of ICT use, there is some concern about whether usage is also harmful (Borzekowski, 2006; Livingstone & Brake, 2010; Plowman, McPake, & Stephen, 2010; Punamäki, Wallenius, Nygard, Saarni, & Rimpelä, 2007; Shaw & Gant, 2002; Wartella & Jennings, 2000), particularly as it relates to psychological well-being (Borzekowski, 2006; Jackson, Zhao, et al., 2008; Kraut et al., 2002; Rideout et al., 2010). Most of the research on the impact of ICTs on well-being has focused on the relationship between Internet usage and depressive symptomatology in young and older adults among majority white samples (Gross, Juvonen, & Gable, 2002; Kraut et al., 2002; Kraut et al., 1998; Morgan & Cotten, 2003; Sanders, Field, Diego, & Kaplan,

2000). Among the studies that focus on younger children, findings regarding the impact of the Internet on aspects of psychological well-being have been mixed with studies reporting both positive and negative findings (Attewell, Suazo-Garcia, & Battle, 2003; Jackson, Zhao, et al., 2008).

To better inform the literature on the effects of ICT usage on children, we must move beyond a focus on the relationship between Internet usage and depression to examine a variety of types of ICT use on dimensions of psychological well-being (Cotten, Goldner, Hale, & Drentea, 2011; Jackson, Zhao, et al., 2008; Valkenburg & Peter, 2007). Although depression is a prevalent mental illness with vast implications for public health (National Alliance on Mental Illness, 2010) and life outcomes (Bessière, Kiesler, Kraut, & Boneva, 2008), other measures of psychological well-being, such as self-esteem, have been shown to predict future health, criminal behavior, and economic capital (Trzesniewski et al., 2006). We must also move beyond the construction of children as vulnerable victims of a world inundated by ICTs (Livingstone, 2007) since ICTs are only one mechanism by which children are socialized. Children are agentic pursuers of ICTs whose interactions with family, friends, and community, not just ICTs, impact their short- and long-term functioning (Dubow, Huesmann, & Greenwood, 2006). Finally, we also need to focus on two populations that are often ignored in studies of ICT use and impacts: minority youth from urban backgrounds and elementary school children. Few studies focus on either of these populations and those that do are primarily descriptive in nature (e.g., Rideout et al., 2010). The purpose of the current study is to examine the association of hours using various types of ICTs and the frequency of Internet activities on depression, hopelessness, self-esteem, and belonging in a large sample of minority children ages 10–13 years old.

LITERATURE REVIEW

Youth and ICT Usage

Studies illustrate that youth use ICTs at very high levels, compared to other groups. Recent surveys of teens conducted by Pew Internet & Society find that 95% of 12–17 year olds use the Internet (Lenhart et al., 2011; Madden, Lenhart, Duggan, Cortesi, & Gasser, 2013) and 80% use social networking sites (Lenhart et al., 2011). Of those teens who go online, 73% use email, 57% watch videos online, and 78% play online games (Zickuhr,

2010). Total daily hours of ICT use among youth (ages 8–18 years) has increased markedly during the past decade, from 6:19 (hours:minutes) in 1999 to 7:38 in 2009 (Rideout et al., 2010, p. 11). Age is an important predictor of ICT usage. Youth ages 8–10 use ICTs for fewer hours each day (5:29) than do 11–14 year olds (8:40) (Rideout et al., 2010, p. 43). In addition to age, race and gender are also associated with various aspects of ICT usage.

African American youth use ICTs and media more hours each day than White youth (9:44 vs. 6:22, respectively) (Rideout et al., 2010, p. 43). African American youth make greater use of television (5:54 vs. 3:36) and game systems (1:25 vs. 0:56) per day than White youth (Rideout et al., 2010, p. 37), and spend more time using a cell phone (2:49 vs. 1:47) (Rideout et al., 2010, p. 18). There is no substantial difference between African American and White youth in daily hours of computer use (1:24 vs. 1:17) (Rideout et al., 2010, p. 23). However, African American children are less likely than white children to have Internet access (Jackson, von Eye, Fitzgerald, Witt, & Zhao, 2011; Madden et al., 2013; Rideout et al., 2010). In 2009, 78% of African American youth had home Internet access compared to 88% of White youth (Rideout et al., 2010, p. 23). Jackson, Fitzgerald, et al. (2008) found that White children had been using a computer longer than African American children and that African American males used computers and the Internet for less time when compared to other groups. They also found consistent gender differences with regards to video game playing and cell phone usage. Males reported playing video games more than females and females used cell phones more than males. African American children used cell phones more than White children, with African American females using them the most (Jackson, Fitzgerald, et al., 2008).

While these differences show a narrowing digital divide as it relates to computer ownership, the issue of Internet access remains hidden among higher usage of some ICTs by African American youth. African American youth spend more time watching television, playing video games, and using a cell phone (Rideout et al., 2010), two of which have been linked to negative academic outcomes (Jackson, Zhao, et al., 2008; Rideout et al., 2010). There isn't much beyond speculation as to why African American youth use those particular ICTs more than White youth since there do not appear to be differences in parental monitoring or media rules (Rideout et al., 2010) and since other aspects of ICT usage have been found to contribute to more positive outcomes. Early access to ICTs, especially computers and the Internet, have been linked to more positive academic outcomes

(Jackson, Zhao, et al., 2008). Issues of access have been attributed to parents' socioeconomic status (Jackson, Zhao, et al., 2008; Rideout et al., 2010). Since White youth have been found to have earlier access to computers when compared to African American youth, the divide may potentially contribute to long term inequality in education and quite possibly psychological well-being. Consequently, we must explore how digital inequality might influence psychological well-being as the noted variation in usage among African American and White children may result in differences in psychological well-being outcomes which have yet to be examined in a large minority sample.

Psychological Well-Being and ICT Usage

Depression is one of the most common psychological disorders among children and adolescents (Merry et al., 2012) and has been linked to poor academic performance and negative cognitive functioning (Lundy, Silva, Kaemingk, Goodwin, & Quan, 2010). Research from a nationally representative sample of youth reveal a lifetime prevalence rate of 11.7% for Major Depressive Disorder (MDD) or Dysthymia among youth ages 13–18 (Merikangas et al., 2010). Levels of depression have been found to vary among racial/ethnic groups with whites having lower levels of depression when compared to African American, Hispanic, and Asian adolescents (Moon & Rao, 2010). Other aspects of psychological well-being, such as hopelessness, sense of belonging, and self-esteem, have been linked to academic achievement, subjective well-being, and risky behaviors (Bolland, Lian, & Formichella, 2005; Faircloth & Hamm, 2005; Vacek, Coyle, & Vera, 2010). These findings provide support for the importance of determining factors which may contribute to negative psychological well-being among children and adolescents such as ICT usage.

Despite research which confirms that youth utilize ICTs more than any other group, much of the research examining ICTs and psychological well-being has been on white college students and adults. This may be attributed to the fact that early Internet usage began as a tool for scientists and engineers (Kraut et al., 1998) and rapidly found its way to adults and then youth. The proliferation of ICTs, especially computers, among youth is fairly recent, as home computer ownership increased and more appealing content was developed. Early researchers thus focused on the heaviest users of technology. Noteworthy research, such as that of Kraut et al. (1998) and Sanders et al. (2000) who found that higher levels of Internet usage were related to poorer psychological well-being were among the earliest studies of the relationship between ICTs and depression. As technology changed, studies began to examine more applications but continued to focus on populations who owned and had access to ICTs resulting in fewer studies examining the impact of ICTs on youth, especially minority youth. Still, findings from adults and college students provide some support for the relationship between ICTs and psychological well-being.

Unlike findings among adults, several researchers have found ICTs to have a positive impact on psychological well-being among younger samples of college students. For example, Shaw and Gant (2002) used an experimental design to examine the impact of using chat rooms on depression, loneliness, self-esteem, and social support in a sample of college students. Participants engaged in five semi-structured chat sessions with an anonymous partner. The authors found that loneliness and depression decreased and that social support and self-esteem increased from pre- to post-test. Morgan and Cotten (2003) found that use of the Internet for email, participation in chat rooms, and instant messaging (IM) was associated with decreased depression among college students. Rohall, Cotten, and Morgan (2002) found that time spent using Internet chat rooms and IM had a small but significant positive effect on self-esteem among college students. Others have found that some students utilize ICTs specifically to improve their psychological well-being. Barker (2009) investigated the motivation to use social networking sites (SNS) in a sample of college freshmen. She discovered a moderate relationship between negative collective self-esteem and SNS use suggesting that students who felt more negatively about their current peer group may have compensated by seeking out other groups to communicate with online. While these studies may seem dated in terms of ICT research, they are some of the seminal studies in this area investigating the effects of ICT use on well-being among young adults.

These studies also highlight the need for additional investigation of the relationship between ICTs and the psychological well-being of youth since, comparatively, youth use ICTs at higher rates than other groups and they are in earlier stages of development which might result in different outcomes. Consequently, research related to the impact of ICTs is usually framed as having an effect on or being an aspect of the process of adolescent development. Effect related research tends to focus on the impact of ICTs on a problem (i.e., youth violence or poor grades) such as to suggest that ICT usage contributes to such problems (Livingstone, 2007). Process-related research focuses on how youth respond to media and how it benefits them

(Livingstone, 2007). These two types of research provide the foundation for two perspectives and hypotheses related to ICTs and well-being.

Jackson, Zhao, et al. (2008) describe two perspectives that have been used to explain how children's use of ICTs, most often in studies investigating Internet use, affects their psychological well-being. The "dystopian" perspective posits that the more time children spend on the Internet, the less time they spend engaging in more vital activities such as reading and face-to-face interaction with family and friends while concurrently being exposed to potentially damaging content or situations online. In contrast, the "utopian" perspective posits that the Internet affords children the opportunity to express themselves and communicate with others in ways that are novel while encouraging the development of the technological skills necessary for 21st century employment. These two perspectives have led to several hypotheses in the literature.

Research investigating the relationship between the Internet and psychological well-being has typically been guided by three hypotheses, the *social augmentation hypothesis*, the *social displacement hypothesis*, and the *social compensation hypothesis* (Bessière, Sara, Robert, & Bonka, 2008). Consistent with the dystopian perspective, the *social displacement hypothesis* argues that online interactions replace offline interactions so more time online may result in negative effects on well-being. Time online is thought to take the place of meaningful face-to-face interactions, which are valuable for positive psychological well-being. Conversely, the utopian perspective draws on the *social augmentation hypothesis*, stating that Internet usage has a positive impact on psychological well-being. The Internet provides the opportunity for individuals to increase their social resources and social networks by providing additional outlets for social interaction. This in turn results in positive psychological outcomes. Finally, the *social compensation hypothesis* suggests that this positive effect will be particularly strong for people who are socially isolated and lack access to social resources important to psychological well-being (Bessière et al., 2008).

Prior research on the effects of ICT use on youth's psychological well-being is mixed but along with studies among adults, provides some support for all three hypotheses. Findings from research on adults, such as Shaw and Gant (2002) provide support for the social augmentation hypotheses, while Barker (2009) provides support for the social compensation hypothesis. Other research, specifically among youth, finds that some types of communicative uses of technology are associated with negative psychological well-being – supporting the social displacement hypothesis.

Using a nationally representative sample of youth ages 10–17 years, Ybarra, Alexander, and Mitchell (2005) found that using email and chat rooms was positively associated with depressive symptomatology. However, Ybarra et al. (2005) also found that IM use and general Internet use was negatively associated with depression. Another study of Internet communication and well-being suggests that the negative impact on psychological well-being may be attributed to communication with strangers. Valkenburg and Peter (2007) examined the relationship between Internet communication, well-being, and communication with strangers using closeness to friends as a mediator and loneliness as a moderator. Initial results revealed that Internet communication had a negative influence on well-being among a sample of 687 adolescents from the Netherlands. After adding hypothesized underlying mechanisms, this direct relationship did not hold. This led to the researchers concluding that who adolescents communicate with online matters most in regards to well-being outcomes.

Gross et al. (2002) had similar findings. Using a sample of participants 11–13 years old, they found that those who had higher levels of loneliness and social anxiety were more likely to be communicating through IM with strangers as compared to students with lower levels of loneliness and social anxiety who were more likely to be communicating with friends. Finally, among youth ages 11–16 years, Jackson et al. (2011) found that cell phone use was associated with higher self-esteem. Non-communicative technology use (i.e., general Internet use, playing video games) had no significant relationship or a positive relationship with psychological well-being. Jackson et al. (2011) found that general Internet use and playing video games had no significant relationship to self-esteem among a sample of youth ages 10–17 years.

Despite it being well known that pre-adolescence is a critical period for psychosocial development, studies of the relationship between ICT usage and psychological well-being have primarily focused on college students and adults. This is problematic because findings are often used to draw conclusions about younger populations who are developmentally different (Valkenburg & Peter, 2011). Present literature also fails to consider whether existing differences among racial/ethnic and socioeconomic groups with regards to Internet access and usage (Bickham et al., 2003; Lenhart et al., 2010; Rideout et al., 2010) would result in different well-being outcomes for these groups. Considering such differences, what is the impact of ICT usage on the psychological development and well-being of minority youth? The current study examines the relationship of various types of ICT usage and activities on psychological well-being in a predominantly African

American sample of youth ages 10–13. Based on findings from previous research, which have been mixed, we propose the following exploratory hypotheses using both the dystopian and utopian perspectives:

(1) The more hours spent using ICTs and the higher the frequency of Internet usage, the higher the level of negative psychological well-being as measured by depression and hopelessness (social displacement hypothesis).
(2) The more hours spent using ICTs and the higher the frequency of Internet usage, the lower the level of positive psychological well-being as measured by self-esteem and sense of belonging (social displacement hypothesis).
(3) The more hours spent using ICTs and the higher the frequency of Internet usage, the lower the levels of negative psychological well-being as measured by depression and hopelessness (social augmentation hypothesis).
(4) The more hours spent using ICTs and the higher the frequency of Internet usage, the higher the levels of positive psychological well-being as measured by self-esteem and sense of belonging (social augmentation hypothesis).

METHODS

Participants and Procedures

In late 2008, the Birmingham City School District distributed 15,000 One Laptop Per Child (OLPC) XO computers to all 1st through 5th grade students. The XO laptop is a durable tool designed for, but not limited to, children in developing countries (One Laptop Per Child, 2014). With built-in wireless and open source software, it was designed to be a low cost alternative to traditional laptops. The dissemination of these laptops presented a unique opportunity to survey students before and after receiving the XOs to determine the effect of the XO program on a variety of social, academic, and psychological well-being outcomes. The program also presented an opportunity to collect data about students' use of information and communication technologies, including PCs at home, cell phones, and using the Internet for a variety of purposes.

The Birmingham Youth and Technology Study (BYTS) was conducted during the 2008–2009 school year. During that time the school district consisted of 39 elementary schools, of which 27 schools participated in the study. There were 2,915 students enrolled at the participating schools during the 2008–2009 academic year. All 4th and 5th grade students at participating schools were given consent forms to take home to obtain their parent's consent to participate in the study. Only those students who returned signed parental consent forms and who also assented to participate were surveyed. Participation was voluntary and a small incentive was provided to students participating in the study.

Surveys were administered during school hours, in designated areas (i.e., classrooms, cafeteria, etc.), and at times considered most suitable by principals and teachers. Participating students were given a pre- and posttest survey (i.e., time 1 and time 2). The paper and pencil surveys took approximately 45 minutes to complete. The pretest survey was given just prior to the XO dissemination and the posttest surveys about three to six months later. The response rate for the pretest survey was 52% ($N = 1,583$). Only students who participated in the pretest survey were eligible for the posttest survey. About 76% of participants completed the posttest survey ($N = 1,202$). We use data from the pre- and posttest surveys to evaluate the relationship between our independent variables on psychological well-being while controlling for baseline (time 1) values in these variables. The school district is predominately African American (97%) and 83% of students qualify for free or reduced price lunch (Alabama State Department of Education, 2010). The study sample is similar to students in the school district with 83% of the sample being African American and 84% qualifying for free or reduced price lunch.

Dependent Variables

To examine the broad impact of ICT use on psychological well-being, we use four outcome measures: depression, hopelessness, belonging, and self-esteem. The use of multiple measures allows us to identify statistically significant factors across outcomes as well as factors unique to each measure of psychological well-being. Measures of depression and hopelessness are used to represent two factors that contribute to poor psychological well-being, and self-esteem and sense of belonging are used to represent two factors that contribute to better psychological well-being.

Depression

Depression is measured using the Center for Epidemiologic Studies Depression Scale for Children (CES-DC) (see Faulstich, Carey, Ruggiero,

Enyart, & Gresham, 1986). All 20 items that comprise the CES-DC were used in the pretest survey. To reduce response burden on students on the posttest the number of items was reduced to six that were identified as contributing to the internal reliability of the scale. The six-item scale values range from 0 to 18 with higher scores indicating greater levels of depression (alpha = 0.74).

Hopelessness

Hopelessness is measured using three items drawn from the Mobile Youth Study (MYS) 2006 (Bolland, 2007), which used a modified version of the Hopelessness Scale for Children (HSC) (Kazdin, French, Unis, Esveldt-Dawson, & Sherick, 1983). The original scale used a true/false response format and was revised for use in the MYS as 0 = disagree, 1 = not sure, and 2 = agree. Although the MYS documentation does not explain the rationale for the change in response format we chose to use the same questions to enable future comparisons with the data collected in Mobile, Alabama. The scale was created by summing the three items, with a higher score representing a greater level of hopelessness. Factor analysis shows the three items load on a single factor that accounts for 71% of the variance. The Chronbach's alpha for the scale is 0.79.

Belonging

Belonging is measured using three items drawn from the eight items that comprise the Organization for Economic Co-operation and Development's (OECD) Program for International Student Assessment (PISA) measure of student engagement and belonging (Willms, 2003). Due to the young age of students in our sample, response options were modified from a four item Likert options to three options (0 = disagree, 1 = not sure, 2 = agree). All three items are first reverse coded and then summed to create the scale with higher values representing a greater sense of belonging (alpha = 0.80). Factor analysis indicates a single common factor that accounts for 71% of the variance.

Self-Esteem

Self-esteem is measured using five items drawn from the MYS (Bolland, 2007). The original source of the items is the 36-item Self-Perception Profile for Children (SPPC) (Harter, 1985) that codes responses on a five item Likert scale. Taking into account the reading levels of children who participated in the MYS, items were reworded and presented as two contrasting statements that participants were asked to identify one as being "most like you." The score is created by summing the statements that represent positive self-esteem (coded as 1) and statements of negative self-esteem (coded as 0) with a higher score representing higher self-esteem. Factor analysis shows the five items load on a single factor that explains 63% of the variance. Chronbach's alpha for the index is 0.72.

Independent and Control Variables

Hours Using ICT

Hours using ICT is measured using six items that ask respondents how many hours per day they spend using various types of ICT, including: a computer (not the XO), the Internet, the XO, a game system, watching TV, and a cell phone. Responses are coded as one of five ordinal categories (0 = I don't use this at all, 1 = 1–2 hours, 2 = 3–4 hours, 3 = 5–6 hours, 4 = 7–8 hours, and 5 = more than 8 hours). Higher scores represent more hours using that particular type of technology.

Frequency of Internet Usage

Frequency of Internet usage is measured using seven items that ask respondents how often they do each of the following things using a computer on the Internet: (1) send or receive email, (2) send or receive instant messages, (3) visit social networking sites, (4) surf the web, (5) play games, (6) listen to music, and (7) watch videos. Responses are coded as one of five ordinal categories (0 = never, 1 = less than once a week, 2 = at least once a week, 3 = several times a week, and 4 = every day). Higher scores represent more frequent Internet usage.

Demographics and PC Ownership

Demographic and PC ownership were measured at the pretest and are time invariant. Demographics include gender (0 = female, 1 = male), grade level (0 = fifth grade, 1 = fourth grade), race (0 = other, 1 = African American). We do not include age in our analyses due to the high correlation between age and grade level. All students were given an XO to use at home and school. A measure of home PC ownership (0 = don't own, 1 = own) was included to control for differences in access to other computers besides the XO.

Analytical Design

This study examines the impact of hours using ICTs and frequency of ICT usage on four measures of psychological well-being: depression, hopelessness, self-esteem, and belonging. Descriptive statistics are presented in Table 1 and include results using paired *t*-test and Wilcoxson signed-rank sum *z*-test for ordinal variables of significant changes from time 1 to time 2. Ordinary Least Squares (OLS) regression is used to test models that assess the relationship between: types of ICT use and frequency of ICT usage on psychological well-being at time 2 (i.e., posttest) while controlling for these measures at baseline (i.e., time 1). The models also include time invariant demographic variables and a control variable indicating PC ownership in the home. Robust standard errors are computed using the "cluster" command in Stata 13 to adjust for the nonrandom distribution of students, grouped by homeroom teacher. Listwise deletion for missing data yielded a final analytic sample of 870 participants. No independent variable used in the models is missing values on more than 3.3% of cases. However, the analytic sample is comprised of fewer African Americans and more PC owners than students with incomplete surveys. Additionally, students in the analytic sample have better psychological well-being on three of the four measures (lower levels of depression and hopelessness, greater sense of belonging) and made more frequent use of the Internet for three types of activities (i.e., email, surfing, games) at time 2. Post-hoc tests were conducted to examine patterns in the missing values using the "misstable" command in Stata, and pairwise correlations between each dependent (i.e., psychological well-being outcomes) and independent variables. We did not find patterns in the missing values or correlations. This leads us to believe that the differences observed in the analytic compared to the full sample are not likely to substantially affect the findings.

We choose to present results using OLS regression to ease the comparison between models for each outcome. We conducted several alternative statistical models to test the robustness of the OLS regression models presented in Tables 2 and 3 (results available from authors). With few exceptions[1] the results did not differ substantially across statistical models.

RESULTS

Descriptive statistics are presented in Table 1. The mean age of students participating in the study is 10.4 years old and 45% are male. Forty-five percent of the students are in the 4th grade and 55% are in the 5th grade. African Americans make up 83% of the sample and other race 17%. Eighty-two percent of students report having a PC in the household.

Table 1. Descriptives (n = 870).

	Time 1		Time 2	
	Mean	SD	Mean	SD
Demographics and controls				
Age	10.445	0.663	NA	NA
Male	0.453	0.498	NA	NA
Grade 4	0.451	0.498	NA	NA
African American	0.834	0.373	NA	NA
PC owner	0.823	0.381	NA	NA
Mental well-being				
Depression, CESD-C[a]	4.832	4.124	4.134***	3.957
Hopelessness	1.091	1.444	1.076	1.589
Belonging	4.478	1.692	4.684***	1.740
Self-esteem	3.611	1.505	3.651	1.535
Hours using ICT				
Computer	1.466	1.331	2.255***	1.826
Internet	1.857	1.463	2.929***	1.828
XO	NA	NA	1.726	1.487
Game system	2.487	1.743	3.337***	1.786
TV	2.746	1.296	2.648*	1.233
Cell phone	1.656	1.684	2.400***	1.958
Frequency of Internet use				
Send or receive email	1.574	1.596	1.555	1.576
Instant message	1.232	1.511	1.357*	1.531
Visit SNS	2.061	1.723	2.085	1.722
Surf the web	2.213	1.639	2.259	1.561
Play games	3.293	1.097	2.932***	1.304
Listen to music	2.828	1.389	2.716	1.467
Watch videos	2.609	1.411	2.628	1.490

Notes: Demographic and control variables are time invariant. Wilcoxon signed-rank sum test used to assess change between time 1 and time 2 for all other variables.
[a]Paired t-test used to assess change in mean between time 1 and time 2.
$^{*}p<0.05$; $^{**}p<0.01$; $^{***}p<0.001$

The mean of the 6-item CES-DC scale decreased significantly from 4.832 to 4.134 at time 2 ($t = -4.643$, $p < 0.001$). Sense of belonging increased significantly from a mean of 4.478 to 4.684 ($z = 3.950$, $p < 0.001$). There was no significant change in hopelessness or self-esteem.

The mean usage of computers, the Internet, game systems, and cell phones increased significantly from time 1 to time 2 while use of TV decreased. Mean usage of the Internet increased from 1.857 to 2.929 (z = 14.635, $p < 0.001$) meaning that on average, students were using the Internet 5–6 hours per day at time 2 (coded as 3) as opposed to 3–4 hours per day (coded as 2). Mean usage of computers increased from 1.466 to 2.255 (z = 10.641, p < 0.001). Game system use increased from a mean of 2.487 to 3.337 (z = 12.173, p < 0.001) and cell phone use increased from a mean of 1.656 to 2.400 (z = 10.383, p < 0.001). Hours of TV use decreased slightly from a mean of 2.746 to 2.648 (z = −2.175, p < 0.050). Students were provided XO computers after the pretest survey and the mean usage for the XO at time 2 is 1.726 or about 3–4 hours per day (coded as 2).

Use of instant messaging increased significantly from a mean of 1.232 to 1.357 (z = 2.076, p < 0.050) and playing games on the Internet decreased from a mean of 3.293 to 2.932 (z = −7.033, p < 0.001). There was no significant change in the frequency of other Internet activities. At time 2, students report accessing email, visiting social networking sites and surfing the web about once a week (coded as 2), and listening to music and watching videos several times a week (coded as 3).

Does Hours of ICT Usage and Frequency of Internet Usage Impact Psychological Well-Being?

Tables 2 and 3 present the results from OLS regression models estimating the relationship of ICT use at time 2 on each of the four outcome measures of psychological well-being at time 2. For each outcome we present two models. The first model includes time invariant demographic variables, a control variable measuring PC ownership, and a variable to control for baseline level in the outcome of psychological well-being at time 1. The second model adds variables measuring ICT use at time 2 and includes variables measuring ICT use at time 1 to control for baseline levels. We include the R-square for each model and use a Wald test to evaluate if adding the ICT use variables significantly improves model fit. The coefficients for the time 1 variables are not interpreted and although present in the models have been omitted from the tables.

Table 2 presents results for two measures of negative psychological well-being, depression and hopelessness. In models for depression and hopelessness, none of the demographic or control variables are significant. Hours using the XO was significantly related to greater depression (b = 0.260, p < 0.010) and use of instant messaging was significantly related to increased hopelessness (b = 0.103, p < 0.010). However, the change in R-square was small (about 3%) and not significant in either model at the p = 0.05 level, indicating that ICT use did not have a substantial negative impact on depression or hopelessness.

Table 2. Depression and Hopelessness Regressed on Technology Ownership, Use, and Demographic Factors Using OLS Regression ($n = 870$).

	Depression		Hopelessness	
	(1)	(2)	(1)	(2)
Demographics and controls				
Male	−0.103	−0.287	0.099	0.090
	(0.255)	(0.308)	(0.106)	(0.111)
Grade 4	−0.047	−0.099	0.200	0.155
	(0.261)	(0.263)	(0.132)	(0.136)
African American	−0.418	−0.299	−0.229	−0.244
	(0.350)	(0.358)	(0.133)	(0.141)
PC owner	−0.321	0.019	−0.176	−0.195
	(0.324)	(0.372)	(0.131)	(0.146)
Hours using ICT				
Computer		−0.023		−0.009
		(0.109)		(0.045)
Internet		−0.092		−0.033
		(0.105)		(0.044)
XO		0.260**		0.079
		(0.091)		(0.041)
Game system		0.070		0.028
		(0.095)		(0.040)
TV		−0.030		−0.001
		(0.113)		(0.055)
Cell phone		−0.038		0.021
		(0.095)		(0.033)
Frequency of Internet use				
Send or receive email		0.096		−0.069
		(0.119)		(0.041)
Instant message		−0.162		0.103*
		(0.117)		(0.044)
Visit SNS		0.002		0.060
		(0.094)		(0.045)
Surf the web		0.076		0.002
		(0.089)		(0.039)
Play games		−0.239		−0.080
		(0.122)		(0.053)
Listen to music		−0.037		−0.038
		(0.135)		(0.051)
Watch videos		−0.071		0.028
		(0.132)		(0.050)
Intercept	2.990***	3.361***	0.893***	0.654**
	(0.480)	(0.729)	(0.179)	(0.237)
R^2	0.162	0.188	0.121	0.149
ΔR^2		0.026		0.028

Notes: Models include variables for each dependent variable and for independent variables at time 1. Demographic and PC ownership are time invariant and included only as time 1 variables. XO was not available to students at time 1 and is included only as a time 2 variable.
$^{*}p < 0.05$; $^{**}p < 0.01$; $^{***}p < 0.001$.

Table 3. Belonging and Self-Esteem Regressed on Technology Ownership, Use, and Demographic Factors Using OLS Regression (n = 870).

	Belonging		Self-Esteem	
	(1)	(2)	(1)	(2)
Demographics and controls				
Male	−0.060	−0.069	−0.231*	−0.207*
	(0.118)	(0.131)	(0.096)	(0.099)
Grade 4	−0.047	−0.027	−0.220*	−0.266**
	(0.124)	(0.126)	(0.101)	(0.090)
African American	0.599***	0.548**	0.096	0.165
	(0.163)	(0.167)	(0.104)	(0.121)
PC owner	−0.049	−0.148	0.196	0.236
	(0.134)	(0.152)	(0.114)	(0.127)
Hours using ICT				
Computer		−0.019		0.009
		(0.045)		(0.034)
Internet		0.076		−0.055
		(0.049)		(0.035)
XO		−0.095*		0.014
		(0.041)		(0.032)
Game system		−0.027		−0.013
		(0.038)		(0.031)
TV		−0.019		−0.026
		(0.052)		(0.040)
Cell phone		0.047		0.050
		(0.032)		(0.030)
Frequency of Internet use				
Send or receive email		−0.027		0.018
		(0.048)		(0.041)
Instant message		−0.053		0.056
		(0.051)		(0.042)
Visit SNS		0.030		−0.122***
		(0.046)		(0.036)
Surf the web		−0.138***		0.003
		(0.039)		(0.037)
Play games		0.024		0.053
		(0.057)		(0.048)
Listen to music		0.041		−0.033
		(0.064)		(0.047)
Watch videos		0.106*		0.046
		(0.051)		(0.046)
Intercept	2.500***	2.383***	1.781***	1.894***
	(0.244)	(0.339)	(0.197)	(0.257)
R^2	0.170	0.205	0.281	0.315
ΔR^2		0.035		0.034

Notes: Unstandardized OLS regression coefficients and robust standard errors in parentheses. Models include variables for each dependent variable and for independent variables at time 1. Demographic and PC ownership are time invariant and included only as time 1 variables. XO was not available to students at time 1 and is included only as a time 2 variable.

$^*p < 0.05$; $^{**}p < 0.01$; $^{***}p < 0.001$.

Table 3 present results for two measures of positive psychological well-being, belonging and self-esteem. African American race was positively associated with sense of belonging (b = 0.599, p < 0.001). Males (b = −0.231, p < 0.050) and students in the 4th grade (b = −0.220, p < 0.050) on average reported lower self-esteem than females and students in the 5th grade. In contrast to models for depression and hopelessness, measures of ICT use had a slightly larger impact on R-square and significantly improved model fit (belonging, $F(25, 120) = 2.46$, $p < 0.001$; self-esteem $F(25,120) = 1.99$, $p < 0.010$). Frequency watching videos were positively associated with sense of belonging (b = 0.106, p < 0.050). Hours using the XO and frequency surfing the web were associated with lower sense of belonging (b = −0.095, p < 0.050 and −0.138, p < 0.001, respectively). Frequency visiting social networking sites (b = −0.122, p < 0.001) was associated with lower self-esteem.

DISCUSSION

This is the first study to measure the impact of hours using various types of ICTs and frequency of ICT usage on a range of psychological well-being outcomes in a predominantly African American sample. Little research has focused on the positive and negative impacts of ICTs for urban minority youth. Our research suggests that in many cases ICT usage has no impact on psychological well-being. The current research supports studies with adults which note that the impact of ICT usage on psychological well-being vary depending upon the type, timing, and amount of ICT usage (Cotten, Goldner, et al., 2011). However, the results are modest at best, with the variables measuring ICT usage explaining about 3% of the variation in our psychological well-being outcomes. In fact, addition of ICT usage did not significantly improve model fit for depression or hopelessness; although there was a significant improvement in model fit for belonging and for self-esteem.

Time spent engaged in particular types of Internet activities had negative and positive relationships with the measures of positive psychological well-being. Frequency of surfing the web was associated with lower sense of belonging while frequency of watching videos was associated with higher sense of belonging. The lower sense of belonging among those who spent more time surfing the web can be explained by the social displacement hypothesis. Time spent surfing the web may be taking the place of face-to-face interactions which has a negative impact on positive psychological well-being. The positive relationship between sense of belonging and frequency of watching videos may be explained by the social augmentation hypothesis. In line with the process approach (Livingstone, 2007), we suggest that youth who spent more time watching videos may be able to increase their social networks by connecting with other youth who watch videos, thereby increasing their social interactions offline through a common interest. By engaging in a specific activity, such as watching videos, they may increase their chances of finding peers who do the same. African American race was also associated with higher sense of belonging. Considering the racial makeup of the school district, this is not a surprising finding.

Self-esteem was inversely related to frequency visiting social networking sites, being male, and being in 4th grade. The finding that frequency visiting social networking sites was related to lower self-esteem supports previous research (e.g., Barker, 2009) and possibly the social compensation hypothesis. It is unclear why males in the sample have lower levels of self-esteem than females or why 4th graders have lower levels of self-esteem than 5th graders. This is not consistent with findings from studies of older adolescents where females were found to have lower levels of self-esteem and where self-esteem seemed to decline with age (van der Berg, Mond, Eisenberg, Ackard, & Neumark-Sztainer, 2010). However, the current study uses students who have not yet reached adolescence – thus, factors that emerge during adolescence that specifically affect females' self-esteem may not present the same self-esteem disadvantages compared to males. It might also be related to the unique population in this study – an almost all African American sample of elementary school students. Given that African Americans are the majority population in this school system, cultural beliefs and socialization experiences may also come into play in these relationships.

Although previous findings show that cell phone use was positively associated with self-esteem (Jackson et al., 2011), we found no significant relationship between cell phone use and self-esteem among our study participants. It is possible that cell phone use has a more protective effect for slightly older children from more affluent backgrounds as the current sample is slightly younger and likely less affluent than those in the Jackson et al. (2011) study. This is merely speculation as the current data cannot speak to this possibility.

We also included measures of a unique type of technology, the XO laptop. Hours using the XO were associated with higher depression and hopelessness, and lower sense of belonging. It is not clear why hours using the XO were consistently associated with worse psychological well-being among participants. Some research suggests that implementing large-scale technology disseminations in minority communities may be challenging due to previous experiences using technology and related attitudes, as well as structural factors that may present barriers to adopting new technology (Cotten, Hale, Howell-Moroney, O'Neal, & Borch, 2011b). This may be the case with the XO program in Birmingham, Alabama, as the project was highly politicized from the initial discussions and continues to be to this time period (Warschauer, Cotten, & Ames, 2011). In addition, little support was provided to the school system to help them integrate the XOs into the curriculum and the XOs were given to the students (the school system did not retain ownership of the laptops).

One of the limitations of this study is that not all students in the schools participated. Parents had to return consent forms in order for their children to participate and it may be that parents who took the initiative to return the consent forms live somewhat different lives than those who did not return the forms. Unfortunately our data do not provide information to investigate this possibility. A noted limitation is that our analytic sample differed significantly from those excluded from the analyses. It appears that those who had more positive psychological well-being were more likely to complete the posttest survey. This may have introduced bias in the results. An additional limitation is that we rely on student report of daily hours spent using ICT. Primary caregivers and teachers were not included in this study, however data from both sources might have been beneficial for determining the accuracy of student reports. Also, our analysis examines hours spent using ICT and frequency of Internet usage, but not ICT content. Future studies might find measuring ICT content useful for developing strong theoretical foundations from which to explore the impact of ICT on psychological well-being among elementary school and minority students. Finally, the time between the pre- and post-test survey averaged 141 days (ranging from 105 to 183 days) and may be too short to detect the effect of changes in ICT use on psychological outcomes. Future studies should follow youth over longer periods of time in order to disentangle these relationships.

CONCLUSIONS

This study examined the impact of hours spent using various types of ICTs and frequency of ICT usage on psychological well-being in a large, primarily African American urban school district. This study took place during a dissemination of One Laptop Per Child XO Laptops to elementary school students in this school system. Thus, this seemed to be an appropriate time to investigate these issues given the dissemination of a new ICT to the students. The results suggest, that even given the addition of a new ICT, there were few effects on well-being for the students. We suspect this is due to the pervasiveness of ICT devices in their larger social worlds and their prior socialization experiences with ICTs. While the goal of giving the XO laptops to the youth was to help eliminate the digital divide and to prepare them for our information based society, little attention was paid to whether students and their families owned computers before receiving the XO laptops. Some of our earlier research shows that a majority of the students reported owning or having access to computers in their households (Cotten et al., 2010). Though we do not have data to test this, we suspect that these students, as many youth their age do, had substantial prior experience with ICTs; thus, the addition of the XO laptops did not result in a disruption in this setting.

Previous research focused on predominantly White samples of college students and adults and was used to draw conclusions regarding youth. Considering the differences in developmental trajectories of adults and children as well as the differences in access and usage among these groups and along racial/ethnic lines, it was important to examine the impact of ICTs among the sample used in the current investigation. Although there were not many significant relationships found, this is a case where nonsignificant findings are important since earlier studies suggested that ICTs might have a negative influence on development.

While much of the research in this area, including the current study seems to follow the effects approach of examining the impact of ICTs on psychological well-being, it is necessary to consider the process approach in that children and adolescents have at least some agency regarding how much they use ICTs as well as which ICTs they use. Moreover, with the prevalence of information and communication technologies among present day youth, it is important to be aware of the ways in which ICT usage may not only impede but also facilitate healthy development (Borzekowski, 2006; Livingstone, 2007; Punamäki et al., 2007; Shaw & Gant, 2002; Wartella & Jennings, 2000). Previous findings were inconsistent (Attewell et al., 2003; Jackson, Zhao, et al., 2008; Kraut et al., 2002, 1998) and aside from being taken from predominantly white samples of adults, they only examined one type of ICT use, that of the Internet.

The current study examined the impacts of both types of ICT usage and frequency of Internet usage on depression, hopelessness, self-esteem, and belonging, as prior researchers have noted the importance of focusing on a range of technologies being used (Cotten, Goldner, et al., 2011) and outcomes other than depression (e.g., Jackson, Zhao, et al., 2008) in a large minority sample. The results of the current study suggest that use of ICTs may have both positive and negative implications for psychological well-being but those impacts are minor. Perhaps most importantly, the results suggest that the vast majority of these measures of ICT usage and types had *NO impact* on depression, hopelessness, belonging, and self-esteem levels among the participants. Thus, our results suggest that for this sample, ICT use does not impede healthy psychological development. It may simply be that ICT use is such a major part of the normal lives or the process of socialization for these youth that it is not likely to play a significant role in their psychological well-being.

NOTE

1. For depression and hopelessness, we tested OLS regression with logged dependent variables, and count models (i.e., Poisson and negative binomial regression). The results using logged dependent variables did not differ substantially from those in Tables 2 and 3. Tests showed that the variables depression and hopelessness are over-dispersed and that negative binomial regression is a better fitting model. The results using negative binomial regression were not substantially different from the OLS model for hopelessness. However, in the depression model frequency of playing games had a small negative relationship. For self-esteem and belonging we tested a ordered logit model. The results were substantially the same as the OLS model, but for two exceptions: (1) hours using a cell phone had a small positive relationship to belonging, and (2) frequency of email use had a positive relationship to self-esteem.

ACKNOWLEDGMENT

This research was supported by a grant from the National Science Foundation (DRL-0819063; Shelia Cotten, PI).

REFERENCES

Alabama State Department of Education. (2010). Retrieved from http://www.alsde.edu. Accessed on November 15.

Attewell, P., Suazo-Garcia, B., & Battle, J. (2003). Computers and young children: Social benefit or social problem. *Social Forces*, *82*(1), 277–293.

Barker, V. (2009). Older adolescents' motivation for social network site use: The influence of gender, group identity, and collective self-esteem. *CyberPsychology & Behavior*, *12*(2), 209–213.

Bessière, K., Kiesler, S., Kraut, R., & Boneva, B. S. (2008). Effects of internet use and social resources on changes in depression. *Information, Communication, & Society*, *11*(1), 47–70.

Bickham, D. S., Vandewater, E. A., Huston, A. C., Lee, J. H., Caplovitz, A. G., & Wright, J. C. (2003). Predictors of children's electronic media use: An examination of three ethnic groups. *Media Psychology*, *5*, 107–137.

Bolland, J. M. (2007). *Mobile youth survey overview*. Unpublished report.

Bolland, J. M., Lian, B. E., & Formichella, C. M. (2005). The origins of hopelessness among inner-city African American adolescents. *American Journal of Community Psychology*, *36*(3–4), 293–305.

Borzekowski, D. L. G. (2006). Adolescents' use of the internet: A controversial, coming-of-age resource. *Adolescent Medicine*, *17*(1), 205–216.

Cotten, S. R., Goldner, M., Hale, T. M., & Drentea, P. (2011). The importance of type, amount, and timing of Internet use for understanding psychological distress. *Social Science Quarterly*, *92*(1), 119–139.

Cotten, S. R., Hale, T. M., Howell-Moroney, M., O'Neal, L., & Borch, C. (2011b). Using affordable technology to decrease digital inequality: Results from Birmingham's one laptop per child XO laptop project. *Information, Communication & Society*, *14*(4), 1–21.

Cotten, S. R., Hale, T. M., O'Neal, L. J., Anderson, W., Borch, C., Howell-Moroney, M., & Nelson, B. (2010). The impacts of XO laptops in Birmingham, Alabama Schools: Education, participation, and social connection outcomes. *Digital media and learning conference*. San Diego, CA, February 18–20.

Dubow, E. F., Huesmann, L. R., & Greenwood, D. (2006). Media and youth socialization: Underlying processes and moderators of effects. In J. Grusec & P. Hastings (Eds.). *The Handbook of Socialization* (pp. 404–432). New York, NY: Guilford Press.

Faircloth, B. S., & Hamm, J. V. (2005). Sense of belonging among high school students representing 4 ethnic groups. *Journal of Youth and Adolescence*, *34*(4), 293–309.

Faulstich, M. E., Carey, M. P., Ruggiero, L., Enyart, P., & Gresham, F. (1986). Assessment of depression in childhood and adolescence: An evaluation of the Center for Epidemiological Studies Depression Scale for Children (CES-DC). *American Journal of Psychiatry*, *143*(8), 1024–1027.

Gross, E. F., Juvonen, J., & Gable, S. (2002). Internet use and well-being in adolescence. *Journal of Social Issues* *58*(1), 75–90.

Harter, S. (1985). *Manual for the self-perception profile for children (Revision of the perceived competence scale for children)*. Denver, CO: University of Denver.

Jackson, L. A., Fitzgerald, H. E., Zhao, Y., Kolenic, A., von Eye, A., & Harold, R. (2008). Information Technology (IT) use and children's psychological well-being. *CyberPsychology and Behavior*, *11*(6), 755–757.

Jackson, L. A., von Eye, A., Fitzgerald, H. E., Witt, E. A., & Zhao, Y. (2011). Internet use, videogame playing and cell phone use as predictors of children's Body Mass Index (BMI) body weight, academic performance, and social and overall self-esteem. *Computers in Human Behavior*, *27*(1), 599–604.

Jackson, L. A., Zhao, Y., Kolenic, A., Fitzgerald, H. E., Harold, R., & von Eye, A. (2008). Race, gender, and information technology use: The new digital divide. *CyberPsychology and Behavior, 11*(4), 437–442.

Kazdin, A. E., French, N. H., Unis, A. S., Esveldt-Dawson, K., & Sherick, R. B. (1983). Hopelessness, depression, and suicidal intent among psychiatrically disturbed inpatient children. *Journal of Consulting and Clinical Psychology, 51*(4), 504–510.

Kraut, R., Kiesler, S., Boneva, B., Cummings, J., Helgeson, V., & Crawford, A. (2002). Internet paradox revisited. *Journal of Social Issues, 58*(1), 49–74.

Kraut, R., Patterson, M., Lundmark, V., Kiesler, S., Mukopadhyay, T., & Scherlis, W. (1998). Internet paradox: A social technology that reduces social involvement and psychological well-being? *American Psychologist, 53*(9), 1017–1031.

Lenhart, A., Madden, M., Smith, A., Purcell, K., Zickuhr, K., & Rainie, L. (2011). *Teens, kindness and cruelty on social network sites.* Retrieved from http://pewinternet.org/Reports/2011/Teens-and-social-media.aspx. Accessed on April 3, 2010.

Lenhart, A., Purcell, K., Smith, A., & Zickuhr, K. (2010). *Social media and mobile Internet use among teens and young adults.* Retrieved from http://pewinternet.org/Reports/2010/Social-Media-and-Young-Adults.aspx. Accessed on November 14.

Livingstone, S. (2007). Do the media harm children?: Reflections on new approaches to an old problem. *Journal of Children and Media, 1*(1), 5–14.

Livingstone, S., & Brake, D. (2010). On the rapid rise of social networking sites: New findings and policy implications. *Children & Society, 24*, 75–83.

Lundy, S. M., Silva, G. E., Kaemingk, K. L., Goodwin, J. L., & Quan, S. F. (2010). Cognitive functioning and academic performance in elementary school children with anxious/depressed and withdrawn symptoms. *Pediatric Medical Journal 4*, 1–9.

Madden, M., Lenhart, A., Duggan, M., Cortesi, S., & Gasser, U. (2013). *Teens and technology.* Retrieved from http://www.pewinternet.org/Reports/2013/Teens-and-Tech.aspx. Accessed on July 15.

Merikangas, K. R., He, J., Burstein, M., Swanson, S. A., Avenevoli, S., Cui, L., … Swendsen, J. (2010). Lifetime prevalence of mental disorders in U.S. adolescents: Results from the National Comorbidity Survey Replication-Adolescent Supplement (NCS-A). *Journal of the American Academy of Child and Adolescent Psychiatry, 49*(10), 980–989.

Merry, S. N., Hetrick, S. E., Cox, G. R., Brudevold-Iversen, T., Bir, J. J., & McDowell, H. (2012). Psychological and educational interventions for preventing depression in children and adolescents (Review). *Evidence-Based Child Health 7*(5), 1409–1685.

Moon, S. S., & Rao, U. (2010). Youth-family, youth-school relationship, and depression. *Child and Adolescent Social Work Journal, 27*, 115–131.

Morgan, C., & Cotten, S. R. (2003). The relationship between internet activities and depressive symptoms in a sample of college freshmen. *CyberPsychology & Behavior, 6*(2) 133–142.

National Alliance on Mental Illness. (2010). *Major depression fact sheet.* Retrieved from http://www.nami.org/Content/NavigationMenu/Mental_Illnesses/Depression/Mental_Illnesses_What_Is_Depression.htm. Accessed on November 13.

One Laptop Per Child. (2014). Retrieved from http://www.laptop.org. Accessed on October 10.

Palfrey, J., & Gasser, U. (2008). *Born digital: Understanding the first generation of digital natives.* New York, NY: Basic Books.

Plowman, L., McPake, J., & Stephen, C. (2010). The technologisation of childhood? Young children and technology in the home. *Children & Society, 24*, 63–74.

Punamäki, R., Wallenius, M., Nygard, C., Saarni, L., & Rimpelä, A. (2007). Use of Information and Communication Technology (ICT) and perceived health in adolescence: The role of sleeping habits and waking-time tiredness. *Journal of Adolescence, 30*(4), 569–585.

Rideout, V. J., Foehr, U. G., & Roberts, D. F. (2010). *Generation M^2 media in the lives of 8- to 18-year olds.* A Kaiser Family Foundation Study. Retrieved from www.kff.org/entmedia/upload/8010.pdf. Accessed on November 13, 2010.

Rohall, D. E., Cotten, S. R., & Morgan, C. (2002). Internet use and the self concept: Linking specific uses to global self-esteem. *Current Research in Social Psychology, 8*(1), 1–18.

Sanders, C. E., Field, T. M., Diego, M., & Kaplan, M. (2000). The relationship of internet use to depression and social isolation among adolescents. *Adolescence, 35*(138), 237–242.

Shaw, L. H., & Gant, L. M. (2002). In defense of the Internet: The relationship between Internet communication and depression, loneliness, self-esteem, and perceived social support. *CyberPsychology and Behavior, 5*(2), 157–171.

Trzesniewski, K. H., Donnellan, M. B., Moffitt, T. E., Robins, R. W., Poulton, R., & Caspi, A. (2006). Low self-esteem during adolescence predicts poor health, criminal behavior, and limited economic prospects during adulthood. *Developmental Psychology, 42*(2), 381–390.

Vacek, K. R., Coyle, L. D., & Vera, E. M. (2010). Stress, self-esteem, hope, optimism, and well-being in urban ethnic minority adolescents. *Journal of Multicultural Counseling and Development, 38*, 99–111.

Valkenburg, P. M., & Peter, J. (2007). Internet communication and its relation to well-being: Identifying some underlying mechanism. *Media Psychology, 9*, 43–58.

Valkenburg, P. M., & Peter, J. (2011). Online communication among adolescents: An integrated model of its attraction, opportunities and risks. *Journal of Adolescent Health, 48*(2), 121–127.

van der Berg, P. A., Mond, J., Eisenberg, M., Ackard, D., & Neumark-Sztainer, D. (2010). The link between body dissatisfaction and self-esteem in adolescents: Similarities across gender, age, weight status, race/ethnicity, and socioeconomic status. *Journal of Adolescent Health, 47*(3), 290–296.

Warschauer, M., Cotten, S. R., & Ames, M. (2011). One laptop per child Birmingham: Case study of a radical experiment. *International Journal of Learning and Media, 3*(2), 61–76.

Wartella, E. A., & Jennings, N. (2000). Children and computers: New technology, old concerns. *The Future of Children, 10*(2), 31–43.

Willms, J. D. (2003). *Student engagement at school: A sense of belonging and participation, results from PISA 2000.* Paris: OECD. Retrieved from http://www.oecd.org/dataoecd/42/35/33689437.pdf. Accessed on May 16, 2010.

Ybarra, M. L., Alexander, C., & Mitchell, K. J. (2005). Depressive symptomatology, youth Internet use, and online interactions: A national survey. *Journal of Adolescent Health, 36*(1), 9–18.

Zickuhr, K. (2010). *Generations 2010.* Retrieved from http://www.pewinternet.org/~/media//Files/Reports/2010/PIP_Generations_and_Tech10.pdf. Accessed on April 2, 2012.

A LONGITUDINAL EXAMINATION OF THE RELATIONSHIP BETWEEN TECHNOLOGY USE AND SUBSTANCE USE DURING ADOLESCENCE

Christine McCauley Ohannessian

ABSTRACT

Purpose – *The primary goal of this longitudinal study was to examine whether technology use predicts substance use* and/or *whether substance use predicts technology use during adolescence.*

Methodology/approach – *The sample included 1,031 10th and 11th grade students from the Mid-Atlantic United States. The students completed surveys in school in the spring of 2007 and 2008.*

Findings – *Gender differences in technology use were observed with girls texting, e-mailing/instant messaging, and working on the computer more than boys, and boys playing video games more than girls. Technology use also predicted later substance use for boys and girls. Importantly,*

Technology and Youth: Growing Up in a Digital World
Sociological Studies of Children and Youth, Volume 19, 293–313

ISSN: 1537-4661/doi:10.1108/S1537-466120150000019010

technology use was observed to have both negative and *positive effects on youth. Substance use also predicted later technology use for girls.*

Research limitations/implications – *The sample only included adolescents from the Mid-Atlantic United States. In addition, the measures were based on self-reports. Nevertheless, results from this study highlight the importance of considering both negative and positive effects of technology on adolescents. Of note, social types of technology (texting and e-mailing) predicted more substance use for both boys and girls. As such, substance use prevention programs should target these types of technology.*

Originality/value – *Findings from this study underscore the importance of examining both directions of influence between technology use and adolescent adjustment.*

Keywords: Adolescence; technology use; media; substance use; alcohol; marijuana

Adolescence is characterized by change. Many changes occur within the individual (e.g., puberty, cognitive development, identity exploration) and within the individual's contexts (e.g., family, school, peers) (Lerner, Lerner, von Eye, Bowers, & Lewin-Bizan, 2011; Smetana, Campione-Barr, & Metzger, 2006). The sheer number of changes that take place simultaneously may be challenging for some youth. As such, it is not surprising that during adolescence, psychosocial problems such as substance use and abuse increase dramatically (Johnston, O'Malley, Bachman, Schulenberg, & Miech, 2014).

Contemporary theories of human development (e.g., relational developmental systems theories; Lerner et al., 2011; Overton & Lerner, 2014) such as Bronfenbrenner's bioecological theory (Bronfenbrenner & Morris, 2006) and Elder's life-course theory (Elder & Shanahan, 2006) purport that development unfolds through mutually influential relations between the individual and the contexts in which the individual is embedded (Lerner et al., 2011). These theories emphasize that the individual always is "contextually situated" (Overton & Lerner, 2014). Therefore, it is imperative to consider all contexts applicable to adolescents when considering their adjustment. One context that has become increasingly central in the lives of contemporary youth is the world of technology. Therefore, the goal of the present study was to examine the relationship between this key context and adolescent psychosocial maladjustment, as indicated by substance use.

THE PREVALENCE OF TECHNOLOGY USE DURING ADOLESCENCE

Today, 93% of adolescents have a computer at home and 78% have their own cell phone (with 47% owning a smartphone) (Madden, Lenhart, Duggan, Cortesi, & Gasser, 2013). Impressively, 95% of adolescents now are online (Madden et al., 2013). This easy access to technology translates to a substantial amount of time adolescents spend using technology. Research suggests that adolescents now are engaged in technology for more than 7.5 hours a day (Rideout, Foehr, & Roberts, 2010). Of note, approximately 25% of the time that adolescents are engaged in technology is spent using multiple forms of media simultaneously or "media multi-tasking" (Brown & Bobkowski, 2011). Given that technology use has become so central to the daily lives of adolescents, it is important to consider the manner in which this context is associated with adolescent adjustment.

TECHNOLOGY USE AND ADOLESCENT ADJUSTMENT

Most research conducted on technology use in youth has indicated that technology use is negatively related to adjustment. For instance, many studies have found that playing violent video games may be related to aggressive behavior (Anderson & Bushman, 2001; Anderson, Gentile, & Buckley, 2007; Gentile, Lynch, Linder, & Walsh, 2004; Greitemeyer & Mügge, 2014). Watching television, playing video games, and electronic media use also have been linked to physical inactivity, weight, body fat, and sleep difficulties during adolescence (Lemola, Perkinson-Gloor, Brand, Dewald-Kaufmann, & Grob, 2014; Marshall, Biddle, Gorely, Cameron, & Murdey, 2004). The use of the Internet and electronic media has been found to be associated with adolescent psychological problems (loneliness, depression, anxiety) as well (Kraut et al., 1998; Lemola et al., 2014). In addition, television viewing, playing video games, playing computer games, and talking on the phone have been found to be negatively related to academic performance (Durkin & Barber, 2002; Gentile et al., 2004).

However, it is important to realize that technology use also may be linked to positive adjustment. According to the Positive Youth Development perspective (PYD; Lerner et al., 2011), developmental assets present in an adolescent's ecologies or contexts may positively influence the adolescent's development. Consistent with PYD, some research has indicated that the use of certain types of technology may offer benefits to youth. For example, playing video games and watching television have been associated with lower levels of anxiety and depression for adolescent boys (Ohannessian, 2009a). These findings are in line with the tenet that using certain types of technology (e.g., playing video games) may be beneficial for youth because the technology provides young people with a means to psychologically disengage from their problems (Ohannessian, 2009a). Such disengagement may be psychologically adaptive according to coping and motivational theories (Klinger, 1975). Playing video games has been found to be related to better visual processing and cognitive skills in youth as well (Green & Bavelier, 2007; Subrahmanyam, Greenfield, Kraut, & Gross, 2001). Some types of technology (e.g., texting, e-mailing) also may protect youth from experiencing problems by increasing perceived peer support, which may be adaptive by enabling the individual to vent their feelings, and to receive reassurance and advice (e.g., Carver, Scheier, & Weintraub, 1989). Indeed, research has shown that online communication is positively associated with reported closeness to friends (Valkenburg & Peter, 2007). Given that different types of technology may serve different functions, it is essential to separately examine their associations with adolescent adjustment.

It also should be noted that although technology use has been shown to be associated with externalizing problems (Anderson & Bushman, 2001; Anderson et al., 2007; Gentile et al., 2004), less is known about the relationship between technology use and substance use during adolescence. However, theory and recent research suggest technology use may be related to substance use. Consistent with socialization explanations and social modeling (Arnett, 1995; Slater, 2007), in a study assessing middle school students (Slater & Henry, 2013), music-related media exposure was found to predict the onset of alcohol and cigarette use. In addition, in line with the deviance proneness vulnerability model of substance use (Iacono, Malone, & McGue, 2008), which underscores the importance of peer influence, texting and e-mailing/instant messaging with peers have been found to predict alcohol and cigarette use (Ohannessian, 2009b).

Of note, the focus on both the Slater and Henry and the Ohannessian studies was on younger adolescents. The relationship between technology use and substance use in older adolescents is not clear. It also is important to note that both studies were more rigorous than many studies that have examined adolescent technology use because the adolescents were assessed over time. However, only one direction of effect was examined – whether technology use predicts substance use. According to contemporary developmental theories (e.g., relational developmental systems theoretical models of human development; Lerner et al., 2011; Overton, 2010; Overton & Lerner, 2014), characteristics of the individual and the context influence one another in order to produce development. That is, the link between the individual and the contexts in which the individual is embedded is *bidirectional* (Lerner et al., 2011). Clearly, the adolescent is influenced by the contexts in which he/she is immersed. However, at the same time, the adolescent influences these contexts. As such, examining only the effect that technology has on the individual yields an incomplete picture. The influence that the individual's characteristics and behavior has on their use of technology also must be considered. In a recent study focusing on both directions of effect between adolescent technology use and self-competence (Ohannessian, 2014), technology use was found to have a minimal effect on adolescent self-competence. However, adolescent self-competence consistently predicted technology use. These findings underscore the need to examine both directions of effect between technology use and adolescent adjustment.

GENDER DIFFERENCES IN TECHNOLOGY USE

Importantly, research on children and younger adolescents suggests that technology use may differ by gender. For example, many studies have found that boys spend more time playing video games and computer games than girls (Gentile et al., 2004; Marshall, Gorely, & Biddle, 2006; Ohannessian, 2009a). Boys spend more time watching television than girls as well (Gentile et al., 2004; Marshall et al., 2006). Research on gender differences in the use of other types of technology has been more limited. For instance, Hunley et al. (2004) and Ohannessian (2009a) found that girls spend more time talking on the phone than boys. In contrast, Gross, Juvonen, and Gable (2002) found no gender difference in phone use. This difference may be a result of the samples assessed (e.g., Hunley et al.'s and Ohannessian's samples consisted of 9th and 10th grade students, whereas Gross and colleague's sample included 7th grade students). Research on gender differences in Internet use during adolescence also is mixed. For example, Subrahmanyam et al. (2001) found boys to use the Internet more than girls during adolescence, yet Gross (2004) did not observe gender differences in Internet use in her sample of adolescents. However, it should be noted that the Subrahmanyam et al. sample included a much wider age range (10–19 year olds) in comparison to the Gross (7th and 10th grade students) sample. The difference in sample characteristics may have accounted for the discrepancy in results across studies.

LIMITATIONS OF THE CURRENT LITERATURE

Although research on technology use in youth is growing rapidly, many of the studies conducted to date have been limited by sample constraints and research design issues. Cross-sectional designs, the neglect to examine bidirectional influences between technology use and adjustment, and relatively limited information available on specific types of technology (e.g., texting) and their association with adolescent adjustment (including overlooked indicators of adjustment such as substance use) hinder the field. The present study sought to address these limitations in a large, diverse sample of adolescents. In sum, the following research questions were examined: (1) Does technology use predict adolescent substance use? (2) Does adolescent substance use predict technology use? and (3) Does the relationship between technology use and substance use differ by gender?

METHOD

Participants

The sample included 1,031 10th and 11th grade students (490 boys and 541 girls) from the Mid-Atlantic region of the United States (Delaware, Maryland, and Pennsylvania). During the spring of 2007 (Time 1) and the spring of 2008 (Time 2), students were administered surveys in school. The mean age of the adolescents was 16.15 (SD = .75). The majority (58%) of the participating students were Caucasian. However, 23% were African-American, 12% were Hispanic, and 2% were Asian (the remainder responded "other"). These percentages closely mirror the area from which the sample was drawn (71% Caucasian, 23% African-American, 4% Asian, 7% Hispanic; U.S. Census Bureau, 2008). Most of the participants (72%) lived with both of their biological parents (96% lived with their biological mother, 73% lived with their biological father). Of note, the majority of mothers (96%) and fathers (95%) were high school graduates. Some of the parents (26% of mothers and 24% of fathers) had graduated from a four-year college as well. A minority of the parents (10% of mothers and 7% of fathers) had attended graduate school.

Procedure

The study protocol was approved by the University of Delaware's Institutional Review Board. Public high schools in northern and central Delaware, northern Maryland, and southeastern Pennsylvania (within a 60 mile radius of the University of Delaware) were invited to participate in the study. The administration from seven public high schools agreed to have their school participate. In the spring of 2007, 10th and 11th grade students from participating schools, who provided assent and had parental consent, were given a self-report survey in school by trained research personnel (all of whom were certified with human subjects training). Seventy-one percent of the students attending the study schools participated. The majority of students that did not participate were absent on the day of data collection. Only 3% of the students present on the day of data collection did not participate.

The students were told that participation was voluntary, that the data collected were confidential, and that they could withdraw from the study at any time. In addition, they were told that a Certificate of Confidentiality from the U.S. government would protect their privacy. The survey took approximately 40 minutes to complete. In exchange for their completed survey, participants were given a movie pass. All of the students were invited to participate again the following spring (Time 2). The same protocol was followed at Time 2.

Measures

The survey included a demographic questionnaire which included questions such as the age of the adolescent, the gender of the adolescent, and parents' educational attainment (1 = *elementary school* to 6 = *graduate or medical school*). The survey also included measures of technology use and substance use.

Technology Use

The participants were asked how much time they spent "on an average/typical day" text messaging, e-mailing/instant messaging, playing video games (PlayStation, Nintendo, Game Boy, Xbox, etc.), listening to music on an iPod/MP3 player, and working on the computer. The response scale was 1 = *none*, 2 = *less than 1 hour*, 3 = *about 1 hour*, 4 = *about 2 hours*, 5 = *about 3 hours*, and 6 = *4 or more hours*.

Substance Use

In order to assess alcohol quantity, the participants were asked to report how much, "on the average day," they usually drank (beer, wine, or liquor) in the last six months (separate questions were used for beer, wine, and liquor). The response scale ranged from 0 = *none* to 9 = *more than 8 drinks*. A drink was defined as a can/bottle of beer, a glass of wine, or a drink containing liquor (1 shot = 1 ounce). To assess alcohol frequency, they were asked to report how often they usually had a drink (beer, wine, or liquor) in the last six months. The response scale for this question was 0 = *never*, 1 = *a few times*, 2 = *about once a month*, 3 = *2–3 days a month*, 4 = *about once a week*, 5 = *2–3 days a week*, 6 = *4–5 days a week*, and 7 = *every day*. Consistent with convention (Sobell & Sobell, 1995), sums were calculated across the beverage types for quantity and frequency. The quantity and frequency scores were then multiplied to reflect an estimated total volume consumed (QF) (refer to Sobell & Sobell, 1995, for additional information). To assess binge drinking, participants were asked how many times they drank six or more drinks (cans/bottles of beer, glasses of wine, or drinks of liquor) on one occasion during the last six months. The following question was used to assess marijuana use "How often have you used marijuana or hashish in the last six months (non-medical use only)". The response scale ranged from 0 = *no use* to 7 = *every day*. The logarithmic transformation was used for each of the substance use variables because they were skewed.

RESULTS

Multiple Group Analyses

Multiple group comparison models were conducted to determine whether separate analyses should be carried out for boys and girls. Importantly, multiple group analyses enable a more rigorous examination of gender differences than simply including gender as a covariate or separately conducting models by gender. The latter methods do not formally test whether gender differences are statistically significant. Consistent with the recommendations of Vandenberg and Lance (2000), an unconstrained model with freely estimated parameters was compared to models constraining path coefficients, variances, and covariances to be equal across groups (boys and girls).

For the multiple group analyses, a baseline model with parameters unconstrained by gender was estimated for each direction of effect (technology use predicting substance use and substance use predicting technology use). The unconstrained model was compared to a series of constrained models, imposing parameter invariance for boys and girls. When the models predicting substance use from technology use were examined, the chi-square difference test statistic was found to be significant for the structural intercepts, means, covariances, and residuals ($\Delta\chi^2(27) = 43.66$, $p < .05$; $\Delta\chi^2(37) = 341.81$, $p < .001$; $\Delta\chi^2(60) = 455.70$, $p < .001$; $\Delta\chi^2(66) = 496.65$, $p < .001$, respectively), suggesting that these parameters significantly differed by gender. Similarly, when the models predicting technology use from substance use were examined, the chi-square difference test statistic was significant for the structural intercepts, means, covariances, and residuals ($\Delta\chi^2(35) = 107.05$, $p < .001$; $\Delta\chi^2(45) = 411.55$, $p < .001$; $\Delta\chi^2(68) = 524.97$, $p < .001$; $\Delta\chi^2(83) = 602.03$, $p < .001$, respectively), again indicating that these parameters significantly differed by gender. As such, the parameter estimates that follow were derived from the unconstrained model.

Independent samples *t*-tests were conducted to examine gender differences in technology use and substance use. As shown in Tables 1 and 2, at both Time 1 and Time 2, girls reported more text messaging ($t(992) = -4.69$, $p < .001$, $t(635) = -6.12$, $p < .001$, respectively), and e-mailing/instant messaging ($t(995) = -4.51$, $p < .001$, $t(669) = -2.18$, $p < .05$, respectively) in comparison to boys. In addition, at Time 1, girls reported working on the computer more than boys ($t(990) = -2.73$, $p < .01$). In contrast, at both times of measurement, boys reported playing video games more than girls ($t(889) = 16.12$, $p < .001$, $t(447) = 13.20$, $p < .001$, respectively).

Boys also reported more frequent marijuana use than girls at both Time 1 and Time 2 ($t(877) = 3.15$, $p < .01$, $t(538) = 2.24$, $p < .01$, respectively). In addition, at Time 2, boys reported more frequent alcohol use in comparison to girls ($t(386) = 3.15$, $p < .01$).

Table 1. Correlations, Means, and Standard Deviations among Study Variables at Time 1.

Variable	1	2	3	4	5	6	7	8	Boys' Mean	SD
1. Text messaging	–	.46***	.07	.35***	.26***	.18***	.09*	.18**	2.25***	1.53
2. E-mailing/IMing	.33***	–	.13**	.27***	.43***	.12*	.03	.17**	2.76***	1.66
3. Playing video games	.15**	.20***	–	.11*	.24***	−.05	−.07	−.08	3.25***	1.59
4. Listening to music	.25***	.36***	.20***	–	.29***	.00	−.02	.04	3.11	1.70
5. Working on the computer	.18***	.50***	.27***	.37***	–	.03	−.04	.05	3.05**	1.53
6. Alcohol use	.11*	.14**	−.07	.08	−.04	–	.46***	.74***	5.60	13.89
7. Marijuana use	.00	.02	−.03	.07	−.05	.50***	–	.54***	.71**	1.81
8. Binge drinking	.09	.08	−.06	.03	−.02	.72***	.53***	–	1.78	6.04
Girls' Mean	2.74***	3.24***	1.76***	3.12	3.32**	4.14	.44**	1.67	–	–
SD	1.80	1.71	1.28	1.79	1.61	11.59	1.30	6.11	–	–

Notes: Correlations for boys and girls are presented above and below the diagonal, respectively. The means of both boys and girls are marked with * to indicate significant gender differences.
*$p < .05$; **$p < .01$; ***$p < .001$.

Table 2. Correlations, Means, and Standard Deviations among Study Variables at Time 2.

Variable	1	2	3	4	5	6	7	8	Boys' Mean	SD
1. Text messaging	–	.33***	.08	.24***	.12*	.21***	.03	.04	2.81***	1.80
2. E-mailing/IMing	.31***	–	.20**	.32***	.45***	.01	−.12*	−.04	2.57*	1.51
3. Playing video games	.08	.21***	–	.31***	.36***	−.05	−.09	−.14*	3.04***	1.66
4. Listening to music	.25***	.21***	.24***	–	.34***	−.04	−.08	−.12	3.17	1.71
5. Working on the computer	.17**	.37***	.26***	.29***	–	−.15*	−.17**	−.11	3.18	1.55
6. Alcohol use	.15**	.06	−.03	.03	−.09	–	.52***	.87***	11.42***	26.69
7. Marijuana use	.05	−.04	−.03	.03	−.16**	.45***	–	.52***	.87*	2.02
8. Binge drinking	.12*	.03	−.04	.04	−.09	.78***	.49***	–	1.85	6.35
Girls' Mean	3.69***	2.83*	1.57***	3.14	3.26	5.90***	.56*	2.17	–	–
SD	1.92	1.55	1.04	1.76	1.44	13.59	1.57	7.82	–	–

Notes: Correlations for boys and girls are presented above and below the diagonal, respectively. The means of both boys and girls are marked with * to indicate significant gender differences.
*$p < .05$; **$p < .01$; ***$p < .001$.

Bivariate Analyses

Pearson product-moment correlations were calculated to examine the bivariate relations between the study variables. As shown in Tables 1 and 2, text messaging was associated with more alcohol use and binge drinking for both boys and girls. Similarly, e-mailing/instant messaging was related to more alcohol use for boys and girls. In addition, e-mailing/instant messaging was associated with more marijuana use and binge drinking for boys. In contrast, playing video games was related to less binge drinking for boys and working on the computer was related to less marijuana use for boys and girls.

Longitudinal Analyses

Path analysis was conducted to examine whether technology use (at Time 1) predicted substance use one year later (Time 2) and/or whether substance use (at Time 1) predicted technology use one year later (Time 2). The endogenous variables assessed at Time 1, adolescent age, and an SES proxy (parental education) were included as covariates in the models. Full information maximum likelihood (FIML) was used to handle missing data. FIML uses all data available (the covariance matrix and a vector of the means) to produce maximum likelihood-based sufficient statistics. Importantly, FIML has been observed to yield unbiased parameter estimates (Wothke, 2000).

Does Technology Use Predict Adolescent Substance Use?

The model predicting substance use from technology use provided a good fit to the data ($X^2(76) = 182.31$, $p = .00$; CMIN/DF = 2.40; CFI = .96; RMSEA = .04).

Results for Boys

As shown in Fig. 1, text messaging predicted more frequent marijuana use for boys ($\beta = .18$, $p < .01$). In contrast, playing video games predicted lower alcohol consumption ($\beta = -.14$, $p < .05$) and less binge drinking ($\beta = -.12$, $p < .05$) for boys.

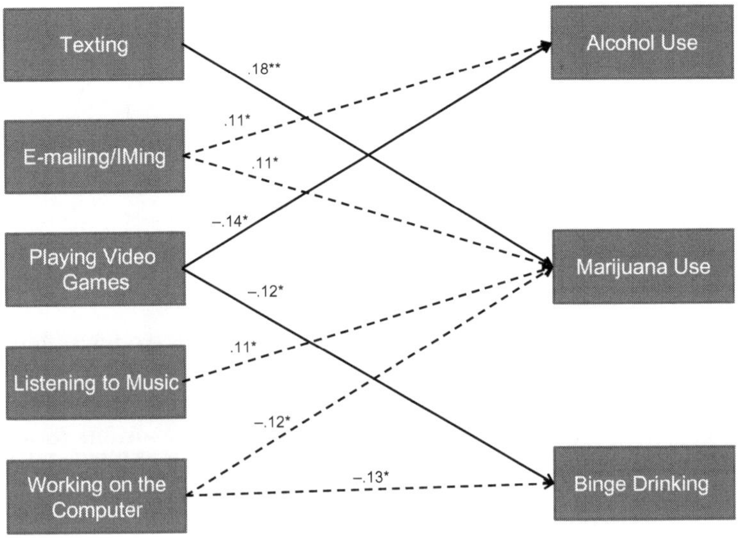

Fig. 1. Model Predicting Substance Use from Technology Use. For Ease of Interpretation, Only Significant Paths are Shown. Boys = Solid Lines, Girls = Dashed Lines. Standardized Regression Coefficients are Presented. Control Variables, Covariances, and Disturbance Terms Are Not Displayed. $*p < .05$; $**p < .01$.

As expected, the dependent variables assessed at Time 1 predicted the dependent variables assessed at Time 2 ($\beta = .41$, $p < .001$ for alcohol use, $\beta = .40$, $p < .001$ for binge drinking, and $\beta = .48$, $p < .001$ for marijuana use). Of note, age significantly predicted alcohol use ($\beta = .11$, $p < .05$) and binge drinking ($\beta = .14$, $p < .05$), indicating that alcohol use increased with age.

Results for Girls

For girls, e-mailing and instant messaging predicted more alcohol use ($\beta = .11$, $p <.05$) and marijuana use ($\beta = .11$, $p <.05$) (see Fig. 1). In addition, listening to music from an iPod, MP3 player, etc. predicted more frequent marijuana use ($\beta = .11$, $p <.05$). In contrast, working on the computer predicted less marijuana use ($\beta = -.12$, $p <.05$) and less binge drinking ($\beta = -.13$, $p <.05$) for girls.

The dependent variables assessed at Time 1 significantly predicted the dependent variables assessed at Time 2 as well ($\beta = .45$, $p < .001$ for alcohol use, $\beta = .49$, $p < .001$ for binge drinking, and $\beta = .51$, $p < .001$ for marijuana use). Of note, age and parental education did not predict any of the variables in the girls' model.

Does Adolescent Substance Use Predict Technology Use?

The model predicting technology use from substance use similarly fit the data well ($X^2(104) = 206.50$, $p = .00$; CMIN/DF = 1.99; CFI = .96; RMSEA = .03).

Results for Boys

Substance use did not predict any of the types of technology examined for boys. Not surprisingly, however, the dependent variables assessed at Time 1 predicted the dependent variables assessed at Time 2 ($\beta = .44$, $p < .001$ for text messaging, $\beta = .31$, $p < .001$ for e-mailing/instant messaging, $\beta = .49$, $p < .001$ for playing video games, $\beta = .43$, $p < .001$ for listening to music, and $\beta = .35$, $p < .001$ for working on the computer). Parental education also predicted working on the computer ($\beta = .12$, $p < .05$), indicating that adolescents with more educated parents worked on the computer relatively more frequently.

Results for Girls

As shown in Fig. 2, alcohol use predicted more text messaging ($\beta = .16$, $p <.05$) for girls. In contrast, marijuana use predicted less text messaging ($\beta = -.14$, $p <.05$) and less time working on the computer ($\beta = -.12$, $p <.05$) for girls.

In addition, the dependent variables assessed at Time 1 predicted the dependent variables assessed at Time 2 ($\beta = .42$, $p < .001$ for text messaging, $\beta = .45$, $p < .001$ for e-mailing/instant messaging, $\beta = .49$, $p < .001$ for playing video games, $\beta = .45$, $p < .001$ for listening to music, and $\beta = .43$, $p < .001$ for working on the computer). Age significantly predicted e-mailing and instant messaging ($\beta = .09$, $p < .05$) as well, with e-mailing and instant messaging increasing with age.

DISCUSSION

The primary goal of the present study was to examine bidirectional relations between technology use and substance use during adolescence. Consistent with prior research suggesting technology may negatively

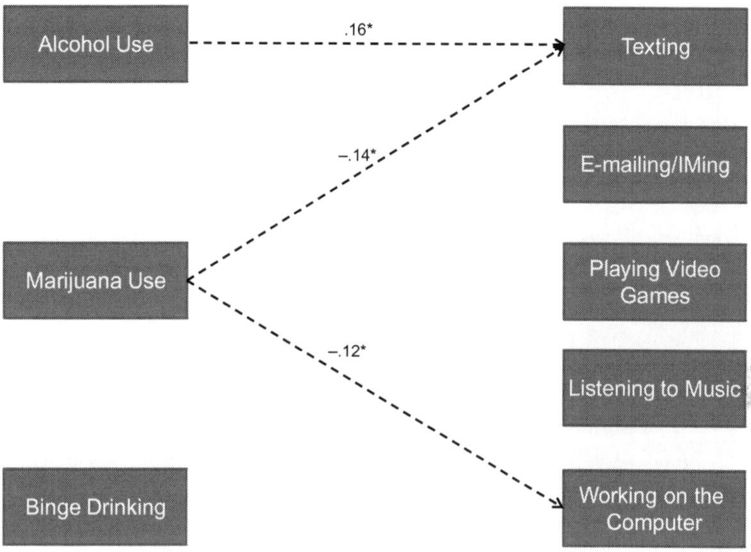

Fig. 2. Model Predicting Technology Use from Substance Use. For Ease of Interpretation, Only Significant Paths Are Shown. Boys = Solid Lines, Girls = Dashed Lines. Standardized Regression Coefficients are Presented. Control Variables, Covariances, and Disturbance Terms are Not Displayed. $*p < .05$.

influence adjustment (Anderson et al., 2007; Greitemeyer & Mügge, 2014; Lemola et al., 2014; Marshall et al., 2004), technology use was found to predict later substance use. More specifically, text messaging predicted more frequent marijuana use for boys and e-mailing/instant messaging predicted more alcohol use and marijuana use for girls. Although relatively few studies have focused on these types of technology specifically, the findings relating to texting and e-mailing/instant messaging are consistent with prior research on younger adolescents. In a previous study focusing on 14–16 year-old adolescents (Ohannessian, 2009b), texting and e-mailing/instant messaging were related to alcohol use and e-mail/instant messaging was related to the initiation of cigarette use. Of note, texting and e-mailing/instant messaging are social types of technology that adolescents primarily use with their peers. As such, it is not surprising that these findings are in line with broader research that has focused on the influence that peers may have on adolescent substance use. Many studies have shown peer substance use to be associated with adolescent substance use (Gibbons et al., 2010). Even during early adolescence, adolescents tend to select friends with similar levels of substance use (Knecht, Burk, Weesie, & Steglich, 2011). Such friends are likely to further influence adolescent substance use via social communication (e.g., texting, e-mailing).

In the present study, listening to music on an iPod/MP3 player predicted marijuana use for adolescent girls. This finding is consistent with Slater and Henry's (2013) research focusing on younger adolescents (7th and 8th grade students). In their work (Slater & Henry, 2013), music-related media exposure was found to predict the onset of alcohol and cigarette use. Taken together, findings from this study and the Slater and Henry study are consistent with socialization explanations and social modeling (Arnett, 1995; Slater, 2007). Research has shown that contemporary music glamorizes substance use (Primack, Dalton, Carroll, Agarwal, & Fine, 2008; Primack, Land, & Fine, 2008). Many musical artists overtly promote substance use behavior and many song lyrics encourage substance use (Primack, Dalton, et al., 2008). Given that some types of music and artists glamorize substance use more than others, an important next step would be to assess whether type of music mediates the relationship between listening to music and adolescent substance use.

Of note, technology use was found to positively influence adjustment as well. For girls, time spent working on the computer and substance use were negatively related. More specifically, computer use predicted less marijuana use and less binge drinking. One hypothesis relating to this finding is that youth who spend more time working on the computer are more academically and achievement oriented. This hypothesis is consistent with our prior research showing more frequent computer use to be related to higher levels of scholastic competence during adolescence (Ohannessian, 2014).

Positive effects of technology use also were observed for boys. More specifically, for boys, playing video games predicted lower alcohol consumption and less binge drinking. These findings are not consistent with the majority of studies that have examined technology use during adolescence. However, some research has shown that playing video games may positively influence youth. For example, playing video games has been found to be associated with better visual processing and cognitive skills (Green & Bavelier, 2007; Subrahmanyam et al., 2001). In addition, we have found playing video games to be related to lower levels of anxiety and depression (Ohannessian, 2009a) and higher levels of athletic competence (Ohannessian, 2014) for adolescent boys. Of note, these findings were especially pronounced in boys from dysfunctional families (Ohannessian, 2009a). One hypothesis is that playing video games may protect youth from developing problems because the games provide them with a means to psychologically disengage from their problems (Ohannessian, 2009a).

It should be noted that playing video games did not predict less substance use for girls. However, girls were significantly less likely to play video games in comparison to boys. In addition, the difference in findings for boys and girls may be related to the different types of games played by boys and girls. For example, boys may be more likely than girls to play games that require them to become immersed in the game, enabling them to easily mentally disengage from their problems (e.g., fantasy type games, sports games). Future studies should examine whether type of game mediates the relationship between video game play and adolescent adjustment. Taken together, results from this study emphasize the importance of considering potential positive effects of technology use, as well as potential negative effects, during adolescence.

As noted, an important contribution of the present study was the examination of bidirectional relations between technology use and substance use. Most studies examining technology use during adolescence have been cross-sectional or have examined only one direction of effect – whether technology use predicts adolescent adjustment. However, contemporary developmental theories (e.g., relational developmental systems theories; Lerner et al., 2011; Overton, 2010; Overton & Lerner, 2014) purport that characteristics of the individual and the context, and the levels within each, influence each other in order to produce development. As such, it is important to examine whether indicators of adjustment influence technology use as well. Of note, in a recent study (Ohannessian, 2014), we found indicators of adolescent adjustment to predict technology use more consistently than the reverse. More specifically, technology use was found to have a minimal effect on adolescent self-competence; however, self-competence consistently predicted later technology use.

In the present study, substance use did not predict technology use for boys. However, alcohol use predicted more frequent text messaging for girls. In contrast, marijuana use predicted less text messaging and less time working on the computer for girls. The findings relating to marijuana are consistent with marijuana's inhibitory effect on behavior. Research has shown that marijuana negatively influences motivation (Lane, Cherek, Pietras, & Steinberg, 2005). As such, adolescents who use marijuana may simply be less motivated and more apathetic in general, which may result in less frequent use of technology. Future research should test this mediational hypothesis.

Although this study contributed to the literature by longitudinally examining both directions of effect between technology use and substance use in a large, diverse sample, limitations of the study should be considered. As noted previously, the sample included adolescents from the Mid-Atlantic United

States. Therefore, the findings may not be generalized to adolescents living outside of this area. In addition, self-report measures were used in this study. However, it is important to note that research has found youth to be accurate reporters of their own behaviors (Deković et al., 2006). Nevertheless, it would be informative for future studies to replicate the results from this study using other types of methodology (e.g., parent reports).

Despite these limitations, results from this study underscore the importance of considering both negative *and* positive effects of technology on adolescents. Of note, social types of technology (texting and e-mailing) predicted more substance use for boys and girls. As such, substance use prevention programs should specifically target these types of technology. In contrast, playing video games appeared to protect boys from substance use; perhaps by helping them to mentally disengage from their problems. Importantly, substance use also predicted technology use. More specifically, alcohol and marijuana use predicted later text messaging. Taken together, these findings point to the need to examine both directions of influence between technology use and adolescent adjustment. Clearly, the examination of only one direction of effect yields an incomplete picture of the complex, dynamic relationship between technology use and adolescent adjustment.

ACKNOWLEDGMENTS

This research was supported by grant 5K01A015059 to Christine McCauley Ohannessian from the National Institutes of Health. The involvement of all of the students who participated in the study is greatly appreciated. Special thanks go to members of the project staff, especially Kaitlin Flannery, Magdalena Owczarska, Kelly Cheeseman, Lisa Fong, Alyson Cavanaugh, Jessica Schulz, Laura Finan, Sara Bergamo, Ashley Malooly, and Ashley Ings.

REFERENCES

Anderson, C. A., & Bushman, B. J. (2001). Effects of violent video games on aggressive behavior, aggressive cognition, aggressive affect, physiological arousal, and prosocial behavior: A meta-analytic review of the scientific literature. *Psychological Science*, *12*(5), 353–359.

Anderson, C. A., Gentile, D. A., & Buckley, K. E. (2007). *Violent video game effects on children and adolescents*. New York, NY: Oxford University Press.

Arnett, J. J. (1995). Adolescents' uses of media for self-socialization. *Journal of Youth and Adolescence*, *24*(5), 519–533.

Bronfenbrenner, U., & Morris, P. A. (2006). The bioecological model of human development. In R. M. Lerner & W. Damon (Eds.), *Theoretical models of human development* (6th ed., Vol. 1). Handbook of Child Psychology. Hoboken, NJ: Wiley.

Brown, J. D., & Bobkowski, P. S. (2011). Older and newer media: Patterns of use and effects on adolescents' health and well-being. *Journal of Research on Adolescence*, *21*(1), 95–113.

Carver, C. S., Scheier, M. F., & Weintraub, J. K. (1989). Assessing coping strategies: A theoretically based approach. *Journal of Personality and Social Psychology*, *56*(2), 267.

Deković, M., ten Have, M., Vollebergh, W. A., Pels, T., Oosterwegel, A., & Wissink, I. B., ... Ormel, J. (2006). The cross-cultural equivalence of parental rearing measure: EMBU-C. *European Journal of Psychological Assessment*, *22*(2), 85–91.

Durkin, K., & Barber, B. (2002). Not so doomed: Computer game play and positive adolescent development. *Journal of Applied Developmental Psychology*, *23*(4), 373–392.

Elder, G. H., Jr., & Shanahan, M. J. (2006). The life course and human development. In R. M. Lerner & W. Damon (Eds.), *Theoretical models of human development* (6th ed., Vol. 1). Handbook of Child Psychology. Hoboken, NJ: Wiley.

Gentile, D. A., Lynch, P. J., Linder, J. R., & Walsh, D. A. (2004). The effects of violent video game habits on adolescent hostility, aggressive behaviors, and school performance. *Journal of Adolescence*, *27*(1), 5–22.

Gibbons, F. X., Pomery, E. A., Gerrard, M., Sargent, J. D., Weng, C.-Y., Wills, T. A., ... Yeh, H.-C. (2010). Media as social influence: Racial differences in the effects of peers and media on adolescent alcohol cognitions and consumption. *Psychology of Addictive Behaviors*, *24*(4), 649–659.

Green, C. S., & Bavelier, D. (2007). Action-video-game experience alters the spatial resolution of vision. *Psychological Science*, *18*(1), 88–94.

Greitemeyer, T., & Mügge, D. O. (2014). Video games do affect social outcomes: A meta-analytic review of the effects of violent and prosocial video game play. *Personality and Social Psychology Bulletin*, *40*(5), 578–589.

Gross, E. F. (2004). Adolescent internet use: What we expect, what teens report. *Journal of Applied Developmental Psychology*, *25*(6), 633–649.

Gross, E. F., Juvonen, J., & Gable, S. L. (2002). Internet use and well-being in adolescence. *Journal of Social Issues*, *58*(1), 75–90.

Hunley, S. A., Evans, J. H., Delgado-Hachey, M., Krise, J., Rich, T., & Schell, C. (2004). Adolescent computer use and academic achievement. *Adolescence*, *40*(158), 307–318.

Iacono, W., Malone, S., & McGue, M. (2008). Behavioral disinhibition and the development of early onset addiction: Common and specific influences. *Annual Review of Clinical Psychology*, *4*, 325–348.

Johnston, L. D., O'Malley, P. M., Bachman, J. G., Schulenberg, J. E., & Miech, R. A. (2014). *Monitoring the Future national survey results on drug use, 1975–2013: Volume I, Secondary school students*. Ann Arbor, MI: Institute for Social Research, The University of Michigan.

Klinger, E. (1975). Consequences of commitment to and disengagement from incentives. *Psychological Review*, *82*(1), 1.

Knecht, A. B., Burk, W. J., Weesie, J., & Steglich, C. (2011). Friendship and alcohol use in early adolescence: A multilevel social network approach. *Journal of Research on Adolescence*, *21*(2), 475–487.

Kraut, R., Patterson, M., Lundmark, V., Kiesler, S., Mukophadhyay, T., & Scherlis, W. (1998). Internet paradox: A social technology that reduces social involvement and psychological well-being? *American Psychologist*, *53*(9), 1017.

Lane, S. D., Cherek, D. R., Pietras, C. J., & Steinberg, J. L. (2005). Performance of heavy marijuana-smoking adolescents on a laboratory measure of motivation. *Addictive Behaviors*, *30*(4), 815–828.

Lemola, S., Perkinson-Gloor, N., Brand, S., Dewald-Kaufmann, J. F., & Grob, A. (2014). Adolescents' electronic media use at night, sleep disturbance, and depressive symptoms in the smartphone age. *Journal of Youth and Adolescence*, *44*(2), 405–418.

Lerner, R. M., Lerner, J. V., Lewin-Bizan, S., Bowers, E. P., Boyd, M. J., Mueller, M. K., ... Napolitano, C. M. (2011). Positive youth development: Processes, programs, and problematics. *Journal of Youth Development*, *6*(3), 40–64.

Lerner, R. M., Lerner, J. V., von Eye, A., Bowers, E. P., & Lewin-Bizan, S. (2011). Individual and contextual bases of thriving in adolescence: A view of the issues. *Journal of Adolescence*, *34*, 1107–1114.

Madden, M., Lenhart, A., Duggan, M., Cortesi, S., & Gasser, U. (2013). *Teens and technology 2013*. Washington, DC: Pew Research Center. Retrieved from http://www.pewinternet.org/Reports/2013/Teens-and-Tech.aspx

Marshall, S. J., Biddle, S. J., Gorely, T., Cameron, N., & Murdey, I. (2004). Relationships between media use, body fatness and physical activity in children and youth: A meta-analysis. *International Journal of Obesity*, *28*(10), 1238–1246.

Marshall, S. J., Gorely, T., & Biddle, S. J. (2006). A descriptive epidemiology of screen-based media use in youth: A review and critique. *Journal of Adolescence*, *29*(3), 333–349.

Ohannessian, C. M. (2009a). Media use and adolescent psychological adjustment: An examination of gender differences. *Journal of Child and Family Studies*, *18*(5), 582–593.

Ohannessian, C. M. (2009b). Does technology use moderate the relationship between parental alcoholism and adolescent alcohol and cigarette use? *Addictive Behaviors*, *34*(6–7), 606–609.

Ohannessian, C. M. (2014). Anxiety and substance use during adolescence. *Substance Abuse*, *35*(4), 418–425. doi:10.1080/08897077.2014.953663

Overton, W. F. (2010). Life-span development: Concepts and issues. *Cognition, Biology, and Methods Across the Lifespan*, *1*, 1–29.

Overton, W. F., & Lerner, R. M. (2014). Fundamental concepts and methods in developmental science: A relational perspective. *Research in Human Development*, *11*(1), 63–73.

Primack, B. A., Dalton, M. A., Carroll, M. V., Agarwal, A. A., & Fine, M. J. (2008). Content analysis of tobacco, alcohol, and other drugs in popular music. *Archives of Pediatrics & Adolescent Medicine*, *162*(2), 169–175.

Primack, B. A., Land, S. R., & Fine, M. J. (2008). Adolescent smoking and volume of exposure to various forms of media. *Public Health*, *122*(4), 379–389.

Rideout, V. J., Foehr, U. G., & Roberts, D. F. (2010). *Generation M^2: Media in the lives of 8-to 18-year-olds*. Henry J. Kaiser Family Foundation. Retrieved from http://kff.org/other/poll-finding/report-generation-m2-media-in-the-lives/

Slater, M. D. (2007). Reinforcing spirals: The mutual influence of media selectivity and media effects and their impact on individual behavior and social identity. *Communication Theory*, *17*(3), 281–303.

Slater, M. D., & Henry, K. L. (2013). Prospective influence of music-related media exposure on adolescent substance-use initiation: A peer group mediation model. *Journal of Health Communication*, *18*(3), 291–305.

Smetana, J. G., Campione-Barr, N., & Metzger, A. (2006). Adolescent development in interpersonal and societal contexts. *Annual Review of Psychology, 57*, 255–284.

Sobell, L. C., & Sobell, M. B. (1995). Alcohol consumption measures. In J. P. Allen & M. Columbus (Eds.), *Assessing alcohol problems: A guide for clinicians and researchers*. NIAAA Treatment Handbook Series 4. Washington, DC: U.S. Department of Health and Human Services.

Subrahmanyam, K., Greenfield, P., Kraut, R., & Gross, E. (2001). The impact of computer use on children's and adolescents' development. *Journal of Applied Developmental Psychology, 22*(1), 7–30.

U.S. Census Bureau. (2008). *U.S. Census Bureau: State and county QuickFacts 2008*. Retrieved from http://quickfacts.census.gov/qfd/states/10/10003.html. Accessed on May 16, 2010.

Valkenburg, P. M., & Peter, J. (2007). Preadolescents' and adolescents' online communication and their closeness to friends. *Developmental Psychology, 43*(2), 267.

Vandenberg, R. J., & Lance, C. E. (2000). A review and synthesis of the measurement invariance literature: Suggestions, practices, and recommendations for organizational research. *Organizational Research Methods, 3*(1), 4–70.

Wothke, W. (2000). Longitudinal and multigroup modeling with missing data. In T. D. Little, K. U. Schnabel, & J. Baumert (Eds.), *Modeling longitudinal and multilevel data: Practical issues, applied approaches, and specific examples* (pp. 219–240). Mahwah, NJ: Lawrence Erlbaum Associates Publishers.

SEXTING, DIGITAL DISSENT AND NARRATIVES OF INNOCENCE – CONTROLLING THE CHILD'S BODY

Brian Simpson

ABSTRACT

Purpose – *The purpose of this paper is to explore the narratives that construct the practice and regulation of 'sexting', the sending of sexualised images via text message, when engaged in by young people. The aim of this discussion is to better understand the extent to which those narratives recognise young people's agency in relation to their sexuality and the role that new media plays in enabling youth to explore their sexual identity.*

Methodology – *The methodology employed is that of discourse analysis. This approach is used to deconstruct the dominant narrative of sexting contained in the literature, a narrative that constructs it as a problem to be contained and controlled, either through the application of the criminal law or through education and guidance approaches. This paper then investigates an emerging counter narrative that gives greater emphasis to the autonomy rights of youth. A case study involving a*

Technology and Youth: Growing Up in a Digital World
Sociological Studies of Children and Youth, Volume 19, 315–349

ISSN: 1537-4661/doi:10.1108/S1537-466120150000019011

Parliamentary Inquiry in one Australian State into sexting is also employed to further this analysis.

Findings – *This paper concludes that the dominant narrative remains the strongest influence in the shaping of law and the practice of sexting, but that young people may be better served by the counter narrative that recognises their agency in ways that may empower and grant them more control over their bodies.*

Originality/Value – *The paper thus provides an alternative approach to developing new law and policy with respect to the regulation of sexting by youth that should be of value to lawmakers and child and youth advocates.*

Keywords: Sexting; childhood; autonomy; innocence; sexualisation; agency

The phenomenon that has become known as 'sexting' – essentially the digital transmission, usually via mobile technology, of a 'sexualised' image – while practised by all age groups has gained most of its notoriety from it being engaged in by children and youth. Sexting may be of images transmitted without the agreement of the person depicted, but much of the focus of sexting debates tends be on what is often described as 'consensual sexting' – where a person agrees to have their image transmitted or that person may even be the sender. This latter form of sexting attracts much discussion because in law consent may transform an otherwise unlawful act into a lawful one. However, in the case of sexting by young people who are not yet legally adults, their legal incapacity may render any factual consent on their part void. This status – often loosely referred to as a child's incapacity to consent – is borne out of a concern to protect them from perceived harm and exploitation. The sexual nature of sexting rapidly leads to it being viewed as potentially harmful and so discussions about a young person consenting to their image being transmitted in such a context soon becomes a discussion about whether it can be an activity that children are capable of making decisions about.

This activity has as a consequence given rise to many fears and moral panics about the inappropriateness of such behaviour when engaged in by young people, especially on the basis that sexting sexualises the image of the child and so hastens a loss of childhood innocence. The practice also leads to many other debates about sexting and childhood. For example, many commentators distinguish between consensual and non-consensual sexting with the former being regarded as less culpable if no less problematic in terms of how this nevertheless sexualises young people. This suggests that at the core of the 'problem' of sexting is not just the 'wicked' behaviour of young people but also the damage done to the concept of childhood itself. Of course, for some, the distinction between consensual and non-consensual sexting is illusory as in their eyes, the capacity of children and youth to consent to the sexualisation and transmission of their body image is limited if not non-existent. Such a view of young people's ability to make decisions about their own body and the transmission of its image underscores how views about what is to be regarded as an 'appropriate' childhood lie at the foundation of debates on sexting.

Legal responses to sexting have so far tended to focus on the need to protect the child from even self-inflicted harm caused by sexting. There are all the usual 'law failing to keep pace with technology' narratives in legal debates around sexting, but while explicit references to sexting in legislation are relatively recent and for that matter in only some jurisdictions, there has been no failure on the part of the law to provide an ongoing narrative of childhood innocence which has been used by some campaigners eager to reduce sexting activity. 'Law reform' in this sense in relation to sexting is not about articulating a new role for law in terms of advancing the capacity of children to control their body image, but instead is usually about conceding the innocence or vulnerability of young people in a way that reinforces the notion that children and youth must be protected from their own misguided behaviour through education, surveillance and if necessary the blunt instrument of the law. There is an alternative legal narrative that advances the autonomy and agency of young people, but this narrative has so far gained little traction in most debates about sexting.

What this paper seeks to do is to challenge the 'dominant narrative' that surrounds acts of sexting by children and youth and explore the alternative narrative. It will be contended that the law as a tool of cultural practice contains within it the framework for an analysis of sexting that supports the practice as an expression of young people's self-expression and identity formation, and that such an approach to sexting would form a much sounder basis for the establishment of laws that wish to address the harmful effects of sexting while also recognising the autonomy rights of the young.

In this regard, the conceptual framework of this paper acknowledges that legal responses that assume the innocence and vulnerability of young people are both dangerous to them and harmful to their proper development. We should be learning from the history of child protection, juvenile justice, institutional abuse of children, and the removal of indigenous children from their families that to construct children and young people as lacking agency, capacity and the right of self-determination exposes them to the bad judgments of incompetent adults. To explore these issues, this paper utilises the case study of a recent attempt to reform the law on sexting in one Australian State – Victoria. While some may label this reform as 'progressive', on closer scrutiny the reform process reveals all the traditional narratives that abound in this area. This case study demonstrates the manner in which debates about sexting and young people mask wider anxieties about childhood sexuality, obscure intellectual shortcomings in the formulation of law and policy and conceal various disciplinary biases as to what constitutes legitimate knowledge in addressing sexting. It is through a close analysis of the 'evidence' – the submissions made by interested parties to the Parliamentary Committee inquiring into the law that applies to sexting – that one turns traditional debates on sexting away from children and youth and onto how adults view children and youth in a digital environment. In doing this, I find the work and approach of Robbie Duschinsky particularly useful. He formulates an analytical approach (discursive policy analysis) that 'treats the language of policy as both shaped by broader cultural and political structures, such as institutional forms and social practices, and as a fundamental mechanism through which these forms and practices operate and achieve legitimacy' (Duschinsky, 2012, p. 716). Such structures and forces then shape policy in a way that results in various subjective understandings of the world becoming represented in such texts 'as true, discrete, natural and inevitable' (Duschinsky, 2012, p. 716, citing Mottier, 2008). The discussion of sexting, which took place in the Victorian Inquiry into its practice, is a proper subject of such discursive policy analysis.

'DOMINANT NARRATIVES' OF SEXTING AND YOUNG PEOPLE

Policy debates and narratives surrounding the practice of sexting follow a standard approach. For the most part, the concern is with the young people engaging in the behaviour, with only occasional reference to it also being engaged in by older people. Implicit in all discussions is that it is 'bad' or 'inappropriate' behaviour that can only lead to problems later in life for young people. There is also a focus on the amount of sexting taking place among young people, with much research devoted to attempting to determine how prevalent the practice is. Finally, the 'appropriate response' to the practice is then discussed, which usually discusses the role of law, education, and parental control or guidance. To the extent that sexting is often associated with cyberbullying or harassment, the narratives that surround those additional behaviours will at times also be drawn into the discussion. The emerging 'cyber-safety industry', which includes current and former law enforcement personnel and consultants, regulatory agencies and educators, drives much of these sexting narratives. Each of these narrative strands are worthy of some separate discussion.

Sexting as 'Risky' Youthful Practice

It cannot be denied that sexting occurs across all age groups. Nancy Willard notes that adults over 50 are a group that also engage in sexting as they have in common with teenagers both access to the technology and the time to participate in it (Willard, 2010). A number of high-profile cases involving politicians in both Australia (McKenna, 2013) and the United States (Barbaro, 2011) also indicate that this is not a behaviour only indulged in by young people. Indeed, the Victorian Inquiry identified the sexting by a high-profile professional Australian cricketer as one of the first examples of the use of the term (Law Reform Committee, Parliament of Victoria, 2013, p. 2). This might, as Willard contends, suggest that it is behaviour that 'should be considered within the range of normative human sexual behaviour' (Willard, 2010). However, this is soon passed over to emphasise the risks to young people that sexting represents, in particular, the possibility of wide dissemination resulting in damage to reputation and the risk of exploitation (Willard, 2010, p. 542).

Such references to risk and danger abound in most discussions of sexting both in academic literature and in the general media (Ahern, 2013; D'Antona, 2010; Young People Face Online Safety Timebomb, 2013). Many of those who write about sexting proceed from the starting point that sexting is a problem and feel little need to evaluate precisely which values and views of childhood inform their stance. It is then relatively easy to slide from sexting to related ills, and so by association paint a picture of sexting that is almost demonic in terms of its connection with 'raging adolescent hormones and attitudes of invincibility, digital technological advancements and peer pressure' that connects with 'earlier sexual behaviour, promiscuity, sexually transmitted diseases, unwanted pregnancy, substance abuse, harassment and cyberbullying, guilt and shame, arrest and incarceration, depression and even suicide' (Ahern, 2013, p. 23).

There is rarely any attempt in such discussions to explain how sexting is to be explained when engaged in by adults (presumably, they do not have the justification of 'raging adolescent hormones') nor how the development of a sexual identity is just that, a developmental process that will inevitably involve both good and bad decisions (where young people learn from their mistakes). Such discussions debate the sexual behaviour of youth as if youth are a group apart, with little connection to adults and what they do. Instead sexting is presented as highly risky behaviour that can only lead to even riskier behaviour and harm. In many ways, this mirrors, for example, how drug taking was perceived in earlier times, and the claim that taking 'soft' drugs will lead to the taking of 'hard' drugs. It is notable that Ahern, cited above, even claims a connection between sexting and substance abuse. It fits the classical definition of a moral panic (Cohen, 1972).

A particular feature of the 'risky practice' sexting narrative is that of case examples that serve, among other things, to create a particular 'reality' around sexting. These are often stories that illustrate the negative aspects of sexting and how it can lead to harmful outcomes. Wood provides the example of Jessica Logan, an 18-year old who while in high school sent nude images of herself to her boyfriend at his request. He sent them on to other high school girls when their relationship ended, causing her to be labelled as a 'slut and a whore'. Eventually, she suicided (Wood, 2009/2010, p. 152). Such examples are often recycled in the literature and built upon. One commentator on the case remarked, '[w]hile Jesse's story is tragic, the actions of her former boyfriend are not uncommon (Barkacs, 2010, p. 23)'.

It is of course this last sentence that demands further analysis. This slippage from the anecdotal to an assertion of the scope of the problem avoids any pretence of context or discussion of the evidence. Instead, the anecdotal becomes the reality, the 'fact' of its occurrence becomes the evidence that this is a significant social problem worthy of our attention. Thus, the narrative that defines sexting as risky practice is intertwined with and highly dependent on the notion that sexting has reached substantial, if not 'epidemic' proportions, for, if the anecdotal evidence reveals harmful events, then the 'real' fear is that such events are happening in numbers that should scare us.

Prevalence of Sexting and the Loss of Childhood Innocence

I have discussed elsewhere the studies that have attempted to identify the extent of sexting (Simpson, 2013). Given the nature of the activity, it is clearly difficult to identify the actual amount of sexting that occurs. Most of the suggested rates are extrapolations from samples to the larger population of young people, or rely on self-reporting to popular teenager publications. These include rates in one study of 22% of teen girls and 18% of teen boys sexting and 11% of young teens (aged 11–16) having sent their nude or semi-nude image (Wood, 2009/2010, p. 154). Although Wood acknowledges that 'although the study makes interesting reading and is imminently [sic] quotable by the media, it does not reflect an accurate or scientific reporting of the magnitude of the problem' (Wood, 2009/2010, p. 154). Much of the discussions of rates of sexting appear to adopt a position that it is much more prevalent than many imagine, almost as if they are engaging in 'shock and awe', despite some studies citing relatively low rates (Parliament of Australia, 2011, pp. 137–138). Of course, the problem with studies that cite low rates is that these can be dismissed on the basis that many young people will not admit to the behaviour (Parliament of Australia, 2011, p. 137; Simpson, 2013, p. 693).

It is this latter point that hints at the reason for such concern with the prevalence of sexting in the first place. Duschinksy explains why this narrative develops in the context of the childhood sexualisation debates that have occurred in the United Kingdom and globally (Duschinsky, 2012). For Duschinksy, the childhood sexualisation debate is framed in terms of a discussion about the loss of 'purity' or 'innocence' of the female child, as he explains in his analysis of the United Kingdom Sexualisation of Young People Review in 2010 (UK Home Office, 2010):

> The APA Report and the UK Review assume that the very being of teenage girls is unfamiliar with and thus pure of sexuality and desire, though the text observes that in practice it is clear that they behave otherwise. This behaviour, contrary to essence, is caused by 'the current environment', which 'encourages' girls to take on the artificial appearance of sexuality and desire. They therefore come to 'look sexy' in a way that disturbs the prior match between the behaviour of girls and how they should, in truth, 'be'. As a result, the UK Review argues, 'being "sexy" is no longer about individuality' or about girls being true to their 'authentic voice' (UK Home Office, 2010, pp. 34–58). Young people have been displaced from the natural form of 'who they are' by the intrusion of sexualisation. (Duschinsky, 2012, p. 717; UK Home Office, 2010, p. 4)

Much of this debate relies on notions that media portrayals of young people, in particular, young females, furthers this process to the detriment of their development, for example, by making them feel inadequate about their body image (Duschinsky, 2012, p. 719; UK Home Office, 2010, pp. 56–64). Much of this narrative thus relies on the acceptance of a particular set of values that frame the discussion of childhood sexuality. This framework is gendered, subjective and rejects children's agency. Instead, childhood sexuality is framed in terms of the overwhelming nature of the social forces that endanger the (female) child's purity. Girls are sexualised into objects of desire, while boys become 'hyper-masculinised' and 'sexually callous' (Duschinsky, 2012, p. 721; UK Home Office, 2010, pp. 31–33, pp. 68–69).

What is immediately striking in this analysis is the gendered nature of the childhood sexualisation debate. The danger is said to be to girls who are overwhelmed by their passive consumption and acceptance of a hyper-sexualised femininity as portrayed in the media and in advertising. They apparently consume this imagery with neither agency nor resilience. Young males on the other hand must be constrained in their sexual aggression. The manner in which this narrative can be applied to sexting practice is clear – where the young female is the passive or pressured victim of males driven by their more aggressive hyper-masculinity. Although such statements appear to be driven by value positions about the notion of childhood sexuality, what it means to be 'sexualised' and the 'proper' role of males and females, the proponents of this worldview are also driven by the righteousness of their position. The need for 'scientific' data to support their take on the process of sexualisation then becomes overwhelming:

> The UK Review therefore claims that there is a strong imperative to search 'behind' distorted appearances to the central truths, which take the form of quantifiable measures of the frequency of sexual harm: 'behind the social commentary and headlines about inappropriate clothing and games for children, there are real statistics, on teenage partner violence, sexual bullying and abuse' (UK Home Office, 2010, p. 3). As a result, the UK Review is able to state that the 'arguments presented within this document are not based on conjecture', or 'opinion', 'but on empirical data'. This objective data are, in turn, 'presented as objectively as possible so that a public debate could ensue' (UK Home Office, 2010, p. 3) The reason for this need for objectivity is that public understanding has deviated from reality, due to the distorting effects of cultural representations. (Duschinsky, 2012, p. 726)

Against all of this is the argument that what constitutes the sexualisation of childhood must always be about values and competing discourses of childhood. It is of interest that the childhood sexualisation debate is rarely framed in terms of 'appropriate' versus 'inappropriate' sexualisation, on the basis that the 'proper' sexual development of the child must be one of the responsibilities of parenting. To so frame the discussion would immediately suggest the possibility of different views on such matters and a more relativist conception of the child. This would challenge the certainty of the reality of childhood sexualisation held on to by those who articulate that stance. Instead, dominant narratives of childhood sexuality frame the discussion in terms of the 'reality' of sexual harm as opposed to the 'unreality' of those of a different approach (Duschinsky, 2012, p. 727).

Thus, the concern with the prevalence of sexting is the end product of this narrative, as once the sexualisation of children is accepted in the above terms, the incidence of the practice speaks for itself in terms of the harm to children as a consequence. What is set up is a frame of reference that accepts uncritically the consequences of such behaviour as sexting, based as it is on the immaturity of young people, combined with the special vulnerability of girls and the 'natural' aggression of boys. The need for both protection and control is established accordingly.

Law, Parental Control and Education

What flows from the above is that the response of the law to sexting will be grounded in both punitive and protectionist laws. Parents will not be entrusted to decide how they wish their child to develop sexually. Instead the expectation is that they will subscribe to the dominant narrative and ensure that their children's (meaning their daughters') purity is preserved, while young people's (meaning their sons') unrestrained sexual aggression is controlled. The sexualisation narrative also ensures that the law contains contradictory currents. On one hand, it delivers a legal response that views young people who engage in practices such as sexting not as sexual predators but as immature and overwhelmed by a culture that appears to normalise such behaviour or as the gullible victims of peer pressure. For them, education and counselling will be deemed to be the correct response, and in adopting this stance the underpinnings of the sexualisation narrative – the innocent and thus impressionable child – are at the same time reinforced.

Yet on the other hand, another current that is often observed in the law around sexting is more punitive where the sexting is regarded as malicious or harassing. In those cases, the hyper-masculinised male, in particular, is to be feared and the legal response can be much more punishment oriented. Precisely in what form that punishment is to be cast can be problematic, for while such young people may appear to be committing the same harmful acts towards other young people as other sexual predators, the sense that as children they are immature and lacking in agency suggests that they are not as culpable as adults who commit such acts. For them, the punishment is still present, but somewhat diluted. The conceptual problem is that the rationale for any punishment – agency – is potentially logically inconsistent with the rationale for regarding them as having less blame. This was a dilemma facing the Victorian Inquiry discussed below.

The dilemma flows from the underpinnings of the dominant narrative and its connection with the childhood sexualisation debates. Once it is accepted that children and young people are immature and lack agency, it is difficult to readily apportion blame for their actions. At the same time, the framework of the dominant narrative also constructs some young people – particularly males – to be feared in their predatory actions. This suggests a punitive response, or at least some form of control that will respond to that concern. This readily leads to confused and logically inconsistent stances in terms of what should be done. In particular, it often plays out in the context of a discussion around 'consensual' versus 'non-consensual' sexting, which as we will see does as much to disguise the issues as it does clarify them.

COUNTER-NARRATIVES: DIGITAL DISSENT, CHILDHOOD, AGENCY AND THE FEMALE BODY

If the dominant narrative requires the female child to be vulnerable and pure and the male child to be subject to base urges unrestrained by an undeveloped brain, then the counter narrative requires the child to be constructed as having agency and capacity. In writing about the UK Sexualisation of Young People Review Duschinsky comments that 'the UK Review devalues and ignores feminist qualitative research relevant to the topic and reaffirms dehumanising discourses about young women .... It is not that any appeal to purity is dangerous, but that such discourses make particular constructions of young women or the public sphere appear outside of history or relations of power' (Duschinsky, 2012, p. 728 citing Duschinsky, 2011a, 2011b).

What Duschinsky seems to bemoan here is the manner in which the narrative around the sexualisation of children has been constructed to inhibit a critique of the socialisation of young people – and young women in particular – that is empowering of young people and challenges dominant ideas about gender roles in society. A counter narrative speaks to the sexual expression rights of young people and articulates their power to digitally dissent and reclaim control of their bodies. The autonomy and integrity of the child rather than notions of purity and innocence, which ultimately disempower and restrict young people and young women in particular, drive this alternative narrative.

Lara Karaian has proposed such an alternative reading of sexting (Karaian, 2012). In her construction of sexting, the discourse that frame sexting as child pornography 'employ a "mechanism of censorship" that not only circumscribes "the social parameters of speakable discourse, of what will and will not be admissible in public discourse", but also reifies teenage girls as sexual objects, to be seen and not heard' (Karaian, 2012, p. 59 citing Butler, 1997, p. 132). She calls for the agency of young women to be recognised 'so that cultural and legal actors may better acknowledge the "perpetual spirals of power and pleasure" experienced by girls' (Karaian, 2012, p. 59 citing Foucault, 1978, p. 45). Karaian also employs the notion of the moral panic around sexting in order to show how it operates to protect 'respectable' (white and middle class) girls from sexual predators, through making problematic those young women who transgress what is expected of them by sexting their images (Karaian, 2012, p. 60 citing Hasinoff, 2010, p. 3). Karaian also adopts Judith Butler's notion of 'the regulation of the social domain of speakable discourse' (Butler, 1997, p. 133) that applies implicit forms of censorship [that] may be, in fact, more efficacious than explicit forms in enforcing a limit on speakability' (Butler, 1997, p. 130). In particular, this implicit censorship constructs a narrative around sexting that is constructed for young women and not by them (see Karaian, 2012, pp. 63–64 citing Bauerlein, 2008; Nabokov, 1955; Papadopoulos, 2010; Zurbriggen et al., 2007).

Importantly, Karaian advances the notion that a counter narrative 'may present an opportunity to advance new ways of knowing and being in law' (Karaian, 2012, p. 70). She cites authors who regard the dominant narrative's construction of young people as 'the ultimate objects of desire' (Adler, 2001; Kincaid, 1998), as endangering young people as it casts them in an alluring innocence. Instead, Karaian (2012, p. 70) speaks of the 'digital sexual expression rights' of youth, a language that itself suggests empowerment and autonomy.

The manner then in which the practice of sexting may be recast as digital expression of sexual identity is thus not only a challenge to dominant narratives of sexting but also about what those narratives represent – a particular view of childhood and the feminine child in particular which is essentially disempowering of children. This then is more than a claim that concern with sexting is a moral panic as that both suggests that the phenomenon is exaggerated in its extent and implicitly accepts that it may nevertheless be obviously harmful. The counter narrative claims that it may indeed be occurring but its practice is not so obviously negative.

The counter narrative is thus much more nuanced and engages with a body of literature that is rarely referred to in official inquiries or legal discussions about sexting. Richard Chalfen, for example, discusses the need to broaden the context within which sexting by young people occurs beyond notions of sex and bad judgment (Chalfen, 2009, p. 260). He argues for the need to appreciate where sexting fits into the lives of young people:

> In addition to their home, school and sometimes work cultures, other components of a contemporary teenager's world consist of media culture, techno-culture, visual culture and adolescent culture. We should examine how each subculture contributes to and enhances our understanding of where 'sexting' fits in young people's worlds. (Chalfen, 2009, p. 260)

The meeting of these cultures creates a much more complex picture of sexting. For example, living in a techno-culture makes the availability of mobile technology commonplace, and '[t]his is a world where camera phones are well embedded within everyday life and are used to constitute identity' (Chalfen, 2009, p. 260). When combined with the intense visual culture of young people's existence, where how one looks is central to identity formation sexting appears as an almost logical response to that world (Chalfen, 2009, p. 260).

Dominant narratives generally engage with the matter of childhood sexuality only obliquely. They may mention the need for 'appropriate' sexual development, but there is rarely close analysis of the actual content of sexting and how it fits with that development. Cheri Pascoe discusses how young people use social media to develop their relationships. In the context of a committed relationship sexual content may be transmitted but in a manner that counters the panic surrounding sexting in dominant narratives:

> Alice and Josh do text about sex (a practice some might consider "sexting"), but this sex takes place in the confines of a committed relationship. Both are highly concerned about safer sex practices. The background on Josh's computer even sports a large cartoon condom advocating these practices. This story and others like it provide a counterpoint to the parade of dangerous and irresponsible practices regularly featured in media releases about adolescence, sexuality, and new media use. (Pascoe, 2011, p. 6)

This is not to say that there are no risks in online interaction, but the focus on the negative aspects of sexting blocks out consideration of the broader role of new media in the shaping of young people and the manner in which such practices are actually employed by youth in their lives (Pascoe, 2011, p. 6). Pascoe articulates the manner in which social media enables youth to form their sexual identity. One important example she provides is the manner in which new media empowers LGBT youth by enabling them to connect and meet with other LGBT youth when their off-line environment does not provide those opportunities (Pascoe, 2011, p. 9 citing Hillier, 2007; Hillier & Harrison, 2007).

Ultimately, this is about young people regaining control over their sexual lives as '[n]ew media allows youth who are dating to maintain a digital co-presence, to be connected in a way that they cannot necessarily achieve in the physical world' (Pascoe, 2011, p. 9). Such perspectives emphasise the empowering role of new technologies for youth, here in a context that permits them to pursue a non-conventional sexual identity. We must always question what is meant by 'appropriate' use of new technologies by cyber-safety experts and be critical of any attempt to exclude from such 'appropriate' use youth pursing their own digital self-expression.

Deanna Holtzman and Nancy Kulish also explain sexting in a way that addresses the broader meaning it has for young people's sexual development. Rather than seeing it as something to abhor, control and warn the person against, they recount a case study that engages with the manner in which sexting might explain the struggles in young people's lives:

> 'Sexting' or 'sex-texting' has become very [sic] common societal phenomenon with today's adolescents (CBS News, 2009; Goodstein, 2007; Morris, 2011). This phenomenon exemplifies how cultural factors influence the 'normality' of exhibitionism so that the parameters of permissible behavior vary by time and place. Nevertheless, female genital exhibitionism in sexting seems to be less an expression of penis envy rather than a way to appeal to peers and to be seen as 'hot' and sexy. Whether or not such a widespread phenomenon could be considered perverse is an open question. More detailed psychoanalytic data from such cases than we have available to us at this point might help to answer this question. In a recent article, (Lemma, 2010) observed that the use of cyberspace can become a psychic refuse for adolescents struggling with the challenges of integrating the reality and meaning of their changing sexual bodies into their images of themselves. (Holtzman, 2012, p. 279)

It is difficult for many in society to engage with the sexual in young people and as a result there are often many gaps in our understanding of young people's sexual behaviour. This readily results in casting the child who engages in behaviour that appears to be sexualised as a problem and who is to be labelled as unclean or damaged, when the behaviour may also be seen as an expression of their sexual desire (Holtzman, 2012, p. 289).

The counter narrative constructs children's online engagement in quite a different manner to the dominant narrative of sexting. In that narrative, children are out of control, unthinking, and even defined as the 'victims' of their own bodies caused by slow brain development. The counter narrative suggests young people may exercise agency and make choices – perhaps not always wisely – but their choices nevertheless. Importantly, the counter narrative may sexualise the child, but in much more nuanced and holistic way. This is not in any way to advocate en masse sexting by young people. The fear of those who support the counter narrative is that they will be labelled as supportive of child abuse, as that is how the dominant narrative ultimately constructs sexting, as part of a continuum that leads to sexual abuse of children.

The counter narrative must refute the dominant narrative on the basis that it denies young people their human right to develop sexually, that it constructs as a legal problem something that is inherently a child development issue and does so in a way that perpetuates a romantic view of children as innocent. In doing so, the dominant narrative endangers children by constraining them as innocent beings, a state which is attractive to paedophiles (Kincaid, 1998). It also prevents a focus on the sexual development of the child in a way that recognises their individuality and autonomy, which as a consequence will result in harm to the child. In short, young people cannot simply be viewed as impressionable and immature, they can and do make decisions about how they use social media that are often savvy (Boyd, 2014).

VICTORIAN REFORM

In September 2011, the Law Reform Committee of the Victorian Parliament (Victoria is the second largest state in Australia) was asked to inquire into sexting. Prior to this, there had been some media reports outlining how young people had been charged with child pornography offences because they had in their possession or had sent nude images of themselves to their friends. As a result they had been registered as sex offenders. This led to concern that the law was overreaching and in effect punishing young people with a law that was designed to protect them (Law Reform Committee, Parliament of Victoria, 2013, p. 4). The Committee was asked to inquire into 'the incidence, prevalence and nature of sexting in Victoria;' 'the extent and effectiveness of existing awareness and education about the social and legal effect and ramifications of sexting;' and the laws that addressed sexting in cases where the image is created with the consent of the person as well as in cases where the image is transmitted to others without the person's consent (Law Reform Committee, Victorian Parliament, 2011).

The terms of reference of the Parliamentary Committee thus followed a framework consistent in its focus with the dominant narrative described above. The Committee received submissions from various community groups, professional bodies and individuals. In the result, these submissions framed the narrative of the Committee's final report. Thus, an examination of those submissions is important to undertake to determine the extent to which they reinforced or varied from the competing narratives discussed above.

SUBMISSIONS TO THE INQUIRY: CONSTRUCTING SEXTING YOUTH

As might be expected, the submissions received by the Committee emphasised many of the characteristics that frame discussions of sexting in the literature, and which also responded to the manner in which the Inquiry's terms of reference framed the discussion. That is not to say that such a framework is either logical or coherent. Devoid of a strong position on the agency of young people, such a framework often becomes confused over the extent to which young people are capable of understanding and making choices and the extent to which they are immature and impressionable vessels controlled by prevailing cultural expectations and technology.

Thus, young people were often described as unthinking and naïve in submissions to the Inquiry and for that reason it was felt by some that the law should not label them as pornographers when they sext:

> Adolescents may give little thought to the consequences of creating and sharing a naked picture of themselves with another person using this technology. The consequences include the ease with which such an image may be widely distributed to others and the permanence of the image which once available widely cannot be recalled and deleted.
>
> The law has a role to play in protecting minors from the harms which are associated with sexting.
>
> However, the law should operate flexibly to distinguish various offences according to their seriousness and not to disproportionately punish minor offences nor provide inappropriate defences for serious offences. (Family Voice Submission, 2012)

This approach to sexting creates an immediate difficulty. Adopting a view of young people as lacking in understanding when they sext leads to the contradiction that suggests that on the one hand, they need protection rather than punishment, yet the inappropriate nature of the behaviour supports some punishment in less serious cases and an increased level of punishment in 'more serious' cases. The obvious criticism here is that if young people act without thinking, then on what basis is their behaviour to be criminalised? If we accept the view of young people proposed here for the moment, then this repeats the failings of more punitive approaches to juvenile crime where a focus on the act rather than the needs of young people leads to increased criminalisation of their behaviour. In simple terms, this standpoint wants it both ways.

A lack of understanding on the part of young people informed other submissions too. In some, this was thought to combine with the manner in which 'technological determinism' also plays a part in determining the behaviour of young people

> A national survey in 2010 revealed 59% of teenagers have sent sexually suggestive emails or messages … .This confirms our belief that sexting is becoming more prevalent amongst our youth. It also supports our argument that education in Australia is inadequate and not accurately combating this growing issue. Advances in technology are making this form of communication more accessible and the use of such technology is creating an ideology that this behaviour is acceptable. It is therefore our aim to provide information to the public to both assist and educate them on sexting and the legal and social ramifications involved. (Salvation Army Oasis Hunter Submission, 2012)

> Our discussion will focus on consequences associated with the current lack of education and awareness of sexting. It is our experience that children, teenagers and parents do not understand the dangerous and serious nature of sexting and they are not aware of the repercussions associated with the act. (Salvation Army Oasis Hunter Submission, 2012)

This submission also emphasised the need for parents to warn children about the 'dangers of sexting', and 'to think before they act', the need for parents to monitor what their children are doing online, 'warn them about sexual predators', establish 'clear rules on what they can and can't do with their mobile phone' as well as a broader 'increase in education for young people in terms of the social ramifications of sexting' (Salvation Army Oasis Hunter Submission, 2012).

One of the most problematic discussions for those who frame the sexting debate in terms of young people's incapacities is in relation to 'consensual' and 'non-consensual' sexting. To accept the notion that a young person engages in consensual sexting requires them to be accorded some agency, yet the harmful effects of sexting generally suggest that this is not something young people should be consenting to in any event. This leads to some interesting commentary:

> It seems to me that, notwithstanding the notion of non-consent, where the young people involved in creating and sharing the images do so willingly and knowingly, this should not be seen as equivalent to involvement in the production, possession or distribution of child pornography, and that no crime should be deemed to have been committed.
>
> However, this is not to say that the activity is not without risk. It is in fact a high-risk activity, putting both parties at risk of future harm. This risk comes, of course, from the fact that consensual sexting can readily transform into what might be called 'violent sexting'. (Kane Submission, 2012)

What seems to be occurring here is the very phenomenon that advocates for children's autonomy are criticised for doing, that is, the imputing of an adult frame of reference to the actions of a child. The tension this creates is between the notion that the intent underpinning an act should be considered, but yet the harmful nature of the act – in the case of young people – justifies some form of intervention to prevent it. Other submissions perpetuate this confusion:

> MRLSC urges the Inquiry to make the following important distinction with regard to the nature of 'sexting':
>
> - The act of 'sexting' (at the points of creation, sharing, sending, posting) between adolescents can be an act that is consensual, and at any point the action may have an intent that is not offensive, malicious, menacing, abusive, or exploitative.
>
> - The act of 'sexting' (at the points of creation, sharing, sending, posting) between adolescents can be an act that is not consensual, and can have an intent that could be construed as either/and/or: offensive, malicious, menacing, abusive, or exploitative.
>
> This helps to clarify that the problem with 'sexting' is not the act itself, but the unconsenting act. (Macedon Ranges Local Safety Committee Submission, 2012a, 2012b)
>
> We understand that the distinction between personal enjoyment and exploitation is blurred when dealing with young people and sexting however we do believe the law should maintain a distinction and appropriately punish those who do deliberately exploit children, even if those people are children themselves. (Gippsland Community Legal Service Submission, 2012)

Another submission argued for a two-year age difference to be applied where the parties were under 16 in order to ensure that any consent was not obtained by duress or imposition on the part of the older person. Yet they also submitted:

> that the Law Reform Committee considers a minimum age where a child cannot legally give consent to send or receive a sext message or image to another person, regardless of

the age of the other party/parties involved'. (Family Planning Victoria Submission, 2012)

Yet other submissions suggested that the consensual nature of the act is only a starting point for concern. In those submissions, consensual sexting is identified as inappropriate and in need of a response, even if it is to require parents to control their children and take responsibility for the acts that their children lack the capacity to take responsibility for, as well as the manner in which adults generally provide poor role models for impressionable young people:

> Unfortunately, the primary protectors of children – parents – have also been slow in understanding the risks posed by the abuse of new technologies and, despite years of clear messages being transmitted on such issues, still for example do not by and large supervise their children's use of the Internet or of mobile phones/tablets ….
>
> In this regard, it would be a considerable incentive to parents to take their responsibilities seriously if they were also held liable for violent sexting perpetrated by minor children. I am not sure how this would be written into law (nor whether it is even possible in Australia), however it has been shown to be effective in some countries of northern Europe. (Kane Submission, 2012)
>
> The community as a whole must take some responsibility and blame for the actions of today's youth. The early sexualisation of young people is promoted everywhere and in most cases, popular culture simply endorses it. Many of the females idolised by today's children exude sex and sexuality at every opportunity. Impressionable and vulnerable young people are 'brainwashed' into believing that they need to copy the actions of these 'role models' without giving any thought to the possible consequences.
>
> Parents too must accept some responsibility for this phenomenon as often, young people are handed the technology with little or no guidance or supervision. (Susan McLean Submission, 2012)

In these submissions, young people are accorded little capacity to make their own judgments about which role models to follow or how to conduct themselves. They are constructed as impressionable beings, victims of the various forces around them. Other submissions even contended that this has a physiological basis:

> Dr. Deborah Yurgelun-Todd of Harvard Medical School has studied the relation between these new findings and teen behaviour and concluded that adolescents often rely on emotional parts of the brain, rather than the frontal lobe …
>
> This discovery gives us a new understanding into juvenile delinquency. The frontal lobe is "involved in behavioral facets germane to many aspects of criminal culpability," explains Dr.Ruben C. Gur, neuropsychologist and Director of the Brain Behavior Laboratory at the University of Pennsylvania …
>
> It is clear from this new understanding of adolescent brain development that children, adolescents and the developmentally impaired need extra protection from the dangers present in ICT. (Susan McLean Submission, 2012)

This seems akin to the resuscitation of an almost Lombrosian view of how criminals are to be identified. In arguing for brain development as a cause of bad judgment, this ignores the extent to which values determine what is illegal, immoral or inappropriate. Indeed, the very purpose of the Inquiry being conducted was to recommend whether current laws were appropriate. The only way in which this approach has coherence is to accept that the conduct is unarguably criminal or inappropriate and the only matter to discuss is how the law should sanction the behaviour. This explains how it can be said that such theories provide 'a new understanding into juvenile delinquency' as the framework accepts such behaviour as inherently wrong. Yet, the history of juvenile justice would suggest otherwise as we have defined and redefined what is criminal. Of course, another flaw in this explanation is that if immature brain development explains the sexting of young people, then what explains sexting by adults?

Yet, even the above submission goes on to acknowledge that in some way a young person may have agency in such matters, relegating the overwhelming nature of brain development to a mere lapse:

> Whilst I would have no hesitation in applying the law as it stands in certain circumstances, we must ask if it is a suitable punishment for a teen, guilty of no more than a 'brain fade' moment, or one who has willingly sent an image consensually to his/her partner'. (Susan McLean Submission, 2012)

The other conflicting aspect of the framework constructed by the dominant narrative of sexting is the slippage between the young person as unthinking and the young person as having the capacity to make decisions, albeit bad ones:

> The developmental immaturity of some young people means they may engage in transmitting images of themselves of a sexual nature without due consideration about how widely that material may be disseminated. In other circumstances images of a sexual nature may be transmitted specifically to cause distress. (Royal Australian and New Zealand College of Psychiatrists Submission, 2012)

The idea that children are unthoughtful but capable of being educated about how to choose appropriate pathways also suffers from the dilemma that the child is then to be regarded as both immature and undeveloped yet able – with guidance – to have agency for their actions:

> The challenge for anybody undertaking awareness and education work about sexting is that unfortunately for some young people, capturing the image, and pressing the 'send' button may at first seem like a bit of fun, but is often carried out with the absence of forethought of the consequences of their actions. The realisation that once an image enters cyber space and it cannot be deleted, instead it is likely to be further transmitted locally, nationally and even across the world, is not something some young people comprehend. Young people need to ask themselves the question: "If you would not be

> comfortable showing certain photos at the dinner table or to your grandparents, why would you be comfortable sharing them online, sometimes with people you don't even know?". (Australian Federal Police Submission, 2012)

In effect this sets up the dichotomy of the wild and untamed child versus the thoughtful and educated one. In such a dichotomy, the untamed child will often be seen as even more deviant and worthy of control. It can also be seen how this fits with the gendered approach to sexting – where the innocent and pure female child is posited against the cunning and aggressive male child. The regulation of sexting, according to this view, is thus about defending the purity of the female against the exuberant, hormone charged and almost uncontrollable male. Equally, this excludes the sexually competent female who may be cast as the impure seductress. In the result, the legal responses will shift constantly between punishment and protection, yet both act as control devices.

Sexting was portrayed as part of a sexualised culture in other submissions:

> We are experiencing what has been described as a "sexualisation of culture" with cultural materials such as advertising and music clips becoming increasingly sexualised. What defines this sexualisation is that it is heterosexual and gendered in nature and predominantly presents young women in sexual ways, as objects to be consumed and expectations that young men will be sexually interested in this representation. We are not seeing an increase in sexual imagery where sexual diversity is represented or where women and men are represented as sexual equals and where intimacy is shared rather women are passive objects to be consumed and men are the active consumers. (Women's Health Grampians Submission, 2012)

In such cases, sexting is also seen as part of a broader social malaise:

> To continue to address the issue of sexting outside of a gendered framework is to ignore the gender inequalities and power imbalances that it represents. To do this reinforces the continued sexualisation of young women: as objects of sexual desire to be consumed by men and provided for their sexual pleasure. It also reinforces a social norm that young men are to seek out and consume sexual images of young women. (Women's Health Grampians Submission, 2012)

But this construction of sexting itself relies on the passive female absorbing these various messages about the sexualised role of women, and equally for males to conform to the messages they receive. As with all such explanations they struggle with notions of agency and resilience and, as Duschinksy observes, in fact adopt a restricted feminist analysis that denies the possibility of a more radical and critical feminist critique.

The difficulty with the sexualisation discourse is that it creates a tension with the place of consent in these narratives. Consent ultimately relies on the agency of a person, while the sexualisation narrative rests on the extent to which gender is performed and absorbed by social pressures – in effect creating a form of false consciousness. This results in much confusion as some struggle with the attraction of the sexualisation narrative for what it provides in terms of explaining behaviour that is apparently sexualised, while also desiring acknowledgement of young people's right to control their bodies:

> It is well established that gender norms have a powerful influence on people's sexual identity, practices and behaviour, and the way in which they enact their sexuality. Gender-based power inequities between men and women serve to create and maintain social pressures and constraints through which women, particularly young women, negotiate sexual encounters.
>
> Such social pressures and constraints placed on young women have a direct impact on their ability to negotiate and make decisions about safe sexual practices, including their participation in sexting behaviour (Rosenthal, 2005). Consequently, among young people, it is increasingly expected that young women will engage in sexting as a part of what is deemed 'normal' sexual behaviour and relationships (Powell, 2010) ….
>
> Research suggests that coercion and pressure placed on young women is a significant contributor to their participation in sexting behaviour. (Ringrose et al., 2012; Women's Health West Submission, 2012)

The difficulty here is in understanding whether this suggests that 'coercion and pressure' is of a kind that vitiates consent, or whether it is that form of socialisation that we all may be subject to, that questions the extent to which any of us truly has 'free will', but nevertheless cannot be used to avoid responsibility. To this extent, sexting narratives are often based on confused and contradictory ideas about human behaviour.

This is also evident in one piece of oral evidence provided to the Inquiry where in response to a question on whether a young person under 16, who cannot legally consent to sex should also therefore be prevented from consenting to sexting of their image, a cyber-safety consultant replied:

> Yes. I think that would be a really good way of starting it in relation to it providing consistency so people do not have to get their head around one bit of the law and another bit of the law; that is the same. Whilst some people will argue that kids are sexual beings and this is just what they do and it is part of normality, adults have a requirement to keep kids safe. It is not a good career move by anyone, regardless of age, to have naked pictures of themselves floating around cyberspace. Certainly at least under the age of 16 if you make no consent, it makes it very consistent with the age-of-consent laws, and I think that is very sensible, because then you are protecting vulnerable people. You are going to have to look at what is consent, so you would have to have the same things. I would go exactly the same as when is consent not

> consent — coercion, fear, threat, being under the effects of drugs or alcohol and all those sorts of things. That would need to be there as well, because a lot of girls will say, 'But I said yes', but when you dig deeper, they only said yes because they were tricked, were fearful or were threatened, so that of course cannot be consent. I think you have got a really good basis to start with by looking at those and then aligning this to them'. (Susan McLean Oral evidence, 2012)

There is no mention here of the capacity of the child to make decisions about their own body nor the relevance of the maturity of the child. There is also again the emphasis of the wisdom of 'anyone, regardless of age, to have naked pictures of themselves floating around cyberspace', which suggests that the real concern here is indeed not the consensual or otherwise nature of the act, but the simple 'inappropriateness' of the act itself. That girls do so because they may be tricked or drugged is again casting young women as lacking in agency.

DIGITAL DISSENT AND THE CONTROL OF THE CHILD'S BODY

While most of the submissions to the Inquiry adopted the dominant narrative surrounding sexting, there were aspects of the counter narrative also presented in other submissions. Some connected sexting with young people exploring their sexual identity and articulated the need to include their voice in debates:

> Headspace believes that understanding sexting from the perspective of young people is fundamental to the development of strategies for preventing potential harms. Important to this debate/inquiry are the views and lived experiences of young people as these policies and laws impact on their current rights and their future. Involving young people in decisions ensures appropriate design and delivery of laws and improves young people's experience of public life. In addition, the active and informed participation of young people is likely to result in laws and policies that are more effective and relevant to their needs. This will result in better outcomes for young Australians. (Headspace Submission, 2012)

The Victorian Child Safety Commissioner pointed out the inconsistencies in approaches to sexting by young people, albeit still adopting the sexualisation narrative:

> ... sexting is not an activity limited to young people; there have been many examples of high profile adults who have engaged in sexting. In addition, children and young people

> in Australia live in a society in which they are regularly exposed to sexualised images, including sexualised images of young people ….
>
> As a community we are giving our young people quite mixed messages. On the one hand, we are permitting adults to engage in sexting and in the production and dissemination of sexualised images in a wide range of media and advertising aimed at, or available to, young people. Whereas on the other hand, young people risk significant criminal penalties if they themselves produce or disseminate sexualised pictures of themselves. (Child Safety Commissioner (Victoria) submission, 2012)

The Commissioner also submitted that there was a need to empower youth in these discussions and to not see the law as holding all the answers (Child Safety Commissioner (Victoria) submission, 2012).

The problem is that while the rhetoric of empowerment seems to rest on the capacity of young people to claim their rights, the recourse to the sexualisation discourse here, with its connotations of denying agency and young people as vulnerable to socialisation, undermines this. One must ask the extent to which the idea of the empowered child is actually accepted. Such tensions are evident in the following submission that also expressed the notion of the child's right to sexual expression while also being concerned about protection of the child:

> The practice of sexting requires legislators to balance right to privacy, sexual self expression and sexual assertiveness with the need to protect children. CLA's view is that the existing law is not appropriate, nor is it adequate as it conflates consensual sexting between partners with the malicious distribution of sexts …
>
> However, we submit that young people, who may be exploring their sexuality in a risky manner by sexting, should be distinguished in law from adult producers and distributors of child pornography. The law should also distinguish between the culpability of a person who receives a sext and maintains the confidentiality of the sender by comparison with a person who distributes the sext to a third party without consent. (Civil Liberties Australia submission, 2012)

As the example from the oral hearing above demonstrates that such statements about 'sexual self-expression' are quickly put down by the advocates of the dominant narrative on the basis that those views of young people fail to keep them 'safe'. The consequence of this is that even such statements about the autonomy rights of young people remain guarded as they still refer to sexting as 'risky'. Part of the reason for this confused approach to the area seems to be due to reluctance to impart agency to young people for fear this may suggest that sexting may actually involve legitimate activity in terms of identity exploration. Thus, even concepts that are deemed worthy for the developmental benefits they provide, such as play – now clearly a human right for children (United Nations, n.d.) – are used disparagingly as if in doing so emphasises, the risky and therefore illegitimate nature of such activity:

> Teachers and legal advocates for children and young people note that within school cultures practices of sexting are sometimes part of games people play – for instance, photos of body parts are circulated and young people try and guess to whom the image belongs. The popularity of such games, and the pressure on young people to participate, suggests that they are often oblivious to the legal consequences of such sexual games, or they are unable or unwilling to resist peer pressure to participate. (Dobson, Rasmussen, & Tyson submission, 2012)

This is then compounded by the murky nature of what constitutes consensual activity in this area where young people are said to be subject to irresistible forces:

> The difficulties of distinguishing between wanted and unwanted sexual communication and image distribution is compounded by the fact that 'sexting' is a practice that is linked to mediated social networks that are permeable, unpredictable and viral. So the locations in which young people are engaging with one another sexually easily move beyond their control. In this social and technological context, the difficulties in apprehending 'consensual' and 'non-consensual' participation are great and the ability for context-based police discretion in responding to incidences of sexting is crucial. (Dobson et al., 2012)

One submission did challenge many of the positions of the dominant narrative with respect to notions of agency and consent:

> Young people themselves are often the creators of such images, their private sensual motivations being quite different to those who set out to create pornography or for reasons which may include coercion, harassment or cyberbullying. However, young people who commit what some see as the 'deplorable judgement' (Kreimer, 2011, p. 407) of sexting are sometimes themselves accused of producing pornographic images (Leary, 2008) certainly in the eye of the beholders, with a resultant 'moral panic' (Kreimer, 2011, p. 406) in various circles. However, in the view of other commentators, it is a victimless crime – if crime it be ...
>
> It is easy to think of sexting as aberrant, even abhorrent and commentators from a number of discourses depict it in this way. However, commentators closer to the age of the doers tend to view this behaviour as more benign and part of an image-sharing culture in a sexually permissive society, one in which young people see sexualised images virtually everywhere they look. It is not surprising that they create their own sexual imagery, and perhaps unrealistic to expect that they live up to a higher standard than we set for the rest of society. The question arises whether a sexual image taken by a young person of him or herself can be considered pornographic. Public debate continues over whether the young people who create such images are victims or perpetrators. (The Alannah and Madeline Foundation submission, 2012)

This submission touches upon the question of whether sexting is even a proper subject of concern for the law. Whether or not sexting is inappropriate behaviour may be debated, but should the criminal law be applied at all to it? Of course, the answer to this question may lie in an understanding of what the perceived harm is when sexting occurs. The dominant narrative might indicate that the control of the immoral or wild child and the attack on the notion of childhood innocence that adults hold close may explain more about why the heavy handed application of law takes place than any logical consideration of harm to an individual. In many ways young people who sext are almost beyond redemption to some, so perhaps the aim of the law is to prevent the loss of those young people who are not yet corrupted.

When such strands of the narrative are considered, statements about the sexual rights of young people seem marginal comments at best. For example, a submission from the New South Wales Legal Aid Office stressed the autonomy rights of young people in relation to sexting:

> There are a number of possible approaches to sexting. Obviously, the first is that any law regulating or criminalising it is a breach of children's rights to personal autonomy and sexual exploration. This approach has great merits, protecting the rights of young persons and allowing the law to concentrate on 'genuine' cases of child pornography. However of course, this approach ignores the fact that the creation, possession and transmission may in some instances warrant the involvement of the law. (Legal Aid NSW submission, 2012)

The submission then went on to state that while some sexting may be exploitative 'some sexting behaviour is just a consensual expression of sexuality. To allow sexting behaviour of the second kind for consenting adults but deny it to children may be viewed as paternalistic' (Legal Aid NSW submission, 2012).

Yet as we have often seen, when such statements acknowledge that the practice of sexting connects with expressions of identity there is still an undercurrent of suspicion about its real value:

> MRLSC acknowledges that young people may engage in 'sexting' at multiple and diverse levels. As with other sexual behaviours, there may be aspects of 'sexting' that explore and express self-identity and relationships in a way that has reasonable intent and consent. MRLSC also acknowledges that as with other sexual behaviours, there are associated risks. The nuances of exploration and expression through technology appear to amplify particular risks. (Macedon Ranges Local Safety Committee submission, 2012a, 2012b)

Another submission, cited work (Zhang, 2010) that 'argued that sexting is ordinarily a contemporary manifestation of youthful sexual exploration and experimentation' and that 'the motivations for sexting are likely to be found in romance and socialisation' (Just Leadership Program Submission, 2012). The problem is that the narrative that places young people in the same sentence as romance struggles to find any foothold in the law.

THE REPORT OF THE VICTORIAN INQUIRY

While the Law Reform Committee heard some perspectives that had a basis in the agency of young people, the overall weight of the submissions unsurprisingly adhered to the dominant narrative of sexting which depicts young people as immature, impressionable and lacking in agency. This led to a conclusion that sexting had to be understood as an activity that resulted from a lack of judgment on the part of young people and which was in effect a corruption of childhood:

> Through the Inquiry, the Committee heard that while the range of circumstances in which sexting occurred was diverse, most cases of sexting by young people involved a typical set of circumstances. Mr Bill Byrd, Safe Schools Administrator for the Toronto District School Board, told the Committee that sexting typically occurs when small groups of same-sex (usually upper-middle class) younger children meet. Often sexting occurs when that group of children get bored, and one of the children introduces something different (often alcohol or marijuana) to the group, and an additional motivator is introduced – for example, a boy may send a text to them (or vice versa) asking them to "show us something". While children are often coerced into sexting, at the time it simply "looks like fun" to them. Mr Byrd told the Committee that often the exhibitionist of the group, or the most reclusive person, is co-opted into sexting – and invariably once they have done it they immediately recognise that they have made a mistake. (Law Reform Committee, Victorian Parliament, 2013, p. 21)

The report then acknowledged the role sexting plays in the identity formation of young people, but then placed that in the context of young people's immaturity and lesser brain development:

> In the Committee's view, when considering sexting it is important to bear in mind that adolescence is a period of development and is typically a time of experimenting and risk-taking, when young people explore their identity. A number of submissions made the point that significant changes occur in the brain during teenage years, and adolescents are less able than adults to make informed decisions about personal safety and security:
>
> The evidence now is strong that the brain does not cease to mature until the early 20s in those relevant parts that govern impulsivity, judgment, planning for the future, foresight of consequences, and other characteristics that make people morally culpable … Indeed, age 21 or 22 would be closer to the 'biological' age of maturity. (Law Reform Committee, Victorian Parliament, 2013, p. 49)

This led to the conclusion that such immaturity undermined the 'natural' goodness of the young people involved and led them to take risks based on poor decision-making, their lesser developed brains and short-term focus (Law Reform Committee, Victorian Parliament, 2013, p. 40). The Committee noted one submission that stressed 'that often children that participate in sexting do so in the heat of the moment – they are 'good' kids doing things out of character' (Law Reform Committee, Victorian Parliament, 2013, p. 40).

The Report of the Inquiry also connected sexting to the sexualisation of society and protection of the purity of young women's bodies:

> The Committee heard our society is currently experiencing a "sexualisation of culture" that is heterosexual and gendered in nature ....
>
> This is of note and is indicative of the gendered nature of the problem and how it is reflective of broader social norms and stereotypes. Sexting is gendered in nature; it is the utilisation of a digital forum to reinforce already existing gender stereotypes and power relationships. This is the representation of women as sexual objects to be consumed by men and men as consumers who seek out sexually explicit images of women. (Law Reform Committee, Victorian Parliament, 2013, p. 41)

The denial of young women's agency was reinforced not only in this framework but also in its supposed outcome:

> Teen girls are often pressured by teen boys to send or post sexual images of themselves, yet girls who do so are often perceived by teens of both sexes as promiscuous, as stupid, and as lacking the appropriate degree of self-respect or self-esteem. (Law Reform Committee, Victorian Parliament, 2013, p. 41)

CONSENT AND SEXTING

The sexualisation narrative that became embedded in the whole analysis of sexting by the Inquiry is ultimately the basis upon which the agency of young people is denied. As might be expected the Report discussed the distinction between consensual and non-consensual sexting as this underpins much of the discussion in the literature for it represents the core difference between the exploitative use of another person's body image and those cases where it is done with the consent of the person. The difficulty in the case of young people is that there is in the community angst around the notion of young people consenting to behaviour that is regarded as sexually inappropriate or harmful in their eyes – and it is the sexualisation narrative that 'proves' the harmful nature of sexting. This narrative renders the consensual nature of the act in effect irrelevant:

> Some witnesses suggested that there are good reasons for discouraging children and young people from engaging in sexting, even when they consent to it, as harm can result even where sexting is done voluntarily and kept private ....
>
> As discussed above, sexting also contributes to the highly sexualised culture in which children and young people live, which can have an adverse impact on their development and wellbeing.
>
> ... when sexting reinforces gender stereotypes, unequal gender power relations and coercion, the practice is harmful and a form of violence against women.
>
> Free will and consent can also become blurred when gender stereotypes exist that support and encourage young women to measure their value and worth on their sexual attractiveness and availability:
>
> It is only by advocating for and providing alternative examples of women's sexuality that we can ensure young women are able to make decisions in their best interests and not those overly influenced by negative gender stereotypes. (Law Reform Committee, Victorian Parliament, 2013, p. 45)

Certainly, recourse to notions of the 'best interests' suggests that this is not an area for young people to exercise agency but for them to conform to adult expectations about the proper role of the child. In the end, reliance on claiming that young people need to be guided towards 'appropriate' use of technology seems more of an exhausted concession than a strong view of what that might mean:

> The ubiquitous nature of these technologies means that suggestions for protecting children from abusing them, or being abused through them, by removing the technologies, or monitoring use, will become increasingly futile. The best means by which young people can be protected from the abuse of technology is to provide an education that allows them to discern appropriate use of those technologies, and to identify appropriate use by others. (Law Reform Committee, Victorian Parliament, 2013, p. 69)

Of course, what is deemed 'appropriate' can only be understood by reference to one of the narratives of sexting. The problem lies perhaps in the disconnection between the one that many adults lean towards and the narrative that many young people actively live by.

INQUIRY RECOMMENDATIONS

For the purpose of this paper, the relevant recommendations of the Inquiry came down to two. The first was that children and young people should not be prosecuted for child pornography offences when they sext to their peers (Law Reform Committee, Victorian Parliament, 2013, p. 141). This was subject to some provisos: '(a) that the film or photograph depicts only the accused person; or (b) that, at the time of making, taking or being given the film or photograph, the accused was not more than two years older than the minor was or appeared to be; and (i) The film or photograph depicts the accused person engaged in lawful sexual activity; (ii) The film or photograph depicts the accused person and another person or persons with whom the accused could engage in lawful sexual activity; or (iii) The film or photograph depicts a person with whom the accused could engage in lawful sexual activity, or more than one person, all of whom the accused could engage in lawful sexual activity with' (Law Reform Committee, Victorian Parliament, 2013, pp. 142–143). At the core of this recommendation are notions of consensual activity. In relation to the relative ages of the young people involved is the view that a larger age difference might imply coercion of the older person over the younger (Law Reform Committee, Parliament of Victoria, 2013, p. 141). The notion of the young people being otherwise engaged in lawful sexual activity is also connected to consensual behaviour as part of what makes sexual activity between two people lawful is that it is by consent of both, including in Victoria between two children over the age of 12 where the other child is not more than two years older than the other (Crimes Act, 1958 (Vic), s.45(4)). This recommendation was in effect subsequently enacted by the Victorian Parliament (Crimes Act, 1958 (Vic), s.70AAA).

However, it was in relation to the creation of a new non-consensual sexting offence that the denial of the capacity of young people to consent at all occurred though not necessarily at the Inquiry stage. The rationale for a new offence of sexting was based on the view that 'the Committee believes it is likely that some people in the community are using sexting to harm others, and the Committee believes that where that harm is significant, it is appropriate that those people be prosecuted ...[and] the Committee believes that a specific offence for sexting, that is seen by the community to be prosecuted, will assist to inform the community on expectations around the appropriate use of communications technologies' (Law Reform Committee, Victorian Parliament, 2013, p. 149). The Committee recommended the creation of a new offence of sexting of an intimate image of the other, and specified that it would be a defence to such an offence that the person depicted in the image consented to its distribution (Law Reform Committee, Victorian Parliament, 2013, p. 152). The Report made no reference to the age of the persons involved, even though the discussion in the Report may have provided the justification for differential treatment based on age.

The actual offence of non-consensual sexting passed by the Parliament did differentiate by reference to the age of the persons. While an adult may claim that the sending of a sext was consented to by the person depicted in the image, the consent of a person under 18 to a sext of themselves is irrelevant (Summary Offences Act, 1966, s.41DA). The penalty is a maximum period of two years imprisonment. However, in deciding whether the offence is committed in addition to the distribution of the image, there must also be consideration of whether 'the distribution of the image is contrary to community standards of acceptable conduct' (Summary Offences Act, 1966, s.41DA(1)(b)). These are further explained in the legislation as requiring regard to be had to: the nature and content of the image; the circumstances in which the image was captured; the circumstances in which the image was distributed; the age, intellectual capacity, vulnerability or other relevant circumstances of a person depicted in the image; and the degree to which the distribution of the image affects the privacy of a person depicted in the image (Summary Offences Act, 1966, s.40). Thus, the age of the person sending the sext and the circumstances may be considered but not in a way that precludes the offence due to the consent of the young person depicted in the sext. Clearly, as the submissions and report of the Inquiry would overall suggest the need to control and correct the immature and harmful behaviour that is sexting is unlikely to see community standards read in such a way as to recognise the importance of young people having agency. Instead, it is more likely that the sexualisation narrative will influence what is acceptable and so demonise both the level of harm done to childhood and the person who distributes the image. What is also of note is that after the Inquiry spent most of its time taking evidence about the manner in which young people engage with new media and technology on the basis that sexting was overwhelmingly participated in by young people, it was adults who gained the benefit of the defence of consensual sexting.

LAW AND THE CREATION OF THE DIRTY CHILD

We live in a time when the sexualisation of childhood is being played out in varied and disturbing ways. The challenge is to confront the common narratives in each of them to better understand how children are better protected from harm. As the above discussion indicates, much of the concern surrounding sexting proceeds from the corruption of childhood innocence that it is said to encourage. That some would suggest that sexting may actually demonstrate young people's capacity to express their own sexual identity is advanced by others as just more evidence of that corruption. Rather, if anything, they would suggest that the sender of the sext is part of the problem, the 'dirty child' that corrupts others. In debates about sexting those children are often the young males and the 'fallen' are the young females seduced into having their images collected and sent.

Yet notions of the innocent child play out in other areas of current concern about the treatment of children. In relation to the institutional abuse of children in religious institutions, Harry Ferguson has described how children in institutional care were treated harshly because they were seen as a threat (Ferguson, 2007).

> Such children were socially constructed as a grotesque 'other', seen as a polluting presence and in need of retraining and reforming. (Ferguson, 2007, p. 133)

In this view, the child has to be saved to ensure its ongoing innocence, but this may require brutal treatment:

> Re-moralising children meant returning them to 'innocence' by knocking the devil out of them. Thus what children were supposed to be helped to forget created a rationale for their systematic abuse. (Ferguson, 2007, p. 134)

In other words, the child who had lost its innocence was deemed to be the appropriate recipient of harsh and brutal treatment. When one considers that many of the children who were so treated complained at the time but were not listened to, one can see the importance of the recognition of capacity and agency as an empowering tool. But such empowered children are also a threat, and it is time that we confronted the manner in which children denied agency become more vulnerable to the misuse of adult power. We may also argue over the nature of the harsh treatment that the child who is said to lapse receives – after all treating children who sext as child pornographers is now regarded as a step too far in many jurisdictions – but changing the manner of punishment is barely reform if we continue to deny the agency of young people in relation to their bodies which sets them up as subjects of concern in the first place when they seek to claim that agency in practice.

The discourse on sexting then might be recast as a struggle between competing views of childhood. On the one hand is the vulnerable child, more usually female who must be protected from the demonic child, more usually male. Although, even here some young females will possibly fall into the shameful behaviour and although seduced into it, they will be fallen nevertheless. In this view of children and young people, the battle is over innocence and purity, and the tools are protective programs to guide the still innocent into appropriate behaviour and punishments for the transgressors. The competing view is that of young people who claim agency, and particularly young women, confident in expressing their sexuality. Such young people are a threat to the authority of adults, but also to the romantic notion of childhood. The pathways are not made clearer by the moral panics and confusion that surround sexting. The Victorian experience is instructive in this on another level – that adults are capable of granting a degree of responsibility to young people at times, while holding them to be the impressionable victims of social pressures at other times. That at the end of the law reform process in Victoria, we observe one law amended to prevent young people being treated as pornographers in recognition of the digital world they inhabit, while at the same time creating a new offence that denies them agency in the same digital world seems hard to fathom.

However, it is possible to understand that result if we move away from the idea that the law protects the vulnerable child and instead ask how well the law has articulated the capacity of young people to claim their own identity, including their sexual identity. Such a view of the law must rest on the recognition of young person's agency and in that agency they may also claim true protection from the various harms that have befallen children in the past, and may continue to do so, unless the law recognises the right of young people to define their own identities, including as digital citizens. Law might do more justice to young people by recognising their self-worth in this way, speaking less to the need to control and constrain the 'dirty child' of its imagination, and confronting the reality of children's capacity.

In terms of future directions for understanding and developing policy in relation to young people and digital technology, it is clear that lawmakers would be well advised to broaden their understanding and evidence base. The dominant narrative that has become embedded in policy surrounding young people and sexting, for example, has been articulated without any firm basis in matters that go to the autonomy and empowerment of young people. This is certainly an area that requires far more research to better understand not just how alternative narratives might operate in this area, but how such approaches may also better protect young people, and enhance their capacity to develop their own identity and claim their space in society.

REFERENCES

Adler, A. (2001). The perverse law of child pornography. *Columbia Law Review, 101*, 209–273.

Ahern, N. R. (2013). Sexting: Serious problems for youth. *Journal of Psychosocial Nursing and Mental Health Services, 51*, 22–30.

Australian Federal Police Submission. (2012). *Law Reform Committee, Victorian Parliament, Inquiry into Sexting*. Submission No. 57.

Barbaro, M. (2011, June 6). *Weiner admits he sent lewd photos; Says he won't resign*. Retrieved from http://www.nytimes.com/2011/06/07/us/politics/07weiner.html?pagewanted=all&_r=0

Barkacs, L. L. (2010). Do you think i'm sexty? Minors and sexting: Teenage fad or child pornography? *Journal of Legal, Ethical and Regulatory Issues, 13*, 23–31.

Bauerlein, M. (2008). *The dumbest generation: How the digital age stupefies young Americans and Jeopardizes our future (Or, Don't Trust Anyone Under 30)*. New York, NY: Tarcher/Penguin.

Boyd, D. (2014). *It's complicated: The social lives of networked teens*. New Haven, CT: Yale University Press.

Butler, J. (1997). *Excitable speech: A politics of the performative*. New York, NY: Routledge.

CBS News. (2009). *"Sexting" shockingly common among teens*. Retrieved from http://www.cbsnews.com/stories/2009/01/15/national/main4723161.shtml

Chalfen, R. (2009). It's only a picture': Sexting, 'smutty' snapshots and felony charges. *Visual Studies, 24*, 258–268.

Child Safety Commissioner (Victoria) Submission. (2012). *Law Reform Committee, Victorian Parliament, Inquiry into Sexting*. Submission No. 25.

Civil Liberties Australia Submission. (2012). *Law Reform Committee, Victorian Parliament, Inquiry into Sexting*. Submission No. 27.

Cohen, S. (1972). *Folk devils and moral panics: The creation of the mods and rockers*. London: MacGibbon and Kee.

Community Practitioner. (2013). Young people face online safety timebomb. *Academic Journal, 86*, 6.

Crimes Act 1958 (Vic).

D'Antona, R. (2010). Sexting, texting, cyberbullying and keeping youth safe online. *Journal of Social Sciences, 6*, 523–528.

Dobson, A., Rasmussen, M., & Tyson submission, D. (2012). *Law Reform Committee, Victorian Parliament, Inquiry into Sexting*. Submission No. 34.

Duschinsky, R. (2011a). Ideal and unsullied: Purity, subjectivity and social power. *Subjectivity, 4*, 147–167.

Duschinsky, R. (2011b). Nietzsche: through the lens of purity. *Journal of Nietzsche Studies, 44*, 50–64.

Duschinsky, R. (2012). The 2010 UK home office 'sexualisation of young people' review: A discursive policy analysis, *Journal of Social Policy*, 41, 715–731.

Family Planning Victoria Submission. (2012). *Law Reform Committee, Parliament of Victoria, Inquiry into Sexting*. Submission No. 18.

Family Voice Submission. (2012). *Law Reform Committee, Victorian Parliament, Inquiry into Sexting*. Submission No. 4.

Ferguson, H. (2007). Abused and looked after children as 'Moral Dirt': Child abuse and institutional care in historical perspective. *Journal of Social Policy, 36*, 123–139.

Foucault, M. (1978). *History of sexuality: An introduction* (Vol. 1), New York, NY: Random House.

Gippsland Community Legal Service Submission. (2012). *Law Reform Committee, Parliament of Victoria, Inquiry into Sexting*. Submission No. 17.

Goodstein, A. (2007). What teens and tweens are really doing online totally wired. Retrieved from http://totallywired.ypulse.com/archives/2007/02/the_new_generation_gap_require.php

Hasinoff, A. (2010). *Making sense of sexting: Adolescents as careless, irrational, and sex-crazed.* Unpulished dissertation chapter.

Headspace Submission. (2012). *Law Reform Committee, Victorian Parliament, Inquiry into Sexting*. Submission No. 22.

Hillier, L., & Harrison, L. (2007). Building realities less limited than their own: young people practising same-sex attraction on the internet. *Sexualities, 10*(1), 82–100.

Holtzman, D. H. (2012). Female exhibitionism: Identification, competition and camaraderie. *The International Journal of Psychoanalysis, 93*, 271–292.

Just Leadership Program Submission. (2012). *Law Reform Committee, Victorian Parliament, Inquiry into Sexting*. Submission No. 59.

Kane, J. Submission (2012). *Law Reform Committee, Victorian Parliament, Inquiry into Sexting*. Submission No. 10.

Karaian, L. (2012). Lolita speaks: 'Sexting,'teenage girls and the law. *Crime Media Culture, 8*, 57–73.

Kincaid, J. (1998). *Erotic innocence: The culture of child molesting*. Durham, NC: Duke University Press.

Kreimer, S. (2011). *Pervasive image capture and the first amendment: Memory, discourse, and the right to record.* Research Paper No. 10-06. University Of Pennsylvania Law School.

Law Reform Committee, Parliament of Victoria. (2013). *Inquiry into Sexting.*

Law Reform Committee, Victorian Parliament. (2011, September 1). *Terms of Reference, Inquiry into Sexting.*

Law Reform Committee, Victorian Parliament. (2013). *Inquiry into Sexting Report.*

Leary, M. (2008). *Self produced child pornography: The appropriate societal response to juvenile self-sexual exploitation.* Catholic University of America, Columbus School of law.

Legal Aid NSW Submission. (2012). *Law Reform Committee, Victorian Parliament, Inquiry into Sexting*. Submissoin No. 50.

Lemma, A. (2010). An order of pure decision: Growing up in a virtual world and the adolescent's experience of being-in-a body. *Journal of the American Psychoanalytic Association, 58*, 691–714.

Macedon Ranges Local Safety Committee Submission. (2012a). *Law Reform Committee, Victorian Parliament, Inquiry into Sexting*. Submission No. 54.

Macedon Ranges Local Safety Committee Submission. (2012b). *Law Reform Committee, Parliament of Victoria, Inquiry into Sexting*. Submission No. 54.

McKenna, M. (2013, August 8). *An old-fashioned scandal.* Retrieved from http://www.theaustralian.com.au/news/features/an-old-fashioned-scandal/story-e6frg6z6-1226693031208?nk=71152d7d73c371d9f80c111262237736

Morris, A. (2011). They know what boys want. *New York Magazine*, February 7, pp. 31–36.

Mottier, V. (2008). Metaphors, mini-narratives and foucauldian discourse theory. In T. Carver & J. Pikalo (Eds.), *Political language and metaphor: Interpreting and changing the world* (pp. 282–294). London: Routledge.

Nabokov, V. (1955). *Lolita*. Paris: Olympia Press.

Papadopoulos, L. (2010). *Sexualisation of young people review*. London: Home Office.

Parliament of Australia. (2011). *High-Wire Act Cyber-Safety and the Young: Interim report*, Joint Select Committee on Cyber-Safety Commonwealth of Australia.

Pascoe, C. J. (2011). Resource and risk: Youth sexuality and new media use. *Sexuality Research and Social Policy*, *8*, 5–17.

Powell, A. (2010). Safe sext'? Young people, technology and sexual violence. *DVRCV*, *2*, 1–5.

Ringrose, J., Gill, R., Livingstone, S., & Harvey, L. (2012). *A qualitative study of children, young people and 'sexting': A report prepared for the NSPCC*. London: National Society for the Prevention of Cruelty to Children.

Rosenthal, D. (2005). Sexual wellbeing and life transitions: A focus on adolescence and old age. In M. T.-S. Gifford (Ed.), *Sexual Health: An Australian Perspective*. Melbourne: IP Communications Pty.

Royal Australian and New Zealand College of Psychiatrists Submission. (2012). *Law Reform Committee, Victorian Parliament, Inquiry into Sexting*. Submission No. 13.

Salvation Army Oasis Hunter Submission. (2012). *Law Reform Committee, Victorian Parliament, Inquiry into Sexting*. Submission No. 7.

Simpson, B. (2013). Challenging childhood, challenging children: Children's rights and sexting. *Sexualities*, *16*, 690–709.

Summary Offences Act 1966 (Vic).

Susan McLean Oral Evidence. (2012). Oral evidence to the Committee, *Law Reform Committee, Victorian Parliament, Inquiry into Sexting*,7 August.

Susan McLean Submission. (2012). *Law Reform Committee, Victorian Parliament, Inquiry into Sexting*. Submission No. 12.

The Alannah and Madeline Foundation Submission. (2012). *Law Reform Committee, Victorian Parliament, Inquiry into Sexting*. Submission No. 42.

UK Home Office. (2010). Sexualisation of young people review. London.

United Nations. (n.d.). *United Nations Convention on the Rights of the Child, article 31*.

Willard, N. E. (2010). Sexting and youth: Achieving a rational response. *Journal of Social Sciences*, *6*, 542–562.

Women's Health Grampians Submission. (2012). *Law Reform Committee, Victorian Parliament, Inquiry into Sexting*. Submission No. 14.

Women's Health West Submission. (2012). *Law Reform Committee, Victorian Parliament, Inquiry into Sexting*. Submission No. 21.

Wood, R. H. (2009/2010). The failure of sexting criminalization: A plea for the exercise of prosecutorial restraint. *Michigan Telecommunications and Technology Law Review*, *16*, 151–178.

Zhang, X. (2010). Charging children with child pornography – Using the legal system to handle the problem of 'sexting'. *Computer Law & Security Review*, *26*, 251.

Zurbriggen, E. L., Collins, R. L., Lamb, S., Roberts, T.-A., Tolman, D. L., Ward, L. M., & Blake, J. (2007). *Report of the American psychological association task force on the sexualization of girls*. Washington, DC: American Psychological Association.

TECHNOLOGY AND THE FOSTERED CHILD: A NEW SOCIAL WORK RESPONSE

Jennifer E. Simpson

ABSTRACT

Purpose – *The advent of the Internet and social medial presents major challenges to contemporary social work involving young people and their families particularly in the sphere of fostering.*

Methodology/approach – *This discussion paper begins an exploration of how risk and vulnerability, associated with the use of social networking, are usually managed using a traditional model that is rooted in psychologically informed social work.*

Findings and originality/value – *The discussion then moves to viewing an alternative model of social work that is informed by the sociology of childhood. An argument is made that this model can be effectively used to meet the challenges of safeguarding in a technological age.*

Keywords: Risk; social media; fostered children; fostering; child protection

Technology and Youth: Growing Up in a Digital World
Sociological Studies of Children and Youth, Volume 19, 351–367

ISSN: 1537-4661/doi:10.1108/S1537-466120150000019012

INTRODUCTION

The advent of mobile phone technology and the Internet has brought with it significant challenges for those who work within the arena of fostering, in particular, social work practitioners (Fursland, 2011a, 2011b; Simpson, 2013; Stephenson, 2009). Such challenges began as early as 2009. In a British online forum, 'Care Space', a number of posts were received from social work practitioners and foster carers that related to issues associated with children and young people making contact with birth parents and relatives, to their detriment (Cooper, 2009). This was closely followed by an article on the extent to which young people in foster care should be monitored. Stephenson (2009) in the opening paragraph of her article cited the problem as, '*Regulating contact between children in care and their birth parents has always been a delicate balancing act, but the proliferation of ways of communicating including text messages and social networking sites has made it even more challenging*' (2009, p. 1). Stephenson (2009) goes on to cite examples of vulnerable children being unsettled and their foster care placements being undermined by the manipulative behaviour of birth parents using mobile phones and social media to illicitly contact their children.

The concerns raised by Stephenson in 2009 were taken up far more substantially in 2011 by Eileen Fursland writing under the patronage of The British Association of Adoption and Fostering. Her publication entitled '*Foster care and social networking: A guide for social workers and foster carers*' not only provides guidance and advice to the said audience, but also gives a series of explanations regarding the opportunities and risks afforded by the Internet, and in particular social networking sites. Fursland (2011a, 2011b) emphasises the vulnerability of fostered young people. She also provides a range of case studies that exemplify how they make use of the Internet and social networking sites. There are a total of five case studies in Fursland's publication (2011a), but for the purpose of this discussion paper, only four are referred to as these highlight key issues in relation use, as well as the responses by social work practitioners and foster carers.

The first case study selected is that of a 13-year old adolescent girl in foster care, a victim of cyber-bullying by her fellow students. The cyber-bullying used the medium of a website that a large number of students contributed to and used to post negative comments (Fursland, 2011a, p. 27). The response by the foster carers was to contact the school; but the response they received was that there was little that could be done. The social work practitioner supporting the foster carers reported the matter to The Child Exploitation and Online Protection Centre (CEOP). CEOP communicated with the social networking site and the website was removed.

Another case study concerns a group of siblings who, in their previous placement, had been allowed to use Facebook to stay in touch with friends. Having moved to their new placement, they were banned from using Facebook and their access to the family computer was limited by the foster carers because there were frequent disagreements regarding use and turn-taking among the sibling group. The resulting solution offered by the foster carers was for each of the children to e-mail their friends, even though this was not the preferred method of communication (Fursland, 2011a, p. 30).

A further case study involving a fostered adolescent female aged 13 concerned conditions of her continued contact with her adopted brothers, one of which was not to have contact with her birth mother. The young person got in touch with her mother and other members of the birth family by using Facebook. As a result of the young person's actions, contact with her mother was arranged. But the outcome of this arrangement was that it did not continue and the young person was then not allowed to have contact with her adopted brothers because she had breached one of the conditions agreed upon (Fursland, 2011a, p. 35).

The fourth and final case study highlights a significant divergence of opinion by social work practitioners. The case involved two fostered adolescent females and their boastful communication with each other about promiscuous activities. What is of note is the reaction of the social work practitioner supporting the foster carers. Her expectation was that the fostered children's allocated social workers would take action to manage the situation. Instead, a contrary view that branded the foster carers as 'spying and being controlling' was held by one of the social work practitioners who had case responsibility for one of the fostered adolescent females (Fursland, 2011a, p. 45).

Three out of these four case studies provided by Fursland (2011a) offer an insight into the use fostered children make of the Internet and social networking for the purposes of maintaining friendships and ties with family, despite the potential for negative consequences. One case study in particular demonstrates that fostered young people, like many other adolescents, can also become victims, and that both the risk and threat can be external to the home environment (Lenhart et al., 2011). The majority of case studies give a very strong inference in relation to an aversion of risk and this is demonstrated by the way in which the adults (both foster carers and social work practitioners) responded to the use of social networking sites.

Since 2009 there have been further reports such as that from CEOP which identified key characteristics in terms of use of the Internet by young people. The report by CEOP entitled 'Threat Assessment of Child Sexual Exploitation and Abuse' (2012) highlights that use of the Internet and other mobile technology is now firmly embedded in the everyday lives of young people. Moreover, it is shown that the Internet with its quick and easy links to education, personal social networks and entertainment means that average use has increased among 12–15 year olds from 14.9 hours in 2011 to 17.1 hours in 2012: a rise of 15% (Children and Parents: Media Use and Attitudes Report, Ofcom, 2012, as cited in CEOP, 2013, p. 6). A further significant development is the opportunity provided for children and young people to use the Internet to create separate identities and therefore engage in risk-filled behaviours that would not be acceptable in the offline world (CEOP, 2013).

More recently, research by Sen (2015, p. 2) has highlighted another area of vulnerability for fostered children. That is child sexual exploitation either by an individual or gangs. Sen points out that exploitation and victimisation occur both online and offline. More importantly, he shows that the vast majority of victims are female and that there is an over-representation of children and young people who are fostered (Office of the Children's Rights Director, 2012; The Child Exploitation and Online Protection Centre, 2013 as cited by Sen, 2015, p. 2). Other researches (Carrick-Davies, 2011; Livingstone, Olafsson, O'Neill, & Donoso, 2012; Livingtone & Brake, 2010; Nylander & Larshammer, 2012) illustrate that ease of access to the Internet and social networking sites is now more immediate because of mobile technologies (this term encapsulates the revolution in technology. Smartphones (Android and iPhones) enable users to communicate from anywhere across a variety of platforms that include text, instant messaging, social media and telephoning; Nylander & Larshammer, 2012).

Moreover, research undertaken by Carrick-Davies (2011) involving vulnerable young people being taught in PRUs[1] (some of whom will be fostered young people transitioning from one placement to another) revealed that for many of them their primary gateway to the Internet was via their smartphone. He also found that there were other risks associated with the use of smartphones. These included inappropriate tags of photographs, geolocation and also 'complex' gifting, for example, the sharing of music and photographs that might be of a pornographic nature (sexting). There are a number of findings drawn from this study that echo similar studies about children and young people's use the Internet. For example, the correlation between online and offline vulnerabilities for certain young people is problematic (Livingstone & Brake, 2010, p. 3); that risk is best understood from a young person's perspective and experience (Livingstone & Brake, 2010, p. 4). A key finding was 'the significance of the mobile phone in the often complicated inter-relationships that characterise young people's lives and their perception of risk' (Carrick-Davies, 2011, p. 15).

In summary, the risks experienced by young people, in general, when using the Internet and mobile technologies can be categorised into three main areas: content, contact and conduct (May-Chahal et al., 2014). The first is that of *content*, example of which include pornographic materials in the form of images and videos. The second is *contact*. This may involve an individual who is online endeavouring to persuade a young person to take part in an interaction either online or offline that then leads to risky behaviours. Examples of such risky behaviours include requesting that a young person engage in sexual chat, or send an indecent image of him/herself, or meet the individual in person (CEOP, 2013, p. 10). The third risk is associated with *conduct*. This is where the young person is the perpetrator, which then threatens other children and young people (May-Chahal et al., 2014, p. 599).

However, it is important to note that the level and exposure to vulnerability by young people is not necessarily clear cut. Moreover, Sen (2015) contends that, as yet, there is a not a robust understanding of how the relationship between the offline and online worlds either promote or hinder vulnerability. There is an obvious gap in research regarding vulnerability which has yet to be filled; nevertheless, there is room to consider, as this discussion paper does; how the use of the Internet and mobile technologies by fostered young people is managed by social work practitioners using a traditional model that is rooted in psychologically informed social work. The discussion focuses on an alternative model of social work that is informed by the sociology of childhood. An argument is made that this model can be effectively used to meet the challenges of safeguarding in a technological age. It is also contended that this alternative model provides an opportunity to develop with fostered young people a partnership that promotes a shared goal of protection and also a way of working that allows for their experiences to be heard and validated.

The Need for Change in the Child Protection and Care System

Firmin (2013, p. 2) states that many of the risks faced by young people come from outside the home and are situated in their peer groups, schools and neighbourhoods. She importantly adds that both parents and professionals are struggling to effectively manage these risks and that the existing child protection system is inherently designed to deal with risk emanating from *within* the familial environment. Hanson and Holmes (2014, p. 3) provide further insight by stating, '... *some risks are not simply present in the child's environment, but are created by interactions between that environment and adolescent developmental changes and tasks*'. The way in which vulnerability and risk are effectively managed by practitioners has taken on a greater level of significance because of the major shift in policy direction that is occurring in relation to adolescents and the current care and child protection system. This shift in policy began with a published position statement by The Association of Directors of Children's Services (ADCS) entitled 'What is Care For – Alternative Model of Care for Adolescents' (The Association of Directors of Children's Services, 2013). In particular, the position paper questions whether the current child protection system is fit for purpose and is able to meet the needs of adolescents. This same questioning was taken up by the coalition government as it was keen to develop a range of successful approaches that would achieve higher quality and improved outcomes for young people who are either entering the care system for the first time or who are fostered (Department of Education, 2014, p. 4).

This growing sense of doubt about the existing care system for adolescents can be strongly linked to the increasing body of research about the way in which the child protection and care system responds to adolescents. Gorin and Jobe (2013) have through their research highlighted the varying and complex needs of maltreated young people as they were referred to Children's Social Care Services and how they were responded to. In particular, Gorin and Jobe (2013) identified that child protection processes employed for young people entering the care system were likely to be inappropriate as their immediate circumstances included multi-faceted problems that were difficult to resolve and therefore required a different response. The reason for this was because of the expansive range of risks that were identified as being external to the home environment combined with existing risk-taking behaviours exhibited by the young people themselves. Gorin and Jobe (2013) further substantiate their findings by referring to Rees, Gorin, Jobe, Stein, Medforth and Goswami (2010 as cited by Gorin & Jobe, 2013), who remarked a number of years earlier that as children matured and became adolescents they were invariably more independent and mobile and that this development brings with it access to wider social networks and associated risks linked to different surroundings that are external to the home environment.

It has been established that many of the risks faced by adolescents are external to the home environment. How do these findings contribute to what we know about use of the Internet and social networking by adolescents in general, and then more specifically fostered young people? LaMendola (2011) states that young people engage in social networking that only requires episodic exchanges of data without a knowledge of others who may be looking on (p. 4). Moreover, the effortlessness with which these networks can be accessed means that a fundamental need of adolescent development is by satisfying some requirement for attachment through a variety of relationships. This point is further reinforced by Livingstone (2011), who emphasises that young people, whether they are identified as vulnerable or not, wish to maintain communication not only with friends but also with a wider circle of people (as seen in three out of the four case studies mentioned earlier). boyd and Marwick's (2011a, p. 2) commentary on contemporary teenage life highlights that it consists of relationships and social ties that are played out in the arena of such social networking sites as Facebook, Reddit and Tumblr. In particular, a performance on the social media stage not only draws the attention of peers but also enhances the individual's own personal worth and standing among them and serves as a means to obtain social capital. In summary, key development tasks of this particular life stage are being undertaken which according to John Hill (1983, as cited in Subahmanyam & Greenfield, 2008, p. 124) are identity, autonomy and sexuality. This is substantiated by Livingtone and Brake (2010), who in seeking to explain the use of SNS by adolescents identify the link between opportunities and risks. They argue that as adolescents experience a range of opportunities online these are positively correlated to the risks they encounter. Consequently, the more opportunities adolescents seek to avail themselves of online, the equal number of risks they expose themselves to. This situation, according to Livingtone and Brake (2010), mirrors the adolescent characteristics of ongoing human development which are exploration and risk-taking (p. 79).

boyd and Marwick (2011a, p. 15) add a further dimension to our understanding by asserting that historically our common interpretation of private and public '*has been built upon a set of dichotomies and divisions whether they be spatial (workplace, home) temporal ('on' or 'off' the clock), or object-related work (work Blackberry or parent's car)*'. These dichotomies and divisions cannot be immediately applied to the Internet or social media: an actuality that is understood by young people. Therefore, the situation we have is adolescents who are concerned about privacy and in many cases are clear about when, how and with whom they want to communicate. Nevertheless, adults are alarmed at the way in which the social norms related to privacy are being 'transgressed' on a daily basis (boyd & Marwick, 2011a). More often than not adults are reading the communication by a young person based on their offline understanding of the individual, and are failing to appreciate the '*linguistic and cultural*' cues employed by adolescents when they communicate with each other (boyd & Marwick, 2011b, pp. 17–18).

One might reason that the response by social work practitioners to these risks has mainly relied upon a way of working that is rooted in psychologically informed social work. Psychologically informed social work draws on theoretical approaches that include systems, ecological and attachment theories. The manifestation of psychologically informed social work is best seen in the preoccupation of securing the individual safety of a young person and protecting the childhood that s/he has had (Goodyer, 2013). Within psychologically informed social work, there is a tendency to pathologise the behaviours of fostered young people, highlighting their irresponsibility, immaturity and inability to comprehend the possible consequences of their actions.

At this juncture, it is critical to note that this model of social work is not only concerned with the psychological, but also informed and influenced by the biological development of children and young people. A key exemplar of the dominant narrative of psychological and biological development is Piaget's (1953) theory in relation to cognitive development and the notion that a child has to pass through four successive stages before acquiring the necessary competence for adulthood. The work of Piaget (1953) and others like Susan Isaacs contributed to the normative model of developing cognition in children (Hendrick, 1997). Key messages from Isaac's work include the fact that although children have the same mental ability as adults there are differences in that the brain of a child does not function to the same capacity as that of an adult. Other dominating narratives relate to children being governed by their basic wishes, fantasies and fears (Hendrick, 1997, p. 52). These messages and that of Burt and Issacs (who were influential in the development of Child Guidance Clinics of the 1920s and 1930s where 'troubled' children came to be 'treated') led to specific theories in relation to the psychological and biological development of children, young people and childhood. Children and adolescents were therefore deemed as a category of human being in which the idea of natural growth was comprehensively accepted, and where the inner workings of the child were recognised as having great significance in terms of long term development towards adulthood and the avoidance of deviant behaviour (Hendrick, 1997, p. 53;

James and Prout, 1996, p. 20). Critically, James and Prout (James, Jenks, & Prout, 1998, p. 79) have claimed that the domination of this particular narrative is because it has gained scientific respectability.

Evidence of the psychological model of social work can be found in Fursland's (2011a, 2011b) guide to social networking for social work practitioners and foster carers. She stresses the vulnerabilities of fostered adolescents and their supposed lack of skill to communicate and behave appropriately online. The advice provided by Fursland (2011a) strongly encourages foster carers and social work practitioners to employ online monitoring techniques. It would seem that Fursland's (2011a) motivation for encouraging this approach is based on a likelihood that young people who are fostered have already experienced the deleterious effects of either abuse (physical, emotional or sexual) or neglect. Moreover, as a consequence of such abuse and neglect, and the probability of existing emotional, social and behavioural difficulties, these young people may be unable to instigate strategies for dealing with online threats (Fursland, 2011a). Such a position reflects a pathological model that identifies the problems experienced as inherent within the character of the fostered young person.

The biological and pathological model of child and adolescent development coupled with the supposed inherent vulnerabilities of fostered young people would lead one to assume that psychologically based social work is entirely appropriate. This would be the case if the link between vulnerability and risk was entirely straightforward. Sen (2015) and others (Livingstone, 2011; Livingstone et al., 2012; Livingtone & Brake, 2010; McLeod, 2007) argue that this may not be the case. Sen (2015, p. 3) in particular comments that, '*the exact relationship between online and offline vulnerability still needs to be better understood and the evidence does not support an assumption that young people with care experience are per se, at greater risk online*'. Sen (2015) is not alone in voicing doubt; the research undertaken by Livingstone et al. (2012) suggests that young people's use of the Internet when it was managed by monitoring software, tended to impair not only online skills but also digital skills and opportunities that would provide them with the opportunity to maintain their safety.

AN ALTERNATIVE MODEL

Having considered the characteristics of psychologically informed social work in relation to the use of Internet and mobile technologies by fostered young people, is there an alternative model that could be employed? As shown earlier, there is a growing dissent beginning to emerge regarding the provision of social care services to young people that, it could be argued, is underpinned by psychologically informed social work. It may be that there is a need to engage in a model of social work that can deal with the external risks encountered, and which promotes a way of working with fostered young people that is built on a foundation of relationship (The Care Inquiry, 2013). Goodyer (2013) in a recent paper about understanding the childhoods of fostered children observed that, currently child and family social work has been reliant upon a model of working that views children, young people and their behaviours in a psychological way. Whilst practitioners have engaged with child development and assimilated knowledge that has informed their practice, what they have not done is to critically reflect upon the literature that acknowledges child development has been and can be socially constructed (Goodyer, 2013).

The first challenge to the dominating discourse of Piaget, (1953) and others (Hendrick, 1997; Lavalette & Cunningham, 2002) was made by Aries (1962 as cited in Lavalette & Cunningham, 2002, p. 11), who claimed that childhood had not always existed but rather it had been discovered. Aries' supposition, though criticised and maligned in various quarters (Fuller, 1979; Weekes, 1989; Wilson, 1980; Woodward, 2001 as cited in Lavalette & Cunningham, 2002, p. 15), created the necessary room for children and childhood to be seen beyond the confines of biological and psychological development. Lavalette and Cunningham (2002, p. 18) state, '*It focused the sociological imagination on childhood and emphasised that childhood is not a static life stage but has altered across time and space*' (Qvortrup, 1987). Furthermore, this major shift in sociological thinking has highlighted that childhood is influenced by factors that relate to the social and cultural characteristics of a certain era.

The loosening of the grip of biological and psychological child development since the 1970s has led to the deconstruction of childhood. Now consideration is being given to the experiences of children and the understanding of their reality (Alanen, 1998; Hutchby & Moran-Ellis, 1998; James, Jenks, & Prout, 1998; James & Prout, 1990a; Lavalette & Cunningham, 2002; Neale & Smart, 1998; Woodhead, 1990, 2005). Sociologists such as Qvortrup (1987 as cited in Lavalette & Cunningham, 2002); James and Prout (1990a, 1990b) and James et al. (1998) have postulated that although children, can and should be seen as distinct from adults, they are able to contribute and co-construct our adult understanding of childhood. More importantly, the emphasis put forward by the academics listed above clearly identifies that it is only children, and not adults, who are able to fully articulate childhood. Put another way, '*children occupy and conduct themselves in worlds that are full of meaning to them, but about which adults are, at least partially, ignorant. These are the 'children's childhoods'. Children are best placed to describe and analyse this world, better at any rate than adult outsiders*' (Lavalette & Cunningham, 2002, p. 24).

As has been mentioned above, psychologically informed social work draws on theoretical approaches that include systems, ecological and attachment theories. The approach does not include enabling young people to have a significant influence on the nature and type of protection they receive. Nor does this type of social work allow for the experiences and knowledge of fostered young people to be formally recognised and accepted. Furthermore, the approach does not recognise the wider societal challenges that young people face, for example, institutional and cultural oppression (Dominelli, 1997, p. 238), which is particularly the case where young people (and children) as Mayall (2000) has argued are deemed to be an oppressed minority group.

A model of social work informed by the sociology of childhood is in essence the opposite of psychologically informed social work. Crucially, it acknowledges that children and young people have agency and therefore are not passive recipients of whatever happens to them. Goodyer (2013) states the following, '... *the sociology of childhood offers both macro and micro level explanations of children's and young people's lives and positions children as citizens with rights*'. The concept of citizens with rights can bring about fundamental changes in the way social work practice is carried out with young people. Such changes include working with young people and allowing them to define the problem which is of critical importance. It also allows the social work practitioner the opportunity to understand the young person's '*experience of risk through their own narrative account and to analyse the contextual phenomena found there*' (Carrick-Davies, 2011, p. 3). Furthermore, this approach acknowledges the need not only to consider the wishes and feelings of the young person but also to respect their wishes about actions taken on their behalf (Butler & Williamson, 1994 as cited in Goodyer, 2013, p. 399).

Why is a model of social work that is informed by the sociology of childhood being posited as a replacement at this particular time? For an answer, we have to go back to what has been stated by Gorin and Jobe (2013), Firmin (2013) and Hanson and Holmes (2014), all of whom identify that the risks faced by young people are often external to either the family or foster home environment and therefore are not easily managed. A model of social work informed by the sociology of childhood with its recognition that children and young people are experts of their own lives and are able to '*engage in feasible opportunities to learn and train, to help others, or to participate in decisions that affect them*' (Hanson & Holmes, 2014, p. 22) provides a framework within which they are capable and able to share with social work practitioners and foster carers common goals in relation to their own protection. Hanson and Holmes (2014) argue that in order to reduce the risk faced by young people who are fostered or identified as vulnerable, there is a need to work with them in order to bring liberty from the inhibiting influences of low self-esteem, and move them towards positively channelling their energy and will in the direction of outcomes that are better in the long term. Hanson and Holmes (2014, p. 25) claim that this approach '*....is about empowering young people and is a key mechanism in promoting resilience*'.

Moreover, Hanson and Holmes (2014, p. 5) add that this requires a '*paradigm shift*' in how young people are seen and how the risks they face are understood. Unlike the psychologically informed social work model it is possible, using the sociology of childhood, to acknowledge that certain risks are meeting the needs of a young person in the here and now and that any intervention might be refused on the grounds that his/her immediate needs would not be met (Hanson & Holmes, 2014). Crucially, Hanson and Holmes (2014, p. 25) warn that if practitioners do not work with adolescence and those behaviours associated with choice and engagement, '[these] *will remain a potent force in their various vulnerabilities*'.

Social work informed by the sociology of childhood offers social work practitioners the opportunity to understand the complex lives and histories of fostered young people. It can also promote partnership working in the sense that fostered young people have the room and space to express their views about their own care experiences thereby enabling them to actively engage and participate in decision-making about their own lives (Goodyer, 2013). A further advantage of this model of social work is that it can readily promote safeguarding that is much more child-centred primarily because social work practitioners are able to work in a fashion that allows young people to participate in decision-making. The model can also be effectively used to highlight and inform the development of child protection services that are congruent to the needs of young people who are entering the system, as well as those being fostered. A model of social work informed by the sociology of childhood chimes with the recommendation made by Munro about the need for social work organisations to give credence to the views and experiences of both children and the social work practitioners who work alongside them, as part of supporting effective social work practice (Munro, 2011, p. 105).

The necessity of social work practitioners to engage with the complex and difficult terrain of safeguarding has been brought into sharp focus by mobile technologies that include easy access to the Internet and social networking. The requirement for a different way of working is also supported by the children's rights agenda which is best illustrated by the United Nations Convention on the Rights of the Child, 1989 (UNCRC) (UNICEF, 1989). The UNCRC stipulates children's and young people's rights to express their views in all matters that affect them (Article 12), freedom of expression through any medium of the child's/young person's choice (Article 13) and protection of privacy (Article 16). The children's right agenda as specified by the UNCRC and the duty of social work practitioners to safeguard vulnerable young people would immediately seem to be at odds because, as stated in the opening paragraph of this paper, there is a need to keep in mind the best interests of the young person. Nevertheless, one can contend that the rights of young people, along with ensuring their welfare, can be achieved through the implementation of an approach that entails taking a real interest in their use of the Internet. This position is eloquently stated by The Fostering Network as:

> *You wouldn't expect children to cross a road without teaching them how, and the same applies to the web. However, you also wouldn't ban children from crossing roads because of the dangers, nor make them feel that every time they do they are likely to get run over. A positive and safe approach is what is called for.* (The Fostering Network, as cited in Fursland, 2011a)

CLOSING REMARKS

Evidence has been presented that much of the risk faced by young people in general, as well as fostered adolescents is situated outside of the immediate familial environment. With this growing realisation has come a greater awareness that the current child protection and care system are not effectively meeting the needs of young people who present with multi-faceted problems (Gorin & Jobe, 2013). An argument has been made that the social work response to fostered young people and their use of the Internet and social networking have been influenced by psychologically informed social work that has a tendency to pathologise them and emphasise their vulnerabilities, despite a body of research indicating that the risk and vulnerability in relation to the offline and online worlds are not straightforward (Sen, 2015).

A case has also been made for a model of social work that is informed by the sociology of childhood. This model of social work brings with it the opportunity for practitioners to move beyond the confines of the biological and psychological model of young people which is characterised by immaturity, dependence, physical weakness and the need to protect (Lansdown, 1994 as cited in Morrow & Richards, 1996, p. 97). The model offers the possibility of seeing children and young people as human beings who are able to exercise self-determination and can make better sense of their own lives than can the adults who surround them. For social work practitioners, an equilibrium has to be achieved in terms of supervision, support, independence and protection, and most importantly, choice when working with fostered young people. It is not expected that the features of what can best be described as child-centred practice should be exercised in the order that they are written. Rather, the order is likely to change according to the needs, capabilities and level of vulnerability of the individual fostered young person with whom a social work practitioner is working.

NOTE

1. Pupil Referral Units – an educational provision that supports a wide range of school-aged children in centres outside of mainstream school but often work in close partnership with groups of schools and other agencies (http://prus-ap.org.uk/).

ACKNOWLEDGEMENT

Grateful thanks goes to David Kennelly for his encouragement and support.

REFERENCES

Alanen, L. (1998). Children and the family order: Constraints and competencies. In I. Hutchby & J. Moran-Ellis (Eds.), *Children and social competence: Arenas of action* (pp. 30–47). London: Falmer Press.

boyd, d., & Marwick, A. (2011a, October). Social privacy in networked publics: Teens' attitudes, practices, and strategies. A decade in internet time: Symposium on the dynamics of the internet and society, Oxford Internet Institute, Oxford.

boyd, d., & Marwick, A. (2011b, October). The drama! Teen conflict, gossip, and bullying in networked publics. A decade in internet time: Symposium on the dynamics of the internet and society, Oxford Internet Institute, Oxford.

Carrick-Davies, S. (2011). *Munch, poke, ping: Vulnerable young people, social media and E-Safety*. London: Teacher Development Agency (TDA).

CEOP. (2013). *Threat assessment of child sexual exploitation and abuse*. London: Child Exploitation and Online Protection Centre.

Cooper, J. (2009). Technology and supervised contact between children and young people and birth parents, *Community Care*. Retrieved from http://www.communitycare.co.uk/carespace/forums/technology-and-supervised-contact-between-5758.aspx#21485

Department of Education. (2014). *Rethinking support for adolescents in care or on the edge of care: Department of education's children's social care innovation programme*. Retrieved from https://www.gov.uk/government/uploads/system/uploads/attachment_data/file/342052/Rethinking_support_for_adolescents.pdf

Dominelli, L. (1997). *Sociology for social work*. Basingstoke: MacMillan.

Firmin, C. (2013). We are failing to respond to the risks faced by teenagers. *The Guardian*. Retrieved from http://www.theguardian.com/society/2013/nov/26/child-protection-failing-teenagers-risk

Fursland, E. (2011a). *Foster care and social networking: A guide for social workers and foster carers*. London: Association of Adoption and Fostering.

Fursland, E. (2011b). Social networking and adoption *CW360: A comprehensive look at a prevalent child welfare issue. Child Welfare and Technology, Spring 2011*, 20.

Goodyer, A. (2013). Understanding looked after childhoods. *Child and Family Social Work*, *18*, 394–402 doi:10.1111/j.1365-2206.2012.00858.x

Gorin, S., & Jobe, A. (2013). Young people who have been maltreated: Different needs – Different responses? *British Journal of Social Work*, *43*, 1330–1346.

Hanson, E., & Holmes, D. (2014). *That difficult age: Developing a more effective response to risks in adolescence*. Dartington: Research in Practice.

Hendrick, H. (1997). Constructions and reconstructions of british childhood: An interpretative survey, 1800 to present. In A. James & A. Prout (Eds.), *Contemporary issues in the sociological study of childhood* (2nd ed., pp. 33–60). London: Falmer Press.

Hutchby, I., & Moran-Ellis, J. (1998). Situating children's social competence. In I. Hutchby & J. Moran-Ellis (Eds.), *Children and social competence: Arenas of action*. London: Falmer Press.

James, A., Jenks, C., & Prout, A. (1998). *Theorizing childhood*. Cambridge: Polity Press.

James, A., & Prout, A. (1990a). Re-presenting childhood: Time and transition in the study of childhood. In A. James & A. Prout (Eds.), *Contemporary issues in the sociological study of childhood* (pp. 227–246). London: Falmer Press.

James, A., & Prout, A. (1990b). A new paradigm for the sociology of childhood? Provenance, promise and problems. In A. James & A. Prout (Eds.), *Contemporary issues in the sociological study of childhood* (pp. 7–32). London: Falmer Press.

James, A., & Prout, A. (1996). Strategies and structures: Toward a new perspective on children's experiences of family life. In J. Brannen & M. O'Brien (Eds.), *Children in families: Research and policy* (pp. 41–52). London: Falmer Press.

LaMendola, W. (2011). *Child welfare and technology in child and welfare technology pp. 4 and 12*. Retrieved from http://cascw.umn.edu/wp-content/uploads/2013/12/CW360_2011.pdf

Lavalette, M., & Cunningham, S. (2002). The sociology of childhood. In B. Goldson, M. Lavalette & J. McKechnie (Eds.), *Children, welfare and the state* (pp. 9–28). London: Sage.

Lenhart, A., Madden, M., Smith, A., Purcell, K., Zickuhr, K., & Rainie, L. (2011). *Teens, kindess and cruelty oon social network sites: How American teen navigate the new world of 'digital citizenship*. Washington, DC: Pew Internet and American Life Project.

Livingstone, S. (2011). *Social networking: Risks and opportunities for youth in child and welfare technology pp. 23*. Spring. Retrieved from http://cascw.umn.edu/wp-content/uploads/2013/12/CW360_2011.pdf

Livingstone, S., Olafsson, K., O'Neill, b., & Donoso, V. (2012). *Towards a better internet for children*. EU Kids Online. Retrieved from http://www2.lse.ac.uk/media@lse/research/EUKidsOnline/EU%20Kids%20III/Reports/EUKidsOnlinereportfortheCEOCoalition.pdf

Livingstone, S., & Brake, D. R. (2010). On the rapid rise of social networking sites: New findings and policy implications. *Children and Society*, *24*, 75–83.

Mayall, B. (2000). The sociology of childhood in relation to children's rights. *International Journal of Children's Rights*, *8*, 243–259.

May-Chahal, C., Mason, C., Rashid, A., Walkerdine, J., Rayson, P., & Greenwood, P. (2014). Safeguarding cyborg childhoods: Incorporating the on/offline behaviour of children in everyday social work practices. *British Journal of Social Work*, *44*, 596–614.

McLeod, A. (2007). Whose agenda? Issues of power and relationship when listening to looked after children. *Child and Family Social Work*, *12*, 278–286.

Morrow, V., & Richards, M. (1996). The ethics of social research in children: An overview. *Children and Society*, *10*, 90–105.

Munro, E. (2011). The Munro review of child protection: Final reporta child-centred system, Department for Education, London.

Neale, B., & Smart, C. (1998). *Agents or dependents?: Struggling to listen to children in family law and family research*. Working Paper No. 3. University of Leeds, Centre for Research on Family, Kinship and Childhood.

Nylander, S., & Larshammer, M. (2012). The phone as a tool for combining online and offline social activity: Teenagers' phone access to an online community. *International Journal of Mobile Human Computer Interaction*, *4*(4), 22–36.

Office of the Children's Rights Director. (2012). *Children's views on fostering*. London: Department of Education.

Piaget, J. (1953). *The origin of intelligence in the child*. London: Routledge and Kegan Paul.

Qvortrup, J. (1987). The sociology of childhood. *International Journal of Sociology*, *3*, 3–37.

Sen, R. (2015). Not all that is solid melts into air? Care-experienced young people, friendship and relationships in the 'Digital Age'. *British Journal of Social Work Advance Access*, 1–17. doi:10.1093/bjsw/bcu152

Simpson, J. E. (2013). A divergence of opinion: How those involved in child and family social work are responding to the Internet and social media. *Child and Family Social Work*. doi:10.1111/cfs.12114

Stephenson, J. (2009). How far should young people in care be monitored? *Community Care*. Retrieved from http://www.communitycare.co.uk/Articles/29/10/2009/113007/The-impact-of-technology-on-supervised-contact.htm

Subahmanyam, K., & Greenfield, P. (2008). Online communication and adolescent relationships. *Future of Children*, *18*(1), 119–146. doi:10.1353/foc.0.0006

The Association of Directors of Children's Services. (2013). *What is care for – Alternative model of care for adolescents*. London: The Association of Directors of Children's

Services. Retrieved from http://www.adcs.org.uk/download/position-%20statements/2013/ADCS_position_statement_What_Is_Care_For_April_2013.pdf

The Care Inquiry. (2013, April). *Making not breaking: Building relationships for our most vulnerable children*. London: House of Commons.

UNICEF. (1989). *United Nations convention on the rights of the child*. Retrieved from http://www.unicef.org.uk/Documents/Publication-pdfs/UNCRC_PRESS200910web.pdf

Woodhead, M. (1990). Psychology and the cultural construction of children's needs. In A. James & A. Prout (Eds.), *Contemporary issues in the sociological study of childhood* (pp. 61–82). London: Falmer Press.

Woodhead, M. (2005). Early childhood development: A question of rights. *International Journal of Early Childhood*, *37*(3), 79–98.

"THAT IS SO NOT TRUE": ADOLESCENT PERSPECTIVES OF ADULT MISCONCEPTIONS OF TEEN TEXT MESSAGING

Sarah Tulane, J. Mitchell Vaterlaus and Troy E. Beckert

ABSTRACT

Purpose – *The authors of this paper sought to qualitatively understand the perceived stereotypes associated with the youth culture surrounding texting.*

Methodology – *Drawing on responses from 161 high school students and using qualitative methods as outlined by Bogden and Biklen to analyze data, four main themes of perceived adult misconceptions emerged. Adolescents spoke about adults' general misperceptions about adolescents' motivations to text, practices associated with texting, the content of their messages, and the impact of texting behaviors on adolescents' developmental processes.*

Findings – *The existence of a youth culture that is imperceptible to adults seems evident from these findings. Texting is a primary form of*

Technology and Youth: Growing Up in a Digital World
Sociological Studies of Children and Youth, Volume 19, 369–393

ISSN: 1537-4661/doi:10.1108/S1537-466120150000019013

communication for adolescents and, for this sample, enhances face-to-face relationships. Adolescents indicated they have control over their texting choices and behaviors.

Implications – *Two common limitations existed in this study. These included using a convenience sample from only one high school, and including only adolescent perspectives. Future research should also examine the parent–child dyad in relationship to texting.*

Originality/value – *This research contributes to the understanding of social implications of adolescent text messaging and helps to further clarify the youth culture of texting. Although there are media that have been extensively studied in relation to adolescent development, newer media such as social networking and texting need further examination, as they have become definitive components of youth culture.*

Keywords: Text messaging; adolescent media use; family relationships

Text messaging is a prevalent topic in entertainment and news media. Most of the attention given in media outlets highlights adult concerns about the potential dangers of teen texting. Adult fears about text messaging are multiplied because American adolescents use texting as their primary form of communication (Lenhart, 2012). Indeed, many adults feel that today's youth communicate exclusively via electronic means. Although adolescent texting has increased markedly over the years, research addressing the developmental implications of text messaging lags behind (Mahatanankoon & O'Sullivan, 2008). An overarching negative adult opinion about adolescent texting, stemming primarily from high-profile media exposure, has influenced private and public policy without much input from the adolescent texters.

Adolescents, in this study, are recognized as being a part of the *Millennial generation* (born between 1980 and 2000; Pew Research Center, 2010). Generational theory posits that generational cohorts exist with people who are born within about a 20-year time period and who also share a location in history, beliefs and behaviors, and a sense of membership in the generational group (Strauss & Howe, 1991). The theory allows for comparisons between generational cohorts. The Millennial generation is distinguished from previous generations, including the *Boomers* (born between 1943 and 1960) and *Generation X* (born between 1961 and 1981; Coomes & Debard,

2004), by their technology use (Pew Research Center, 2010). Not just their access to technology devices, but also how they have "fused their social lives into them" (Pew Research Center, 2010, p. 6).

Supporting the technological uniqueness of the Millennial generation, Thurlow and Bell (2009) suggested that adolescents have historically been quick to adapt to new technologies, this includes the adoption of text messaging as a primary communication method (Brown & Bobkowski, 2011). This particular technology calls for social science attention because, from a societal perspective, texting is related to most forms of adolescent interactions including those with parents (Williams & Williams, 2005) and friends (friendship quality, Cupples & Thompson, 2010; relationship formation, Harley, Winn, Pemberton, & Wilcox, 2007; and relationship duration and relationship termination, Holtgraves, 2011).

As a private form of communication, Millennial adolescents prefer to use texting to build face-to-face social networks (Berg, Taylor, & Harper, 2005). Texting is becoming a part of young people's lives at earlier ages, even 10 and 11 year olds enjoy cell phones, in part, because of the opportunity to text (Davie, Panting, & Charlton, 2004). The appeal of texting is to maintain social relationships and schedules with short messages, which can be sent with enough delay for thinking and processing while enjoying instantaneous transmission. Perhaps the ease and speed with which adolescents adapt to new technologies contribute to adult concerns and misunderstandings over teens' texting behaviors.

As texting becomes a more common component of adolescent life, media have highlighted parental concern and educator apprehension over the potential negative developmental impacts of adolescent text messaging (Kemp, 2010; Thurlow & Bell, 2009). Many adults are concerned about the impact of text messaging on adolescent school performance. Adults have also expressed concerns about other socially dangerous outcomes from texting including concern over message content (Draper, 2011). For example, heightened concern centers on "sexting" (sending messages sexual in nature or that contain sexual images; Draper, 2011). Adults have also expressed a fear that face-to-face communication is diminishing as teens are losing abilities necessary for personal, face-to-face relationships (Thurlow & Bell, 2009).

News broadcasts and popular media depictions provide some justification for adult concerns about teen text messaging. Nevertheless, it is apparent that there has been limited research concerning positive aspects of text messaging and that adolescents report several positive perceptions of their preferred method of communication. As the prominent users of the medium, teenagers should be given a voice regarding the actual adolescent experience of text messaging. These stakeholder opinions can then be used to either add accuracy to adult arguments or clarify adult misgivings surrounding the behaviors. As a necessary first step toward accomplishing this purpose, the current study was designed to gain a grounded understanding of adolescent perceptions about adult misconceptions (stereotypes) of adolescent texting behaviors.

ADULT VERSUS ADOLESCENT PERCEPTIONS

With few exceptions, adolescent sociality is influenced by family relationships, especially by parents. During adolescence, most teenagers learn to balance their quest for autonomy while maintaining positive relationships with their parents (McElhaney, Allen, Stephenson, & Hare, 2009). Thus, the cell phone, specifically text messaging, is a potentially important element in our understanding of adolescent development. On the one hand, cell phones have been described as "symbolic umbilical cord(s)" (Ling, 2005, p. 175) aiding in the constant connection between adolescents and their parents. On the other hand, many adolescents see it as a device that enhances opportunity for autonomy and individuality (Blair & Fletcher, 2011).

Adults and adolescents have both perceptual differences and perceptual commonalities regarding text messaging – highlighting generational differences in texting (Vaterlaus & Tulane, 2015). Understanding the complexities of this phenomenon is challenged by the striking contrasts within and between the two populations of adults and adolescents. At one extreme, some adults view texting as a complete waste of time (Cupples & Thompson, 2010). This view is often accompanied by the difficulty many adults have with working texting into their day-to-day social interaction Adolescents text more than adults (Lenhart, 2010), and have a much easier time seamlessly working texting into their social lives than do adults (Bryant, Sanders-Jackson, & Smallwood, 2006). Commonalities in opinions about texting between adults and adolescents also exist. Both parents and adolescents agree that cell phones are important as safety measures, helping adolescents to keep connections with family and friends (Lenhart, Ling, Campbell, & Purcell, 2010). Some parents even consider cell phones as facilitators toward increasing social status for adolescents (Blair & Fletcher, 2011).

ADULT INVOLVEMENT IN ADOLESCENT TEXTING

Some measure of adult involvement in texting activity can be beneficial for adolescents. Researchers have found that parents' limiting and monitoring of text messaging has a positive effect on their adolescents' texting behaviors, such as decreasing both the likelihood of adolescent sexting, and the likelihood of teens reporting having sent regretful text messages (Lenhart, 2010). Additional research suggests that parental authority boundaries are changing through cell phone use (Williams & Williams, 2005). For example, two-thirds of parents indicate they have taken away cell phones as a punishment (Lenhart et al., 2010). Unfortunately, this punishment creates a problem for parents by eliminating the option for continual contact with adolescents who are otherwise constantly available through cell phones (May & Hearn, 2005). In a study of parents and teenagers, Williams and Williams (2005) found that parents are extending their presence into adolescent space by tracking their child's whereabouts and behaviors through text messaging.

YOUTH CULTURE AND STEREOTYPES

Within generational theory, it is not uncommon for the rising generation to create shared beliefs and behaviors that are perceptually different from previous generations (Howe & Strauss, 2003; Strauss & Howe, 1991). Although people from several generations use technology (Duggan, 2013; Lenhart et al., 2010), the Millennials' behaviors and beliefs surrounding technology have been marked as different from other generations (Pew Research Center, 2010). For example, adults and adolescents report different purposes and uses for texting (Vaterlaus & Tulane, 2015). However, caution is needed when making declarations about all people within a generation because "there are as many differences in attitudes, values, behaviors, and lifestyles within a generation as there are between generations" (Pew Research Center, 2010, p. 5). As such, we acknowledge the value of identifying major differences between generations through a generational theory lens, but deviate to focus on a youth cultural lens. Scholars have posited that a youth culture has become apparent surrounding the use of adolescent text messaging (Oksman & Turtiainen, 2004; Tulane, 2012).

A youth cultural perspective provides a framework that allows for commonality and variance within culture, because cultural values, practices, and beliefs exist on continua (Lynch & Hanson, 2011). Acknowledging within-cultural differences, in addition to cross-cultural differences, may decrease the likelihood of promoting stereotypes based on the behavior or values of some cultural members to the majority culture. The proposed youth culture surrounding texting has often been identified to be imperceptible to adults (Oksman & Turtiainen, 2004; Tulane, 2012). Oksman and Turtiainen suggested this youth culture impacts communication patterns and the way adolescents construct their worldview. For example, cell phones allow adolescents to carry their social networks with them. Texting is a quiet way for adolescents to address their social network without bothering others or involving adults in their social communications (Oksman & Turtiainen, 2004). As text messaging has become a preferred form of communication, some problematic behaviors (e.g., sexting, texting while driving) have begun to manifest in a minority of users (Lenhart et al., 2010). Because the youth culture is often imperceptible to adults (Oksman & Turtiainen, 2004; Tulane, 2012), adults may also fail to acknowledge the continua of cultural values, practices, and beliefs. Thus, the shared youth culture around texting creates a group of young people with which adults can generalize perceptions of some teens' texting behaviors to all adolescent texters (Vaterlaus & Tulane, 2015).

These generalizations about adolescent texting can be explored in terms of stereotypes. It is important to note that stereotypes are different from generalizations. Charon (2012) points out that it is the nature of humans to generalize and form categories to help understand behavior. Stereotypes are generalized beliefs shared by a group and are used by individuals to help explain behavior (McGarty, Yzerbyt, & Spears, 2002). Stereotypes function in terms of absolutes, are judgmental, do not open possible understanding of differences, and can block other potential understandings or categories (Charon, 2011). Stereotypes often influence perceptions, emerge because of group differences, impact the way people are treated, and influence the way a stereotyped group behaves (McGarty et al., 2002). The negative labels of stereotypes for an individual or group can lead to "self-fulfilling prophecies" in behavior (Aronson, 2002). Furthermore, the perception of a stereotype can influence the way an individual behaves. An apprehension at conforming to a stereotype, either a personal evaluation of stereotype conformation or in the perception of another, is referred to as "stereotype threat" (Aronson, 2002).

There is a tendency for older generations to criticize younger generations (Gardner & Davis, 2014). Furthermore, for current youth generations, technology has a different meaning than it does for older generations. Gardner and Davis (2014) indicated members of youth generations referring to technology, "to instances of hardware or software that came into existence during their own conscious lives. If already present when they were very young, the form of technology is just part of the background scenery" (pp. 52–53). Because stereotypes are based on categories of people (McGarty, 2002), and texting is the most common communication tool in adolescent daily life, the group that has easily adapted to the technology is more likely to be categorized and stereotyped. The consequences of potential stereotypes and stereotype threats relating to adolescent texting have not been empirically explored.

PURPOSE OF THE CURRENT STUDY

The topic of text messaging has received much attention in the media, and research concerning the social and developmental implications is needed. The current exploratory study was developed to better understand the perceived prevalence of adult stereotypes concerning adolescent texting from an adolescent perspective. The qualitative nature of this study not only opens the exploration of perceived stereotypes but also begins to explore the potential stereotype threats surrounding adolescent text messaging. This investigation into youth culture experience serves a step forward in understanding potential adult misconceptions (stereotypes) adolescents perceive and potential impacts of said stereotypes on adolescent text messaging behavior.

METHODS

Data were collected from high-school-aged adolescents twice during the 2011 school year. The participants were enrolled in multiple information technology courses in a high school located in an urban area in a western state. Information technology courses were selected so the surveys could be administered online (taking up less class time, allowing for password protection of surveys), and because these courses were required for all students. In accordance with IRB requirements, one week before surveys were administered, students in the participating classes were given parental declination forms and asked to return the form if parents did not desire their students to participate in the study. Given the innocuous and anonymous nature of the survey, the IRB approved parental declination in this instance. No forms were returned. On the day of survey administration, students were reassured that survey completion was not mandatory. On the first page of the online survey, students were directed to an informed consent letter explaining the study and indicating survey completion was not mandatory. Acknowledgment of the informed consent was necessary to begin the survey. Surveys were administered on two occasions, once in the fall semester and once in the spring semester. Both survey administration times were near the end of the school semester. All students who were mentally capable completed the survey during class.

Over the two semesters of data collection, 256 participants completed the survey. There were 143 participants the first semester and 113 participants the second semester. No students completed surveys in both semesters. All participants in this sample ($n = 161$) and analysis were included because they self-identified as having text messaging available and responded to the question prompt. Of the participants in this study, there were more female ($n = 91$) texters than male ($n = 70$) texters. Texters were mostly seniors ($n = 89$), followed by juniors ($n = 44$), and sophomores ($n = 27$).

Most of the sample were Caucasian ($n = 86$) and Latino ($n = 51$). There were also African American ($n = 7$), Asian ($n = 4$), and Native American ($n = 4$) participants. A portion of respondents ($n = 8$) identified themselves as "other" ethnicity. A large majority of the participants self-identified their family's income as middle class ($n = 101$). Others identified their family income as upper class ($n = 7$) or lower class ($n = 52$).

To examine potential stereotypes surrounding adolescent text messaging, we presented participants an open-ended question asking if they thought adults had any misconceptions about text messaging and then we invited them to explain their response. The final qualitative data set included the compilation of adolescent responses to the open-ended question organized by gender and year in school. No transcription process was necessary due to online data collection.

Data were analyzed using qualitative methodology as outlined by Bogdan and Biklen (2003). These procedures required researchers to read and reread the data to gain a totality of the data. Coding categories, used to examine the data, were developed and refined based on patterns within the data. In accordance with Bogdan and Biklen's suggestions, two researchers read through the data independently multiple times to identify possible coding categories that would capture the common experience among adolescents in this study. Following the independent immersion process, the two independent researchers met, discussed impressions, and agreed upon commonalities in the data. Both noticed patterns surrounding adolescent discussion of adult misconceptions including time investments, generational differences, and a youth culture surrounding text messaging. These themes were collapsed into one coding category of misunderstandings surrounding motivations and practices. Later, the themes that had been collapsed for coding were decidedly used as subthemes to provide depth to each of the identified major themes.

This process of discussion, synthesis, and agreement between the independent coders was used for the other coding categories as well. For example, when discussing coding categories, the researchers noticed many teens referenced misconceptions regarding sexting or other questionable text message content. As these commonalities emerged, they were synthesized into a theme specific to adult misconceptions regarding message content. Others participants were quick to note adult misperceptions of impacts on developmental components of adolescence such as sociality, communication, and academic performance. This became the coding category regarding developmental impacts. The final coding category of "no misconceptions" did not take much discussion among researchers. Participants who did not feel there were misconceptions were concise in their explanations. Using the established coding categories, each researcher then reread and independently coded the data.

While coding responses, there were instances where the researchers failed to agree on items. To resolve these disagreements the researchers again met together and each researcher explained his/her reasoning for the coding decision. Together the researchers consulted the data to ensure the decided theme accurately represented the adolescent experience and, through discussion, consensus was established. A consensus was reached for each item that was disagreed upon during the independent coding process. After data were coded, one researcher re-organized the qualitative data set by the coded themes established by both researchers.

To increase the trustworthiness of the results of the process, inter-rater reliability, member checking, and theoretical connections were used. A percentage of agreement for inter-rater reliability was calculated at 93.9%. Since members of the sample could not be reached due to school district restrictions, a variation of member checking (Cho & Trent, 2006) was used to recruit members of the population (adolescent high school students) to examine the results of the data to verify that the data reflected an adolescent experience. Two adolescents reviewed the results section and reported that the themes were representative of their experience. Finally, a discussion of theoretical application to the results is included in the discussion section.

RESULTS

As mentioned, the coding categories were collapsed based on identified patterns in the data to form overarching categories. The same categories that were collapsed ultimately served as subcategories for organization within the greater categories. All categories and subcategories are presented in Table 1.

As the researchers immersed themselves in the data, four major themes emerged. In order to present results in a logical way, themes that represented an affirmative perspective are presented initially, with themes and subthemes presented in order of prevalence. First, 60 adolescents (37.3%) spoke about adults' general misperceptions about motivations and practices associated with adolescent texting. Next, 37 adolescents (23.0%) discussed adult misconceptions about the content of their messages. Finally, 22 adolescents (13.7%) talked about adults having misconceptions about the impact of texting behaviors on adolescents' developmental processes.

The majority of the participants commonly agreed that adults did have misconceptions about adolescent text messaging. However, 42 participants (26.1%) responded "no" to this question and explained their position. A general theme entitled "no misconceptions" emerged. Participants supported their perspective by stating common misconceptions are actually true and discussed how adults do not have misconceptions because they also use text messaging.

Table 1. Coding Categories and Subcategories.

Misconceptions of motivations and practices
Youth culture
Stereotypes of beliefs about practices
Consumption
Misconceptions of Message content
General content concerns
Sexting
Misconceptions about developmental impacts
Social impacts
Academic and language impacts
No misconceptions

MISCONCEPTIONS OF MOTIVATIONS AND PRACTICES

Of the participants in this sample, 60 (37.3%) indicated that adults had misconceptions regarding motivations and practices. The majority of the participants who discussed adult misconceptions of motivations and practices focused on texting as being something that teens easily understood, yet adults had a harder time grasping – indicating a potential youth culture. Next, adolescents indicated that many adults take the practices and motivations of a few teenagers and generalize them to all teenagers. Finally, these participants explained that adults had misconceptions about adolescent motivations and practices when it comes to consumption, both actual time and personal investment in texting.

Youth Culture

The majority of the participants who spoke of adult misconceptions about motivations and practices surrounding teenage text messaging behaviors noted something of a youth culture that adults do not comprehend. Some discussed these misconceptions as adults' lack of ability to use texting or as a result of a generational difference. Some participants talked about the inability of adults to see the value of texting that is readily apparent to an adolescent.

Participants frequently highlighted that adults have misconceptions about teenage use of text messaging because adults lack the abilities needed to effectively text that adolescents possess. For example, one participant said, "Just because adults text like 40 wpm (words per minute) doesn't mean they have the right to be jealous of teens texting" (female, senior). Another participant said, "They criticize texting in general. If they knew how then I'm sure they would do it just as much as we do" (male, junior). One participant indicated that because adults are not as adept at texting as are adolescents that it makes adults focus on potential miscommunication. She said:

> They don't feel comfortable with technology or see past the whole "it's awesome to send someone a message instantly" thing and worry about the realities like, what if they misunderstand the text, or what if they say something they wouldn't normally say when, for example, talking to them in person or even on the phone. (female, senior)

These participants also noted that some adults do not understand teenagers' texting behaviors because of their age or the generation to which they belong. One participant simply stated the adults have misperceptions, "... because they are old" (female, junior). A few participants discussed the generational differences between teens and adults. One said, "... when they were young the thing to do was to call, but for us it's to text" (male, senior). Another participant said, "From my experience, my parents feel as if texting is irrelevant, and it is just easier than calling people like the way they did it in the 80s" (female, senior). Speaking to this generational inability to understand the motivations and practices surrounding text messaging, one participant noted:

> I believe it is because their generation did not grow up with the technology of our day so they don't understand our fascination with things that help us communicate more easily and efficiently than a phone call or e-mail. They believe talking in person or calling is easier and a better way of communicating to others, but now-a-days that is considered old school. (female, senior)

For these participants, some adults do not see the ease in communication through texting, nor do they understand how it can be employed appropriately. One participant explained that adults have misconceptions about motivations and practices because they do not see what adolescents value in this form of communication. She said:

> I think adults don't see the value in texting as a form of quick communication and networking. It's easy to text a lot of people and you can "keep in touch" that way while never actually having to see the person. Later in a business or company having these "friendships" can be useful. (female, senior)

Other participants similarly acknowledged that adults do not see what teenagers see, and that adults do not understand appropriate practices surrounding text messaging. One participant summarized this idea well when she said:

> Of course, my parents think it would be easier to just call, but they don't realize how easy it is if you're busy to just text someone to tell them: "I'm running late, I'm on my way, call me when you can, meet for lunch???" Some situations are good for texting. Now arguing, that should be in person, or at least over a phone call, but a text can be a lot quicker and more convenient for the other person too. My parents always say that I answer a text quicker than a phone call, and that's true. Sometimes I can't answer a phone call!! Mom – quit calling me in class, I CAN'T ANSWER :). (female, senior)

Stereotypes of Beliefs about Practices

Many participants explained that the misconceptions adults express about text messaging are based on their stereotypes of the motivations and practices of a select few adolescents to the whole adolescent population. Some of the commonly misconstrued motivations and practices generalized to all included general judgments or ignorance, specific texting instances that could be dangerous or disruptive, and misconceptions about the amount of texting in which adolescents engage.

These participants felt adults use stereotypes of texting to explain all teenagers. One said, "Not all of us are the same. Some people text more than others" (female, senior). Another participant echoed this response and said, "Not everyone is the same and we shouldn't be judged for a few peoples' ignorance" (female, junior).

Some participants spoke about a small group of adolescents that ruin the image of adolescent texting motivations and practices for adults. One participant spoke about texting while driving and texting during class. She said, "They seem to think that all teenagers text while driving and in class, when there are those who don't rely on text messages that much" (female, senior). One participant said, "... it's only the few that don't follow rules or etiquette that puts a bad image of texting into someone's mind" (female, senior). Similarly, one participant said:

> They [adults] have a tendency to think that all teenagers text a lot, but really not all of them do. Those who do text a lot text so much that they make it seem as though the rest of teens are also texting. (female, sophomore)

Consumption

Adolescents who indicated that adults do have misconceptions about motivations and practices of adolescents' texting discussed adults' faulty views of the amount of time, including physical and mental engagement, in which adolescents use text messaging. Some adolescents reported that adults see their practices as a waste of time, and some suggested adults see texting as consuming in the way a drug is consuming.

A prevalent theme from this category of responses was that adults think adolescent motivations and practices with texting are time consuming. One participant said, "They think we are on the phone all the time.

If we were on the phone all the time we would never get anything done, but I always get my stuff done" (male, junior). Another participant said, "Adults feel we are wasting time, but they don't understand that we like to stay in contact" (female, sophomore). One participant defended the practice of spending time with texting. She said, "They think you text too much. Well, duh. You're having a conversation with someone!" (female, sophomore)

A few adolescents indicated that adults' misconceptions take text messaging and compare it to something addictive, like a drug. One participant said, "… they believe that texting is a drug and it takes over people's lives" (male, junior). Similarly, another participant said, "They think we are addicted to it, when in reality, we aren't. Obviously we could quit texting at any time, because cell phones didn't always exist and they didn't always have texting available" (female, sophomore).

MISCONCEPTIONS OF MESSAGE CONTENT

Of the participants who responded to the question, 37 (23.0%) indicated that adults had misconceptions regarding the content of their text messages. Adolescents explained that many adults' misconceptions about texting come from misperceptions about the content of a text message. These participants suggested that adults, in general, assume the content is negative or harmful. For example, many discussed sexting and some talked about parents thinking that adolescents are texting strangers. One participant summed up these adult misconceptions about content by saying, "They may think that we are texting someone we don't know, or texting something dirty, or planning to do something bad with someone else" (female, sophomore). Another participant also addressed all of these concerns when he said, "They [adolescents] might be thinking wrong ideas, [engaging with] bad influences, or sending nude images, or talking to the wrong kind of people" (male, junior).

General Content Concerns

Some adolescents expressed adult misconceptions concerning bad content or behaviors associated with text messages, generally. This subtheme emerged through statements like: "They think we're doing something bad or hiding stuff" (male, senior). "They usually think teenagers are texting bad things" (male, senior). "When teenagers text it is not always about the bad things. Most of the time they are still innocent little conversations going on" (male, senior). A female participant's words further explain adult misconceptions concerning texting content. She said, "Adults feel that teenagers who text are up to no good. That is SO not true. Some things we text are really important and just fun natured. I, personally, am never texting badly" (female, sophomore).

Sexting

Many participants focused on adult concerns about adolescents' sexting. One participant said, "They see on the news about the porn and make wrong assumptions about every text" (female, junior). Some participants defended the real content of adolescents' messages. For example, one adolescent said, "They think every teenager is doing something wrong with their cell phone, like sending pictures of things they shouldn't. Not every teen does that, but it's true some do" (female, senior). Another participant said, "They think all we do is text and sexting. But it is the way we talk to another when we can't see them" (male, sophomore). One participant was very straightforward and said, "Not everyone is sending out texts for girls to show them their boobs" (male, senior).

Other participants suggested there was a misconception about message content associated with texting, but they also supported this adult perspective based on the content of their own messages. One participant said, "They think we are talking about naughty stuff, and sometimes we are and sometimes we are not. I got to admit that I've been caught talking about naughty stuff" (female, sophomore). One participant suggested that whether they were texting or talking on the phone, similar content would emerge. He said, "The news makes it sound worse than it really is. Sometimes we are just talking and yes, sometimes it is more erotic. But isn't it gonna be the same when talking on the phone?" (male, sophomore). Another participant said,

> Since there's been a lot of "sexting" going around the parents/adults will place more cautions on their teens for texting way too much. They could be talking to a complete stranger right in front of their parents and their parents wouldn't even know a thing. It's kinda ridiculous. (female, sophomore)

MISCONCEPTIONS ABOUT DEVELOPMENTAL IMPACTS

Twenty-two adolescents (13.7%) expressed how adults have misconceptions about the developmental impacts of text messaging. Some spoke about social impacts, and others spoke about academic and language impacts.

Social Impacts

Adolescents expressed that adults are very concerned about the social impacts of text messaging. For example, one participant said, "Some adults think teenagers who text too much are losing their social life. In some cases I can agree, but they shouldn't assume it's occurring to every teen" (female, senior). Some participants felt that this social aspect of texting was not damming to social abilities, but a facilitator of social abilities. One participant said, "I believe that [texting] is needed to help with the social part of life when needing to talk about your feelings" (male, junior).

One participant talked about adults' concerns regarding the impact of texting on face-to-face interactions. She said, "I believe adults are worried about the interaction between teens and, therefore, the lack of interaction between teens when texting comes into the bigger picture" (female, senior). Another participant similarly said, "They think that teenagers are losing their ability to speak and act appropriately in a social setting, but we do just fine in both" (female, senior). One participant said:

> Parents believe we only socialize through text, which for me is not true. I still hang out with my friends every weekend, go on lots of dates, I call my cousin and can talk for hours with her about new things in our lives, and I still have school, church, and sport games that I socialize with others. (female, junior)

Academic and Language Impacts

Some of the participants in this study perceived adults having misconceptions about texting because adults believe it is hindering adolescents' academic abilities and their proper use of language. One participant spoke regarding texting impacts on language. She said, "When we speak improperly it's mostly just to piss our parents off" (female, senior). One participant stated that texting has just as much impact on adolescent language abilities as it does on adult language abilities. He said, "It ruins our English just as bad as theirs" (male, senior). Another participant explained, "They think we are stupid for it, that it is lowering our communicating ability, when in fact it is helping us communicate, and if used correctly can be a healthy environment" (male, junior).

Some participants who indicated that this was a misconception that adults possessed about texting also pointed out that there was potential for language impacts. One participant said, "They [adolescents] use it sometimes in school like 'lol,' 'y' instead of 'why,' '4' instead of 'for,' well, you get my point" (female, junior).

NO MISCONCEPTIONS

A portion of the sample (42 participants, 26.1%) indicated that they did not perceive that adults had misconceptions about texting. Responses that indicated a stance of no adult misconception grouped in two ways: (1) misconceptions about texting that are commonly discussed actually represent adolescent texting behaviors and (2) adolescents are not the only people who text, and adults understand texting because they too use this technology feature.

Many of the responses explained that the supposed adult misconceptions about text messaging are actually accurate. Some of the participants even supported adult beliefs about texting. For example, one participant said, "Texting can be very distracting. I do 100% believe no one should text and drive" (female, junior). Another participant said, "They just all think teens text too much and I'm not gonna lie, I do text A LOT (: ha ha" (male, senior). One participant felt it was not an adult misperception that was the problem, but actual teenage practices. He said, "A lot of us who do text are little brats about it and need a reality check" (male, senior).

Others who said they did not believe adults had misconceptions about adolescent texting felt this was a result of adults having the ability to text, as well. One participant said, "Adults text too. Teens aren't the only ones that text" (female, senior). Another teen said, "All the adults I know text as well" (male, junior).

DISCUSSION

Mixed findings have emerged in previous research pertaining to adult ideas about teen texting. Some research findings even suggest an agreement in opinions about texting between adolescents and adults (Lenhart, 2010).

Participants in this study expressed multiple misconceptions surrounding adolescent text messaging they have perceived in adults. Many spoke about adult misconceptions regarding adolescents' motivations and practices, others discussed misconceptions about message content, and some felt that adults had misconceptions about the developmental impact of texting. These themes provide a starting point to understanding adult stereotypes relating to members of the youth culture. Further, these results are important to consider since they are not simply news or popular media reflections of adults' opinions, but a review of the actual adolescent experience.

YOUTH CULTURE

Exploring adolescent perceptions of adult misperceptions relating to teen texting served as an innovative method for generating a clearer picture of the youth culture that has developed around teen text messaging. The question required adolescents to reflect on their personal cultural self-awareness – highlighting "some of the values, behaviors, beliefs, and customs that are identified within one's own [culture]" (Lynch & Hanson, 2011, p. 46). As adolescents pondered the discrepancies between perceived adult misconceptions, they detailed their actual cultural experiences. According to participants in this study, members of youth culture value frequent, quick, and asynchronous communication. This cultural value is supported by previous research on adolescent texting (Bryant et al., 2006). The customs related to texting within youth culture were perceived to provide immediate social benefits (personal connection) and potential professional benefits in the future. Further, cultural rules for appropriate types of cell phone communication emerged for different settings. For instance, an adolescent encouraged her mother to quit calling her during class, because it was not an appropriate cell phone behavior.

Adolescents tried to normalize their cultural practices of texting by comparing them to technological social practices of adults when they were youth – landlines versus texting. However, in general, adolescent responses supported the proposition that the youth culture was incomprehensible or invisible to adults (Oksman & Turtiainen, 2004; Tulane, 2012). This was often supported by proposed generational differences in adolescent and adult ability to use technology in the direction of previous research reports – adolescents have an easier time engaging in this newer media than adults (Bryant et al., 2006).

Framing the study using perceived adult misperceptions highlighted several features within the youth culture. The framework also acknowledged some within cultural differences – adolescents specifically commented on how there were shared rules and practices, but there were also some differences in perceptions between members of this youth culture (e.g., misconceptions vs. no misconceptions). Future research should consider further identifying the unique cultural factors within this youth culture surrounding texting and other new media. Learning accurate information about culture is one way to increase cross-cultural competence (Lynch & Hason, 2011). Thus, creating increased accessibility to empirical and public knowledge about youth culture may decrease any unwarranted/undocumented misconceptions or stereotypes relating to the adolescent cultural experience.

PERCEIVED ADULT MISCONCEPTIONS OF TEXTING

Perceived Misconceptions

Considering the results as a whole, the majority of adolescents reported adult misconceptions about adolescent texting – some of which have been highlighted in previous research. Cupples and Thompson (2010) reported that adolescents perceived that adults viewed texting as a waste of time and respondents in this study validated this finding as they discussed perceived misconceptions about adolescents' over consumption of texting. In addition to the amount and frequency of texting, some adolescents reported adult misconceptions about general content distributed through texting and specifically sexting. A few participants disclosed sending sexual images, which is consistent with Lenhart et al.'s (2010) reported small percentage (4%) of teens sending sexts, while the larger part stated that this was not practiced by the majority of adolescents. Some participants indicated sexting was a misconception, and others reaffirmed adult perceptions that sexting is a part of adolescent texting. Both perceptions reflect potential stereotypes for adolescents. This could be a reflection of (1) a perceived stereotype as well as (2) a "self-fulfilling prophecy" stereotype threat (Aronson, 2002) surrounding texting behavior.

Adolescent participants explained that adults also had misconceptions about the potential developmental impacts of text messaging. Some spoke about misconceptions over social impacts. This misconception is supported by past research. Researchers have shown that texting is often used to build face-to-face relationships and social networks (Berg et al., 2005; Ehrenberg, Juckes, White, & Walsh, 2008; Ling, Julsrud, & Yttri, 2005). For the participants in this study, this was also the case. They felt they were building social networks, not destroying social abilities. This is important to consider that texting may actually be enhancing social skills and social networks – not eliminating an adolescent's ability to communicate in face-to-face situations.

Participants collectively described adult misconceptions about the impact of texting on language development. Both the media and parents have mentioned this as a concern (Thurlow & Bell, 2009). One participant felt this was a misconception, but noted that textese (e.g., acronyms and abbreviations used in texting) is showing up in school. However, the majority of participants in this study indicated that this idea of negative language impacts was a misnomer. In fact, some felt that adults have just as much potential for language impacts. Although participants were not asked the extent their parents participate in texting, adolescents felt that the potential for texting to influence developmental abilities was not targeted solely to their own age group. It is important to recognize that adolescents recognize they are not the only users of text messaging. Furthermore, some felt that a common misconception of their own behavior was not limited by age. These particular results, surrounding developmental impacts of text messaging, indicate that adolescents recognize the adult stereotype surrounding their texting behaviors, but that the stereotype was unfounded and the opposite true.

The final group of adolescents in the study did not report any misconceptions from adults about actual adolescent texting behaviors. There was also a group of participants who reported that adults had an accurate perception of adolescent texting behaviors. These participants felt either that adult perceptions were accurate or that adults text as much as teenagers do, so it is a non-issue. For this group of adolescents, stereotypes were not significant, especially since they were able to group adults into a technology using category with themselves.

A Pathway to Texting Stereotypes

Reflecting on the current findings and previous research, a potential pathway for the development of texting stereotypes emerged. Participants made it clear that adults have obtained information about the risks and problematic behaviors that can occur via texting. It is likely that much of this comes from media representations of teen texting. However, as pointed out by participants, because adults also text (Duggan, 2013) some of their information is probably derived from personal experience, but is not translatable because adults have not acculturated into the youth culture surrounding texting.

Participants in this study indicated that the public information privy to adults is largely based on the behavior of the minority adolescent texter. This claim is supported by previous research showing that only a small percentage of adolescents engage in deviant texting behaviors (Lenhart et al., 2010). Participants do not deny that problematic behaviors occur via texting. In fact, a small portion of the current sample openly admitted to behaviors such as sexting and texting too much. However, participants did deny that the minority texting behaviors were representative of all adolescent texters. The participants explained that perceived stereotypes developed as the minority texting experience are generalized to all adolescent texters. Further, adolescents felt that it was unjust to have their own texting behavior judged by these perceived stereotypes – hinting at the presence of stereotype threat (Aronson, 2002).

LIMITATIONS AND CONCLUSIONS

The qualitative nature of this study allowed for an in-depth understanding of adolescents' perceptions of adult misperceptions of text messaging. Two common limitations also existed in this study. These included using a convenience sample from only one high school, and including only adolescent perspectives. Future research should also examine the parent–child dyad in relationship to texting. As historically noted, parents and adolescents may, at times, have conflictual relationships, with many of the encounters being over mundane issues (Collins & Laursen, 2004). Texting and other communication media are so embedded in youth culture and daily living that it is ordinary (Cupples & Thompson, 2010), not something extraordinary. As such, texting often becomes a possible source of conflict for adolescents and parents. The participants in this study provided their perspectives of this potential conflict with parents and adults surrounding texting, while simultaneously explaining texting is a part of being an adolescent – it is their youth culture.

This research adds a necessary positive examination of media use. As mentioned, parents and news media are quick to point out the potential negative impacts of text messaging, yet this research provided a more positive light for understanding the impacts of texting. According to these adolescents, even though they are engaging in texting as a primary form of communication, it is still contributing to their face-to-face relationships, not necessarily decreasing their abilities to communicate with others in physical social settings. Adolescents are also very aware of themselves, what their behavior looks like, and how adults perceive their actions. Adolescents recognize the potential for harm to themselves through texting. They are also quick to point out the level of control they can exercise in their own lives over their texting behaviors.

Results also indicated the reality of shared perceived stereotypes relating to adolescent texting and hint at the possibility of stereotype threat. Future research is warranted to identify what stereotypes are actually present among adults. Identifying the commonality of the stereotypes can then lead to further investigations into "why they are shared and how they come to be shared" (McGarty et al., 2002, p. 5) ways of viewing adolescent texting. Also, stereotype threat has been associated with poorer or negative outcomes for people in a variety of settings (see Galdi, Cadinu, & Tomasetto, 2014; Silverman & Cohen, 2014). More in-depth exploration of the experience of stereotype threats related to adolescent technology use and the potential influence on behavior (e.g., perceived stereotype threat related to lack of social skills because of technology use – how does this influence actual social skills?) represents an important next step.

Finally, this research contributes to the understanding of social implications of adolescent text messaging. Although more remains to be examined, this research helps to further clarify the youth culture of texting. The youth culture being built through texting seems to be more impactful than just something surrounding a communication device, but a culture that can help researchers in understanding the lived adolescent experience. Adolescent media use, in general, is a way to accurately understand the youth culture that adolescents develop, and, in turn, the youth culture that influences their development. Although there are media that have been extensively studied in relation to adolescent development, newer media such as social networking and texting need further examination, as they have become definitive components of youth culture. As adolescents further engage in social networking, they present themselves socially and publically using both private and public information. As adolescents further engage in developing their youth culture through newer media, conflict and misunderstanding, and living with stereotypes from adult groups is inevitable.

REFERENCES

Aronson, J. (2002). Stereotype threat: Contending and coping with unnerving expectations. In J. Aronson (Ed.), *Improving academic achievement: Impact of psychological factors on education* (pp. 279–301). San Diego, CA: Academic Press.

Berg, S., Taylor, A. S., & Harper, R. (2005). Gift of the gab. In R. Harper, L. Palen & A. Taylor (Eds.), *The inside text: Social cultural and design perspectives on SMS* (pp. 271–285). Dordrect: Springer.

Blair, B. L., & Fletcher, A. C. (2011). The only 13-year-old on planet Earth without a cell phone: Meanings of cell phones in early adolescents' everyday lives. *Journal of Adolescent Research, 26*, 155–177. doi:10.1177/0743558410371127

Bogdan, R., & Biklen, S. K. (2003). *Qualitative research for education: An introduction to theories and methods*. Boston, MA: Allyn and Bacon.

Brown, J. D., & Bobkowski, P. S. (2011). Older and newer media: Patterns of use and effects on adolescents' health and well-being. *Journal of Research on Adolescence, 21*, 95–113. doi:10.111/j.1532-7795.2010.00717.x

Bryant, J. A., Sanders-Jackson, A., & Smallwood, A. M. K. (2006). IMing, text messaging, and adolescent social networks. *Journal of Computer-Mediated Communication, 11*, 577–592. doi:10.1111/j.1083-6101.2006.00028.x

Charon, J. (2012). *Ten questions: A sociological perspective* (8th ed.). Belmont, CA: Wadsworth.

Cho, J., & Trent, A. (2006). Validity in qualitative research revisited. *Qualitative Research, 6*, 319–340. doi:10.1177/1468794106065006

Collins, W. A., & Laursen, B. (2004). Parent-adolescent relationships and influences. In R. M. Lerner & L. Steinberg (Eds.), *Handbook of adolescent psychology* (2nd ed., pp. 331–361). Hoboken, NJ: Wiley.

Coomes, M. D., & DeBard, R. (2004). A generational approach to understanding students. New *Directions for Student Services, 2004*, 5–16. doi:10.1002/ss.121

Cupples, J., & Thompson, L. (2010). Heterotextuality and digital foreplay. *Feminist Media Studies, 10*, 1–17. doi:10.1080/14680770903457063

Davie, R., Panting, C., & Charlton, T. (2004). Mobile phone ownership and usage among pre-adolescents. *Telematics and Informatics, 21*, 359–373. doi:10.1016/j.tele.2004.04.001

Draper, N. R. A. (2011). Is your teen at risk? Discourses of adolescent sexting in United States television news. *Journal of Children and Media, 6*, 221–236. doi:10.1080/17482798.2011.587147

Duggan, M. (2013). *Cell phone activities 2013*. Pew Internet and American Life Project. Retrieved from http://pewinternet.org/~/media//Files/Reports/2013/PIP_Cell%20Phone%20Activities%20May%202013.pdf

Ehrenberg, A., Juckes, S., White, K. M., & Walsh, S. P. (2008). Personality and self-esteem as predictors of young people's technology use. *CyberPsychology & Behavior, 11*, 739–741. doi:10.1089/cpb.2008.0030

Galdi, S., Cadinu, M., & Tomasetto, C. (2014). The roots of stereotype threat: When automatic associations disrupt girls' math performance. *Child Development, 85*, 250–263. doi:10.1111/cdev.12128

Gardner, H., & Davis, K. (2014). *The app generation: How today's youth navigate identity, intimacy, and imagination in a digital world*. New Haven, CT: Yale University Press.

Harley, D., Winn, S., Pemberton, S., & Wilcox, P. (2007). Using texting to support students' transition to university. *Innovations in Education and Teaching International, 44*, 229–241. doi:10.1080/14703290701486506

Holtgraves, T. (2011). Text messaging, personality, and the social context. *Journal of Research in Personality, 45*, 92–99. doi:10.1016/j.jrp.2010.11.015

Howe, N., & Strauss, W. (2003). *Millennials go to college*. Great Falls, VA: American Association of Registrars and Admissions Officers and LifeCourse Associates.

Kemp, N. (2010). Texting versus txtng: Reading and writing text messages, and links with other linguistic skills. *Writing Systems Research, 2*, 53–71. doi:10.1093/wsr/wsq002

Lenhart, A. (2010). *Teens, cell phones and texting*. Pew Internet and American Life Project. Retrieved from http://pewresearch.org/pubs/1572/teens-cell-phones-text-messages

Lenhart, A. (2012). *Teens, smartphones, & texting*. Pew Internet and American Life Project. Retrieved from http://www.pewinternet.org/2012/03/19/teens-smartphones-texting/

Lenhart, A., Ling, R., Campbell, S., & Purcell, K. (2010). *Teens and mobile phones*. Pew Internet and American Life Project. Retrieved from http://www.pewinternet.org/˜/media//Files/Reports/2010/PIP-Teens-and-Mobile-2010-with-topline.pdf

Ling, R. (2005). Mobile communications vis-à-vis teen emancipation, peer group integration and deviance. In R. Harper, L. Palen, & A. Taylor (Eds.), *The inside text: Social cultural and design perspectives on SMS* (pp. 175–193). Dordrect: Springer.

Ling, R., Julsrud, T., & Yttri, B. (2005). Nascent communication genres within SMS and MMS. In R. Harper, L. Palen, & A. Taylor (Eds.), *The inside text: Social cultural and design perspectives on SMS* (pp. 75–100). Dordrect: Springer.

Lynch, E. W., & Hanson, M. J. (2011). *Developing cross-cultural competence*. Baltimore, MA: Paul H. Brooks.

Mahatanankoon, P., & O'Sullivan, P. (2008). Attitude toward mobile text messaging: An expectancy-based perspective. *Journal of Computer-Mediated Communication, 13*, 973–992. doi:10.1111/j.1083-6101.2008.00427.x

May, H., & Hearn, G. (2005). The mobile phone as media. *International Journal of Cultural Studies, 8*, 195–211. doi:10.1177/1367877905052417

McElhaney, K. B., Allen, J. P., Stephenson, J. C., & Hare, A. L. (2009). Attachment and autonomy during adolescence. In R. M. Lerner & L. Steinberg (Eds.), *Handbook of adolescent psychology* (3rd ed., Vol. 1, pp. 358–401). Hoboken, NJ: Wiley.

McGarty, G. (2002). Stereotype formation as category formation. In C. McGarty, V. Y. Yzerbyt & R. Spears (Eds.), *Stereotypes as explanations* (pp. 16–37). New York, NY: Cambridge University Press.

McGarty, G., Yzerbyt, V. Y., & Spears, R. (2002). Social, cultural and cognitive factors in stereotype formation. In C. McGarty, V. Y. Yzerbyt, & R. Spears (Eds.), *Stereotypes as explanations* (pp. 1–15). New York, NY: Cambridge University Press.

Oksman, V., & Turtiainen, J. (2004). Mobile communication as a social stage: Meanings of mobile communication in everyday life among teenagers in Finland. *New Media & Society, 6*, 319–339. doi:10.1177/1461444804042518

Pew Research Center. (2010). *Millennials: A portrait of generation next*. Retrieved from http://www.pewsocialtrends.org/files/2010/10/millennials-confident-connected-open-to-change.pdf

Silverman, A. M., & Cohen, G. L. (2014). Stereotypes as stumbling-blocks: How coping with stereotype threat affects life outcomes for people with physical disabilities. *Personality and Social Psychology Bulletin, 40*(10), 1330–1340. doi:10.1177/0146167214542800

Strauss, W., & Howe, N. (1991). *Generations: The history of America's future, 1584 to 2069*. New York, NY: Morrow.

Thurlow, C., & Bell, K. (2009). Against technologization: Young people's new media discourse as creative cultural practice. *Journal of Computer-Mediated Communication, 14*, 1038–1049. doi:10.1111/j.1083-6101.2009.01480.x

Tulane, S. (2012). *Social implications of adolescent text messaging*. Unpublished Doctoral Dissertation, Utah State University, UT.

Vaterlaus, J. M., & Tulane, S. (2015). Digital generation gaps in parent-adolescent relationships. In C. J. Bruess (Ed.), *Family communication in the digital age*. New York, NY: Peter Lang Publishing.

Williams, S., & Williams, L. (2005). Space invaders: The negotiation of teenage boundaries through the mobile phone. *The Sociological Review, 53*, 314–331. doi:10.1111/j.1467-954X.2005.00516.x

WATCHING TELEVISION AND READING ACHIEVEMENT: A STUDY OF THIRD GRADE LANGUAGE MINORITY STUDENTS

Gregory J. Mills

ABSTRACT

Purpose – *The current paper examines the relationships between watching television for various times of day and reading achievement for a subsample of third grade language minority (LM) students compared to third grade students in general.*

Methodology – *The analysis uses ECLS-K 1998–99 data to first test for significant differences between the two samples, then further explores these relationships using separate OLS multiple regression models, while controlling for past reading achievements and socioeconomic variation.*

Findings – *Building on more nuanced versions of displacement theory, this paper finds a positive relationship between reading achievement and watching television after dinner on weekdays specifically for LM students. For the general sample, watching TV on weekends or weekdays at any time period has no relationship with reading achievement.*

Technology and Youth: Growing Up in a Digital World
Sociological Studies of Children and Youth, Volume 19, 395–419

ISSN: 1537-4661/doi:10.1108/S1537-466120150000019014

Originality/value – *This research suggests the potential for TV or perhaps other media to act as a lingual- or cultural-learning facilitator for LM students, being positively tied to reading achievement. The paper's unique focus on multimedia use and LM students makes it particularly applicable to educators and public policy officials tasked with confronting the reading skills gap for a growing LM student population.*

Keywords: Language minority; multimedia use; immigrant children

INTRODUCTION

The relationship between watching TV and learning for children has been extensively debated, both publically and in various academic fields (Ennemoser & Schneider, 2007; Neuman, 1991; Waters, 1977). There are several reasons this topic draws such wide attention. First, levels of not just TV consumption, but all forms of screen media have been rising considerably in recent years, with estimates that American youth between the ages of 8 and 18 spend over seven hours per day consuming screen-based media (Rideout, Foehr, & Robers, 2010). Also, academic achievement in general is widely perceived to have been dropping in recent decades, with the Programme for International Student Assessment (PISA) consistently ranking the United States in the 20s in terms of reading achievement (OECD, 2014). Lastly, many cultural critics have argued that we are witnessing the more general "dumbing down" of youth, education, and American culture, in general (Gatto, 2002; Sykes, 1995; Washburn, Thornton, & Simon, 1996, Winn, 1977) – and increased multimedia consumption (particularly television) is often argued to be no small part of that process (Rabe, 2010; Stout, 2000).

The scholarly research, however, paints a more nuanced picture. For early elementary school students, past research indicates a slightly negative correlation between watching TV and academic achievement in the aggregate ($r = -.05$), with stronger negative effects for younger children, heavier watchers, and children from higher socioeconomic backgrounds (Razel, 2001; Schmidt & Anderson, 2007; Williams, Haertel, Haertel, & Walberg, 1982), even while the quality of the content watched of course also plays a large role in mediating this process (Anderson et al., 2001). And while this study focuses on TV consumption, past research finds similar relationships with personal computer media consumption as well (Attewell, Suazo-Garcia, & Battle, 2003; Straker, Pollock, & Maslen, 2009).

One group that scholars have largely failed to consider is language minority (LM) students, or students who do not speak English as a first language. This is a significant oversight, as the number of LM students entering the American public school system has dramatically risen in recent decades to about 21 percent (or 10.9 million students) (Aud et al., 2010). These numbers are expected to rise even further in the coming years, thus making literacy and reading achievement among LM students a principle concern among educators and education policy makers.

An additional oversight in this body of research is the way in which watching TV is structured into the family's and child's daily routine. For example, it could be that for younger children, watching children's programming, which may be more educational and is typically broadcasted in the late afternoon, is more beneficially than watching more adult-centered programming (Schmidt & Anderson, 2007). On the other hand, watching TV after dinner might be less likely to displace developmental activities such as after school programs and peer interaction. Thus, the time of day in which watching TV occurs could have significant effects on the potential benefits in terms of academic achievement.

The present study looks to compare the relationships between parent-reported television consumption and reading achievement among third grade LM students and third grade students in general, paying special attention to the time of day in which watching TV occurs, while also controlling for past achievement. These results will help researchers unpack how television and multimedia consumption, in general, might be related to the academic growth trajectories of LM students. Before this, however, I will first review various contextualities of interest concerning watching TV, making the case that the effects of watching TV might be differentiated for LM students.

BEYOND SIMPLE DISPLACEMENT THEORY: A MORE NUANCED VIEW OF THE EFFECTS OF EXPOSURE TO MEDIA

Perhaps, the most widely recognized theory in the study of the effects of watching television on academic achievement is *displacement theory* (Anderson et al., 2001; Neuman, 1991), which hypothesizes that watching TV displaces other activities that would otherwise occupy the child's time, such as reading, playing outside, doing homework, or socializing with friends and family. Typically, the activities that displaced are perceived to be more developmental or beneficial to the child's academic achievement or facilitating pro-social behavior (Williams & Siegel, 1986). And indeed, in an exhaustive research synthesis on the topic, Williams and colleagues did find there to be a slight but negative correlation of hours watching TV and achievement, regardless of sample size, year, and location (Williams et al., 1982). These conclusions had previously gained wide public acceptance in the mid to late 1970s when several reports showed that reading scores across the country had fallen drastically since the mid-1960s (Gaddy, 1986). In 1977, the cover of *Newsweek* magazine ran a story titled "What TV Does to Kids" (Waters, 1977), which painted TV as the definitive reason that an entire generation seemed to be falling behind. When Marie Winn made these claims in her popular 1977 book, *The Plug-in Drug* (now in its 25th anniversary edition), *The Washington Post* argued that she did so "with devastating persuasiveness," suggesting the genuine resonance of these themes in American society.

However, more complex models suggest that rather than a linear negative relationship between watching television and achievement, there exists a curvilinear relationship, where light viewing has little effect or even a slight benefit, while heavier viewing (more than four hours per day) is more likely to lead to significant declines (Beentjes & Voort, 1988; Comstock & Paik, 1991; Razel, 2001; Williams et al., 1982). Razel (2001), for example, proposed a model that is shaped as "an inverted check mark" (373), in that low amounts of viewership can positively affect achievement, with effects becoming negative as viewership becomes heavier. Additionally, she found that the "optimal" amount of time spent watching TV, or the peak of the inverted checkmark, decreases as a child ages, from about two hours per day at the age of nine, decreasing to a half hour by the age of 17. Interestingly, studies on the relationship between computer usage and attainment have shown similar results. First, Attewell and Battle (1999) show that the presence of a computer in the house is associated with educational gains among eighth graders, although the association was highly mediated by SES. Later, Attewell et al. (2003) use time diary data to explore these relationships with the amount of time spent at home on a computer and their original findings stand. In their study, decreases in time spent on other activities such as reading or playing outside only occurred for heavy users. Additionally, time spent computing was even associated with some measures of cognitive gains and lower BMI.

These findings leave us with the question of in a highly contextualized relationship between multimedia consumption and academic achievement that is on the one hand mediated by level of consumption, what other factors are associated with better or worse achievement? Razel (2001) hypothesizes that the relationship is mediated by age and the quality of programing geared toward different age groups – with programming for early elementary students being more educational in character, while middle and high school geared programming is more for entertainment. In line with this hypothesis, Wright, Huston, Murphy et al. (2001) found that children who watched educational programs between the ages of two and three scored higher on several measures of "school readiness," and that these effects were less apparent for children who watched general-audience programming. Likewise, other research shows younger children capable of gaining vocabulary knowledge by watching educational shows such as Sesame Street and others, even with minimal contextual clues (Rice, Huston, Truglio, & Wright, 1990; Rice & Woodsmall, 1988). More recent research using a multi-cohort design has found similar results, where the viewing of children's educational programing was tied to improved reading achievement (Ennemoser & Schneider, 2007). In sum, these lines of research go far to suggest that television can act as a learning facilitator if the content being viewed is cognitively challenging or educational in character.

LM STUDENTS AND WATCHING TELEVISION

It is thus surprising that such little research analyzing the specific impacts of watching television on the scholastic achievement of LM students, as past research does much to suggest that LM students would stand to gain considerably. As noted above, many scholars have pointed out that educational programming will have more positive and/or fewer negative effects on cognitive development (Ennemoser & Schneider, 2007; Rice et al., 1990; Rice & Woodsmall, 1988; Wright, Huston, Murphy et al., 2001). And as many younger LM children are in some stage of language development, it could be that watching target-language television *in general* is inherently educational and conducive to learning. This is because even with entertainment programming, the viewer often must follow complex plotlines, character dialoge, and various cultural scripts (Wierzbicka, 1994a, 1994b), which could conceivably require a high level of attention and concentration, thereby passively facilitating lingual development. It is reasonable to assume that this effect would have beneficial impacts on reading achievement, as found by Ennemoser and Schneider (2007) with children in Germany.

Additionally, Meskill (1998) writes that even entertainment television could serve as a powerful mechanism for language development and cultural assimilation, as it "represents a private window into U.S. culture otherwise not readily observable to nonnative speakers of English" (82). This theory suggests that language and vocabulary development lead to a more thorough adaptation to the classroom environment, which would in turn also have positive effects for academic achievement. Thus, the hypothesis that watching TV would be helpful to LM students is compelling, because these students typically have a great deal more to learn for successful adaptation to the school environment, such as language, culture, and knowledge. And as Meskill (1998) suggests, this might particularly be true if LM children live in more segregated neighborhoods, or otherwise have little chance to interact outside of their residential communities (Iceland, 2004; Sharp & Iceland, 2013).

And indeed, there are findings from the literature on multimedia use and language learning that potentially shed light on these relationships. Multimedia-enhanced learning methods have been shown to be uniquely effective for vocabulary learning among elementary LM students compared to non-LM students. Silverman and Hines (2009) found that LM students who were behind their grade level in terms of vocabulary knowledge showed more progress when given an intervention that included a multimedia component. Additionally, Uchikoshi (2006) investigated vocabulary growth in Spanish-speaking kindergarten students by way of TV viewing at home and school, and found that home-viewing of children's shows was a strong predictor of English language vocabulary growth, which was also tied to a faster pace of lingual and cognitive growth in the classroom.

Thus, there is reason to hypothesize that TV consumption might be positively related to reading achievement, assuming that watching television provides a cognitively or lingually challenging activity for LM students, and that at least a sizable portion of their total viewing time is in English. Additionally, if watching television can be viewed as pathway to cultural assimilation, I might also predict that programming must not need to be educational in character to be beneficial for LM students. LM students might benefit simply from exposure to culture programming that could also facilitate cognitive functioning in the classroom.

TIME OF DAY AND VIEWING CONTEXT

Particularly among young children, the context in which watching television occurs is crucial, as it "occurs in an environmental context that influences what and when viewing occurs, as well as the ways in which viewers interpret what they see" (Huston, 1992, p. 98). The current paper pays particular attention to the time of day in which watching television occurs, as well as the existence of household rules relating to watching TV. The basic premise behind this strategy is that *how* television is utilized might very well offset *how much* television is utilized, as detailed in Attewell's concept of "the second digital divide" (2001) as well as other works (Anderson et al., 2001; Wright, Huston, Vandewater et al., 2001).

Some past studies have looked at weekday television viewing versus weekend viewing, generally finding that children watch more TV on weekend (Bianchi & Robinson, 1997), but that weekday viewing is more influential to academic achievement (Keith, Reimers, Fehrmann, Pottebaum, & Aubey, 1986; Sharif & Sargent, 2006). However, to my knowledge no study has taken into account the *time of day* in which television viewing occurs. This is a considerable oversight. For one, a wide majority of children's programming is in the afternoon, while almost no children's programming occurs after 7:00pm (Woodard IV, 1999). Thus, according to the theory that the effects of television are largely a product of programming (for example, see Anderson et al., 2001), weekday morning and afternoon viewing would be predicted to be more beneficial than in the evening. On the other hand, TV viewing in the afternoon is more likely to be unsupervised by a parent or guardian. In fact, a middle elementary school student watching TV after dinner has been shown to be almost three times more likely to be watching with a parent or other adult, or coviewing (Christenson, 1994). Additionally, children have been shown to prefer noneducational content when the choice of content is left solely up to the child (Giacquinta, Bauer, & Levin, 1993).

Coviewing occurs when children and parental guardians watch television simultaneously. It has been theorized that coviewing can act as a mediator of potential negative effects of television, and as a way of increasing learning through parent–child discussion (Seels, Fullerton, Berry, & Horn, 2004). Indeed, there is supporting evidence that it does in fact facilitate engagement and comprehension of the program itself (Jinqiu & Xiaoming, 2004), such as plot lines and characterization, but very little has been done to directly measure these effects. Particularly for LM students, it could be that the quality of programming has less sway on TV's effect on achievement than family or coviewing context, as academic gains might be more related to general lingual development. While the data used in this analysis does not include information on coviewing or programming quality, looking at the effects of time of day will allow me to theorize about their relationship with academic achievement to be taken up by future research.

Household rules concerning TV have also been shown to be important in terms of cognitive effects (Seels et al., 2004). While rules relating to TV vary in quality, not having any rules has been shown to be particularly detrimental in that it encourages watching of more adult-related content (Wright, St. Peters, & Huston, 1990). Gadberry (1980) performed an experimental study in which one group of six-year old children's viewing time was cut back by half, while a control group was left altered. After just six weeks, the experimental group showed improved cognitive ability as well as increased time reading. Thus, it will also be important to explore the existence of household rules having to do with watching TV and their effect on academic achievement.

HYPOTHESES

Based on the existing literature and arguments presented above, I have formulated several hypotheses about the impact of watching TV on academic achievement:

H1a. I hypothesize that the aggregate amount of time spent watching TV on weekdays will be insignificantly related to reading achievement in the general sample of students.

H1b. I hypothesize that the aggregate amount of time spent watching TV on weekdays will be positively associated with reading achievement among LM students.

H2a. I hypothesize that when decomposed into separate times of day, the amount of time spent watching TV both before and after dinner on weekdays will be unassociated with reading achievement in the general sample of students.

H2b. I hypothesize that when decomposed into separate times of day, the amount of time spent watching TV both before and after dinner on weekdays will be positively associated with reading achievement among LM students.

H3a. I hypothesize that the amount of time spent watching TV on weekends will be unassociated with reading achievement in the general sample of students.

H3b. I hypothesize that the amount of time spent watching TV on weekends will be positively associated with reading achievement among LM students.

METHOD

Sample

The present study will examine data from The Early Childhood Longitudinal Study, Kindergarten Class of 1998–1999 (ECLS-K), sponsored by the National Center of Education Statistics (NCES), which is a large, nationally representative study that focuses on children's early school experiences beginning with kindergarten and follows them through the eighth grade, which was the school year of 2006–2007. The ECLS-K records information from assessments and interviews with the students and their teachers, parents, and school administrators, utilizing a multistage probability sample design. Data was collected from the children by sending a trained assessor to the schools to interview the students. Conversely, the parents' responses were obtained by a trained interviewer via telephone. The sample will include all students had tabulated reading theta scores for either kindergarten or first grade and third grade, and whose parents responded to the third grade phone survey.

I will be comparing a subsample LM students to the larger general sample. The first four rounds of data collection for the ECLS-K, which took place in kindergarten and first grade, notes if the student was from a non-English-speaking household. If it was noted that a particular student was from a non-English-speaking household during any of these rounds, that student was added to the LM subsample. I examine achievement from kindergarten to third grade for several reasons. First, this allows me some control for school effects, as most children do not change schools between these years. Second, by limiting my analyses to these years, I am able to make certain claims about early elementary students and skill development at a time when many researchers claim achievement gaps first begin to arise (Bali & Alvarez, 2004). Lastly, most students who enter elementary school without being able to speak English will have adequately learned to communicate in English by the third grade, so we will likely see the most noticeable effects of not speaking English in the early elementary years.

Dependent Variable

The dependent variable in this study will be operationalized with third grade theta reading scores. Researchers analyzing academic achievement have typically used item response theory (IRT) scores in statistical modeling. However, several education scholars have countered that IRT scores are inappropriate for such analyses, particularly for groups of students with large initial gaps in scores, such as English speakers and non-English speakers (see, for example, Reardon, 2008). Theta scores are based on IRT scores, but are transformed into an interval-scale and normally distributed at each wave, which make them ideal for studying growth over time (Ready, 2010). Additionally, the National Center for Education Statistics has recently stated that ECLS-K theta scores are "ideally suited for measuring growth from kindergarten through eighth grade" (quoted in Ready, 2010).

Independent Variables

The main independent variable of interest is the amount of time spent watching TV in the third grade. Taking advantage of the ECLS-K survey questions that inquire about watching TV in various time periods during the week, this variable will be divided into three different time periods: weekdays before dinner, weekdays after dinner, and weekends. In the analysis, I examine the effect of watching TV on weekdays in the aggregate, separated into before and after dinner, as well as on the weekends in the aggregate.

A second independent variable concerns the existence of rules relating to the TV. The ECLS-K survey asks the parental guardian four questions pertaining to the existence of rules regarding use of the TV: time-use limit per week, time-use limit per weekday, rules regarding content, and rules for how late TV is watched. Possible answers for each of these questions were simply "yes," or "no." Therefore, we cannot assess the quality of the rules, but only if the rule existed or not. A scale was made of these four questions, with a maximum of four when all rules existed and minimum of zero when no rules existed. The scale is internally reliable with a Cronbach's alpha of 0.665.

Two other independent variables which measure the number of days per week a student: (1) does homework and (2) reads non-school-related material, as reported by the student's parental guardian. These are used to control for other behavioral aspects that might influence achievement apart from watching television.

In this study, I control for past achievements by using kindergarten theta reading scores. Previously, it was discussed how theta scores are preferable to IRT scores when measuring growth. Additionally, if the student did not have a recorded theta score for kindergarten, first grade theta scores were substituted. If a student had no kindergarten or first grade theta reading scores, the student was dropped from the study. I control for socioeconomic status (SES), which is a scaled, composite variable consisting of: (1) household income, (2) parents' education, and (3) parents' occupation. Dummy variables were constructed for family structure and gender. For both family status and SES, data from the third grade wave was used, and if it was missing, data first grade or kindergarten was substituted. If no family status or SES level were reported in either of the three waves, the student was dropped from the analysis. In the LM subsample, I control for Hispanic status as over 60 percent of the subsample is Hispanic. On the one hand, while Hispanic children are shown to often come from economically disadvantaged households (Schmeer, 2012), learning English from Spanish might be easily facilitated due simply to large numbers of Spanish speakers in ESL education, as well as television media. Thus, I attempt to control for these complexities.

FINDINGS

Sample Comparisons and Bivariate Correlations

Table 1 shows the distribution of the main variables of concern, comparing the full sample and the LM student subsample. A few points are worth noting. First, the average reading theta score in the LM subsample was slightly lower than for the full sample at each wave. While this makes it imperative to control for past achievements, the sample and subsample theta score means are well within half of a standard deviation from each other. And as discussed above, theta scores are also shown to be more effective than IRT scores in statistical modeling for groups with wide initial gaps in scores (Ready, 2010). Second, LM students on average watch about .15 more hours of television before dinner than the general sample, but about one tenth fewer hours after dinner. This discrepancy could be due to the fact that LM students do homework more frequently than non-LM students, as also shown in the table. Cumulatively, both samples watch on average between 1.8 and 1.9 hours of TV per weekday. On the weekends, LM students watched about one tenth hours more than students, in general, with both reporting watching slightly over five hours. Third, the LM subsample

Table 1. Variable Chart with Means and Standard Deviations for Selected Variables; Full Student Sample and Language.

Variable	Full Sample, Mean (S.D.)	LM Sample, Mean (S.D.)
Dependent variable		
Reading theta, third grade	1.326 (.375)	1.200 (.389)
Independent variables		
Reading theta, kindergarten	−.371 (.546)	−.417 (.604)
TV on weekdays	1.806	1.852
TV before dinner	.835 (.778)	.984 (.853)
TV after dinner	.971 (.754)	.868 (.823)
TV on weekends	5.048	5.177
TV Rules, composite	2.843	3.066
Non-school reading, days per week	3.236 (.817)	3.252 (.857)
Homework, days per week	2.347 (.677)	2.602 (.601)
SES, *z*-scale	.023 (.795)	−.375 (.795)
Hispanic ethnicity (Hispanic = 1)	.155	.595

Data Source: ECLS-K 1998–99.
Note: Full Sample $N = 9{,}777$, LM sample $N = 1{,}170$.

is considerably worse off socioeconomically, with an average standardized SES score of −.375 measured in the third grade, while the average of the full sample is .023. Additionally, about 60 percent of the LM subsample is ethnically Hispanic, compared to just 15.5 percent for the general sample.

Table 2 shows the bivariate correlation matrix, with the LM subsample figures above the diagonal and the general sample below. It is worth noting that the strongest correlate with third grade reading scores for both samples is kindergarten reading scores, again underpinning the necessity to control for past achievements. Also highly related to third grade scores in both samples is household SES, which was expected from past research. Preliminary support for the research hypotheses seems to be mixed at the bivariate level. For the general sample, watching TV at any time period or on weekdays and weekends in the aggregate is negatively associated with reading achievement, and those relationships become stronger between kindergarten and third grade. Additionally, watching TV is negatively associated with non-school reading, which is itself positively associated with reading achievement. Taken together, the bivariate analysis lends more support to the displacement theory of watching TV than it does to the research hypotheses of "no effect." For the LM student subsample, the same holds true for kindergarten, but the relationships weaken in third grade. While TV maintains a negative association with reading scores, the

Table 2. Correlation Table for Dependent and Select Independent Variables, LM Students above the Diagonal.

	Reading Theta, Kind	Reading Theta, Third Grade	Non-School Reading, Days/Week	TV on Week Days	TV Breakfast – Dinner	TV after Dinner	TV on Weekends	TV Rules, Composite	SES
Reading theta, kind	–	.439***	.142***	–.095**	–.060*	–.079**	–.079**	–.012	.255***
Reading theta, third grade	.597***	–	.187***	–.056†	–.063*	–.019	–.053†	.018	.431***
Non-school reading, days/week	.138***	.175***	–	–.159***	–.132***	–.095**	–.031	.054†	.124***
TV on weekdays	–.149***	–.161***	–.097***	–	.741***	.718***	.486***	–.209***	–.131***
TV breakfast – dinner	–.094***	–.119***	–.097***	.747***	–	.065*	.400***	–.158***	–.091**
TV after dinner	–.127***	–.118***	–.105***	.728***	.088***	–	.308***	–.158***	–.132***
TV on weekends	–.093***	–.104***	–.066***	.444***	.330***	.324***	–	–.120***	–.055†
TV rules, composite	–.015	–.014	.064***	–.155***	–.075***	–.153***	–.095***	–	.003
SES	.342***	.444***	.106***	–287***	–.207***	–.217***	–.190***	.018†	–

Data Source: ECLS-K 1998–99.
Note: Full sample $N = 9{,}777$, LM sample $N = 1{,}170$.
Numbers above the diagonal are from the LM subsample.
† $p < .10$; *$p < .05$; **$p < .01$; *** $p < .001$.

correlation loses power and becomes insignificant after dinner on weekdays as well as on weekends. So while trends lack support for the research hypotheses of no effect in the general sample and a positive effect in the LM subsample, watching TV does seem to be less negatively associated with reading scores for the LM subsample, supporting the notion that TV functions somewhat uniquely for LM students.

As expected, the number of days a student reads non-school material is positively related to reading achievement, and the relationship for both the general sample and LM subsample increases in power from kindergarten to third grade. Also not surprisingly, non-school reading was positively associated with household SES. Lastly, the existence of TV rules seems to be very weakly related to theta scores at the bivariate level, but negatively associated with watching TV and positively with non-school reading.

Regression Models

I first run a serious of regressions testing for differences between the full and subsamples. For brevity, these results are shown in full in Appendix. I test for differences by creating interaction variables for LM status and each TV time interval, while also adding the standalone LM status and TV watching variables. The interactions between total weekday TV and TV after dinner were both significant with LM status, leading me to conclude that the relationship between watching TV and reading achievement is statistically different in the subsample of LM students.

The results displayed in Table 3 explore these relationships using OLS regression, where I control for demographic variables to examine the added effects of the independent variables of principle concern to this study, while comparing third grade LM students to students in general. Model 1 for both samples includes controls for kindergarten reading theta scores, household SES, gender, and family structure. In the full sample I control for LM status, while in the subsample, I control for Hispanic ethnicity. In Model 2, I add the aggregate weekday and weekend TV variables, the composite TV rules variable, frequency of non-school reading, and frequency of days with homework.

A few general trends are worth noting about the models overall. First, kindergarten reading scores are a slightly stronger predictor for the full sample throughout all models ($\beta = .489 - .498$, $p < .001$) than for the LM subsample ($\beta = .329 - .337$, $p < .001$), suggesting that reading achievement trajectories for LM students have more variation from kindergarten to

Table 3. OLS Standardized Coefficients for Regression of Third Grade Reading Theta Score s on Selected Control and Independent Measures, Full Sample and Language Minority Subsample (*t*-values in parentheses).

	Full Student Sample			LM Student Sample		
	Model 1	Model 2	Model 3	Model 1	Model 2	Model 3
Independent variables						
Reading theta score,	.498***	.489***	.490***	.337***	.329***	.330***
kindergarten	(60.75)	(59.54)	(59.56)	(13.15)	(12.84)	(12.86)
SES	.249***	.244***	.244***	.330***	.327***	.328***
	(28.93)	(27.52)	(27.54)	(12.39)	(12.21)	(12.26)
Gender	.046***	.037***	.037***	.085**	.073**	.075**
	(6.04)	(4.83)	(4.84)	(3.46)	(2.95)	(3.02)
Hispanic ethnicity	–	–	–	–.060*	–.064*	–.065*
	–	–	–	(–2.34)	(–2.48)	(–2.53)
LM status	–.064***	–.063***	–.062***	–	–	–
	(–8.17)	(–7.99)	(–7.85)	–	–	–
Family structure	–.046***	–.044***	–.044***	.034	.041	.039
	(–5.78)	(–5.47)	(–5.50)	(1.39)	(1.64)	(1.59)
TV on weekdays	–	–.001	–	–	.061*	–
	–	(–0.15)	–	–	(2.12)	–
TV breakfast – dinner	–	–	–.009	–	–	.007
	–	–	(–1.06)	–	–	(0.25)
TV after dinner	–	–	.008	–	–	.076**
	–	–	(0.88)	–	–	(2.91)
TV on weekends	–	.001	.001	–	–.032	–.028
	–	(0.14)	(0.14)	–	(–1.16)	(–1.02)
TV rules, composite	–	–.001	.002	–	.038	.048†
	–	(–0.14)	(0.22)	–	(1.54)	(1.91)
Homework, days per week	–	–.019*	–.019*	–	.018	.022
	–	(–2.46)	(–2.43)	–	(0.72)	(0.88)
Non-school reading, days per	–	.079***	.079***	.088**	.086**	
week	–	(9.98)	(9.98)	–	(3.50)	(3.43)
Intercept	1.45	1.36	1.36	1.33	1.12	1.12
Adjusted *R*-squared	.429	.432	.434	.309	.318	.320

Data Source: ECLS-K 1998–99.
Note: Full sample $N = 9{,}777$, LM sample $N = 1{,}170$.
Numbers in parentheses are *t*-values.
† $p < .10$; * $p < .05$; ** $p < .01$; *** $p < .001$.

third grade. Additionally, SES is a slightly stronger predictor for LM students throughout all models ($\beta = .327 - .330$, $p < .001$) than for the full sample ($\beta = .244 - .249$, $p < .001$). This makes sense if we consider that immigrant families often come from more polarized backgrounds in their countries of origin (Massey et al., 1993), and thus SES would potentially hold more sway on achievement. As expected, LM status has a negative association with reading scores in the full sample, as does Hispanic ethnicity in the subsample, indicating that Hispanic LM students do somewhat worse in reading than non-Hispanic LM students. Lastly, being female has an expected positive effect on reading score outcomes.

The demographic variables in Model 1 explain about 43 percent of the variance in third grade reading scores for the full sample, and 31 percent for the LM subsample, again suggesting more unexplained variability within the LM subsample. Model 2 introduces the aggregate watching TV variables for weekdays and weekends, household TV rules, as well as the frequency of non-school reading and doing homework. Contrary to the bivariate analysis, weekday viewing in the aggregate is shown to have no effect on reading achievement when controlling for demographic variables ($\beta = -.001$), giving support for hypothesis H_{1a}. Contrarily, in the LM subsample, the model shows that weekday viewing is positively associated with achievement ($\beta = .061$, $p < .05$), supporting hypothesis H_{1b}. These results indeed suggest a positive and significant effect of weekday viewing on reading achievement, unique to LM students. Contrarily, the effect of weekend viewing is insignificant for both samples, finding support for hypothesis H_{3a}, but lacking for H_{3b}. Thus, the positive effect of watching TV is uniquely bound to weekday viewing and does not seem to extend to the weekends. Several scholars have found similar effects for weekend screen time for students in general (Keith et al., 1986; Sharif & Sargent, 2006), hypothesizing that the lack of salience for weekend viewing is due to the lack of strong displacement effects, but also typically finds a negative relationship between weekday viewing and achievement due to presence of displacement effects. Conversely in this study, I find that weekday viewing has a uniquely positive effect on reading achievement for LM students, while weekend viewing in the aggregate plays no significant role. The existence of household TV rules is not significant for either sample, while the frequency of doing homework has a slight negative relationship with the general sample only. This is most likely due to the lack of regular homework for third grade students, and the possibility that students who more frequently do homework have more academic trouble than other students. For LM students, homework outside of class might be tied to language learning specifically, rather than general cognitive ability. Lastly, these results control for non-school reading, which is very significant and positively associated with achievement in both the general sample ($\beta = .079$, $p < .001$) and the LM subsample ($\beta = .088$, $p < .01$).

In Model 3, for each sample, I partition weekday viewing into before and after dinner components, while maintaining all other control and independent variables. In the general sample, this manipulation does very little. While the coefficient for TV viewing before dinner tends to lean negative ($\beta = -.009$) and after dinner tends to lean positive ($\beta = .008$), neither come close to reaching significance. Thus, I find support for hypothesis H_{2a}, in that even when disaggregated, weekday TV viewing has no effect on reading achievement in the general sample. In contrast, when I partition weekday viewing in the LM subsample, I find that the bulk of the effect comes from viewing after dinner rather than before. For every standard deviation increase in after dinner TV watching, a student's reading scores increased by .076 standard deviations ($p < .01$), lending some strong support for hypothesis H_{2a}. In fact, Model 3 shows that for LM students, the effect on reading achievement of watching TV after dinner is very similar to that of non-school reading ($\beta = .086$, $p < .01$).

On the other hand, watching TV before dinner on weekdays has surprisingly no discernable effect, and is statistically insignificant. This is indeed a curious finding, as most children's educational programming occurs in the afternoon (Woodard IV, 1999), well before when many third graders would presumably be eating dinner. I do not interpret this to signify that programming type is not an overriding factor in the relationship between watching TV and reading achievement, as prior research has found among elementary students (Ennemoser & Schneider, 2007; Rice et al., 1990; Rice & Woodsmall, 1988; Wright, Huston, Murphy et al., 2001). Contrarily, previous research has shown that when left to their own devices, children will often opt for entertainment media, as educational media is perceived by many children to be boring (Giacquinta et al., 1993).

Unfortunately, the data do not allow me to delve more qualitatively into these relationships, as other than the time of day, no information is given about the context of viewing. However, taking into account the research reviewed earlier, I would hypothesize that the relative significance of watching TV after dinner is related to parents or other individuals being present in the house after dinner compared to before dinner, particularly if one or more parents work full-time during the day. Because of this, it could be that after dinner a parent is more likely to participate in coviewing. As discussed earlier, previous research has found that coviewing can act as a mediator to the effects of television by stimulating parent–child discussion about a particular program or other topics related to the program (Seels et al., 2004). Coviewing has also been found to facilitate engagement and comprehension of the TV program, such as plot lines and characterization (Jinqiu & Xiaoming, 2004). In other words, coviewing could be much more conducive toward lingual development than solitary viewing. An additional factor that could add to the significance of watching TV after dinner as opposed to before is that TV viewing rules might be more likely to be followed as parental guardian figures are likely to be present in the household, particularly, those who work during the day. Indeed, TV rules have been shown in past research to be beneficial in mediating negative cognitive effects of watching TV (Seels et al., 2004). And after partitioning weekday TV viewing the effect of rules on reading achievement increases and becomes marginally significant ($\beta = .048$, $p < .10$), suggesting that the effect of TV has somewhat distinct effects before and after dinner on weekdays.

CONCLUSIONS AND DISCUSSION

The focus of this paper has been to compare the relationships between watching television on the improvement in reading scores between kindergarten and third grade among LM students and students in general. Much of the past research in TV effects has been motivated by displacement theory, which predicts that time spent watching TV displaces other activities that may be more beneficial to academic success. This theory has been amended more recently to include certain contexts when multimedia consumption might be beneficial for academic achievement, particularly when the media content is educational or challenging in nature and when used in moderation. The hypothesis guiding the current research is that these conditions would hold for LM students, as there is some evidence that watching TV is conducive to language learning (Silverman & Hines, 2009), and perhaps cultural learning as well (Meskill, 1998). As such, I hypothesized that these potential gains from watching TV for LM students would serve to offset otherwise negative displacement effects. To assess this hypothesis, I compared a subsample of LM third grade students from the nationally representative ECLS-K 1998–99 dataset to all students in the sample, and tested the relationships between watching TV on third grade reading scores controlling for kindergarten reading scores.

In the bivariate analysis, watching TV either in the afternoon or evening on weekdays or on the weekends had a highly significant negative impact on reading test scores in the general sample, aligned with the expectations of displacement theory. These relationships were less strong on the LM sample, although they still pointed in the negative direction and most were at least marginally significant. In the multivariate analysis, I modeled the effect of watching TV on third grade reading scores while controlling for kindergarten scores as well as a host of demographic factors, in an attempt to gage the effect of watching TV on reading score improvement between kindergarten and third grade. Net of kindergarten reading scores, demographic factors, as well as frequency of doing homework and reading at home, watching TV was shown to have no significant effect on reading scores in the general sample. Additionally, although household TV rules were associated with a reduction in the amount of TV watched, they had no significant effect on reading achievement. These findings are contrary to displacement theory, and favor a "no-effect" explanation of the relationship between watching TV and reading achievement. While this conclusion has been suggested by previous research, the current student examined viewing TV at varying time intervals, used past reading achievement controls, and looked at the simultaneous effect of other academic activities done in the home. While the bivariate relationship between watching TV and reading achievement is highly negative, this relationship disappears completely after the use of thorough controls.

In the LM sample, similar effects were found for watching TV before dinner on weekdays, and on weekends. However, contrary to the general sample, watching TV after dinner had a significant, positive relationship with reading scores. After all controls were added to the model, a standard deviation increase in time spent watching TV after dinner amounted to almost a tenth of a standard deviation increase in third grade reading scores ($p < .01$). In fact, the coefficient for watching TV after dinner had nearly the same size as the coefficient for the frequency of non-school reading. Thus, there is a unique relationship between watching TV after dinner and reading scores for LM students. The main implication of this finding is that use of electronic media, often viewed as harmful to cognitive development, does not seem to be inherently harmful and may actually be beneficial to LM students in overcoming the academic disadvantages associated with coming from a non-English speaking household. To my knowledge, this is a new finding in the literature.

Unfortunately, the limitations of the data preclude me from testing the effects of the type of programming watching, which is perhaps the most commonly cited mediating factor (Anderson et al., 2001). However, provided that most children's educational programming concludes by late-afternoon, it would seem that the hypothesis that educational programming is beneficial while entertainment programming is harmful does not explain these effects. It is possible, on the other hand, that a higher prevalence of parental–guardian presence in the household in the evening compared to the afternoon and weekends is a mediating factor, as has been found in previous research (Giacquinta et al., 1993). After reviewing the relevant literature, parental presence in the household can mediate TV watching in two predominant ways. First, coviewing of television is more likely to occur, which previous research has found to facilitate comprehension of characterization, plot lines, and the engagement with the program overall (Jinqiu & Xiaoming, 2004; Seels et al., 2004). Second, the presence of a parent is in the household is also conducive to monitoring the child's behavior, and thus rules regarding watching TV are more likely to be followed. Indeed, rules regarding TV have been tied to cognitive gains in past research (Seels et al., 2004), and they were a marginally significant predictor of higher reading scores in the LM subsample in this study.

There are several limitations to the data and analysis that deserve mentioning. First, the data I use to measure TV viewing are imperfect as they are based on parents' perceptions about children's TV viewing patterns, which of course will deviate somewhat from viewing patterns in reality. Having a subsample size larger than 1,000 offsets this limitation somewhat, but time-use diary data would of course be more accurate (Attewell et al., 2003; Wright, Huston, Murphy et al., 2001). Second, the data on third grade reading scores and television viewing are cross-sectional, so causality should be interpreted with caution. I attempt to offset this aspect by controlling for past achievements. But to make an interpretation about causality, one must assume that third grade television viewing patterns reported by students' parents are somewhat reflective of previous viewing patterns, which would then cause growth or decline in reading scores over time. Lastly, multimedia use has become much more multifaceted since the collection of these ECLS-K data, and TV viewing may no longer be the dominant media consumed by children. However, I am still inclined to hypothesize that controlled media consumption will be positively related to reading achievement for LM students, which is one of the reasons multimedia is very commonly used in more general language-learning settings (Silverman & Hines, 2009; Uchikoshi, 2006).

Future work could expand on this research in two main ways. First, scholars should do more to parse out the effects of multimedia by capturing the context surrounding its consumption. This study looks at the time of day watching TV takes place, but looking at different types of media devices, the social environment in which media consumption took place, as well as the types of media consumed would add immensely to the usefulness of such analyses. For example, it could be that for LM students watching TV is much more beneficial in the context of coviewing in a family environment; while for students in general, the type of programming is the principle mitigating factor. Future research should also see how these processes play out over time. As child development scholars have pointed out that there is a critical window to language learning that occurs before puberty. Thus, it would be interesting to see if or how these findings sustain over time. With record numbers of immigrant children attending U.S. schools (Suárez-Orozco, Suárez-Orozco, & Suárez-Orozco, 2009), such findings would prove particularly useful to teachers, school administrators, and curriculum developers who work diligently to improve the education of LM students.

REFERENCES

Anderson, D. R., Huston, A. C., Schmitt, K. L., Linebarger, D. L., Wright, J. C., & Larson, R. (2001). Early childhood television viewing and adolescent behavior: The recontact study. *Monographs of the Society for Research in Child Development*, *66*(1), i-154.

Attewell, P. (2001). Comment: The first and second digital divides. *Sociology of Education*, *74*(3), 252–259.

Attewell, P., & Battle, J. (1999). Home computers and school performance. *The Information Society: An International Journal*, *15*(1), 1–10.

Attewell, P., Suazo-Garcia, B., & Battle, J. (2003). Computers and young children: Social benefit or social problem? *Social Forces*, *82*(1), 277–296.

Aud, S., Hussar, W., Planty, M., Snyder, T., Bianco, K., Fox, M. A., ... Drake, L. (2010). The Condition of Education 2010. NCES 2010–028. *National Center for Education Statistics*.

Bali, V. A., & Alvarez, R. M. (2004). The race gap in student achievement scores: Longitudinal evidence from a racially diverse school district. *Policy Studies Journal*, *32*(3), 393–415.

Beentjes, J. W. J., & Voort, T. H. A. V. d. (1988). Television's impact on children's reading skills: A review of research. *Reading Research Quarterly*, *23*(4), 389–413.

Bianchi, S. M., & Robinson, J. (1997). What did you do today? Children's use of time, family composition, and the acquisition of social capital. *Journal of Marriage and Family*, *59*(2), 332–344.

Christenson, P. (1994). Childhood patterns of music uses and preferences. *Communication Reports*, *7*(2), 136–144.

Comstock, G., & Paik, H.-J. (1991). *Television and the American child*. San Diego, CA: Academic Press.

Ennemoser, M., & Schneider, W. (2007). Relations of television viewing and reading: Findings from a 4-year longitudinal study. *Journal of Educational Psychology*, *99*(2), 349–368.

Gadberry, S. (1980). Effects of restricting first graders' TV-viewing on leisure time use, IQ change, and cognitive style. *Journal of Applied Developmental Psychology, 1*(1), 45–57.

Gaddy, G. D. (1986). Television's impact on high school achievement. *The Public Opinion Quarterly, 50*(3), 340–359.

Gatto, J. T. (2002). *Dumbing us down: The hidden curriculum of compulsory schooling* (New ed.). Gabriola Island, BC: New Society Publishers.

Giacquinta, J. B., Bauer, J. A., & Levin, J. E. (1993). *Beyond technology's promise : An examination of children's edcuational computing at home.* New York, NY: Cambridge University Press.

Huston, A. C. (1992). *Big world, small screen: The role of television in American society.* Lincoln: University of Nebraska Press.

Iceland, J. (2004). Beyond black and white: Metropolitan residential segregation in multiethnic America. *Social Science Research, 33*(2), 248–271.

Jinqiu, Z., & Xiaoming, H. (2004). Parent–child co-viewing of television and cognitive development of the Chinese child. *International Journal of Early Years Education, 12*(1), 63–77.

Keith, T. Z., Reimers, T. M., Fehrmann, P. G., Pottebaum, S. M., & Aubey, L. W. (1986). Parental involvement, homework, and TV time: Direct and indirect effects on high school achievement. *Journal of Educational Psychology, 78*(5), 373–380.

Massey, D. S., Arango, J., Hugo, G., Kouaouci, A., Pellegrino, A., & Taylor, J. E. (1993). Theories of international migration: A review and appraisal. *Population and Development Review, 19*, 431–466.

Meskill, C. (1998). Commercial television and the limited English proficient child: Implications for language development. In K. Swan, C. Meskill & S. DeMaio (Eds.), *Social learning from broadcast television.* Cresskill, NJ: Hampton Press.

Neuman, S. B. (1991). *Literacy in the television age: The myth of the TV effect.* Norwood, NJ: Ablex.

OECD. (2014). PISA 2012 results: What students know and can do – Student performance in mathematics, reading and science (Rev. ed., Vol. 1). PISA.

Rabe, L. (2010). What is enough change? *Rhodes Journalism Review, 30*, 9–10.

Razel, M. (2001). The complex model of television viewing and educational achievement. *The Journal of Educational Research, 94*(6), 371–379.

Ready, D. D. (2010). Socioeconomic disadvantage, school attendance, and early cognitive development: The differential effects of school exposure. *Sociology of Education, 83*(4), 271–286.

Reardon, S. F. (2008). *Thirteen ways of looking at the black-white test score gap.* Working Paper. Stanford University, Stanford, CA.

Rice, M. L., Huston, A. C., Truglio, R., & Wright, J. C. (1990). Words from "Sesame Street": Learning vocabulary while viewing. *Developmental Psychology, 26*(3), 421–428.

Rice, M. L., & Woodsmall, L. (1988). Lessons from television: Children's word learning when viewing. *Child Development, 59*(2), 420–429.

Rideout, V. J., Foehr, U. G., & Robers, D. F. (2010). *Generation M2: Media in the lives of 8–18 year olds.* Menlo Park, CA: The Henry J. Kaiser Family Foundation.

Schmeer, K. K. (2012). Early childhood economic disadvantage and the health of Hispanic children. *Social Science & Medicine, 75*(8) 1523–1530.

Schmidt, M. E., & Anderson, D. R. (2007). The impact of television on cognitive development and educational attainment. In E. Wartella, J. P. Murray & N. O. Pecora (Eds.), *Children and television : Fifty years of research*. Mahwah, NJ: Lawrence Erlbaum.

Seels, B., Fullerton, K., Berry, L., & Horn, L. J. (2004). Research on learning from television. In D. H. Jonassen (Ed.), *Handbook of research on educational communications and technology* (Vol. 2). Mahwah, NJ: Erlbaum.

Sharif, I., & Sargent, J. D. (2006). Association between television, movie, and video game exposure and school performance. *Pediatrics*, *118*(4), e1061–e1070.

Sharp, G., & Iceland, J. (2013). The residential segregation patterns of whites by socioeconomic status, 2000–2011. *Social Science Research*, *42*(4), 1046–1060.

Silverman, R., & Hines, S. (2009). The effects of multimedia-enhanced instruction on the vocabulary of English-language learners and non-English-language learners in pre-kindergarten through second grade. *Journal of Educational Psychology*, *101*(2), 305–314.

Stout, M. (2000). *The feel-good curriculum: The dumbing-down of America's kids in the name of self-esteem*. Cambridge, MA: Perseus Books.

Straker, L., Pollock, C., & Maslen, B. (2009). Principles for the wise use of computers by children. *Ergonomics*, *52*(11), 1386–1401.

Suárez-Orozco, C., Suárez-Orozco, M. M., & Suárez-Orozco, M. M. (2009). *Children of immigration*. Cambridge, MA: Harvard University Press.

Sykes, C. J. (1995). *Dumbing down our kids: Why America's children feel good about themselves but can't read, write, or add* (1st ed.). New York, NY: St. Martin's Press.

Uchikoshi, Y. (2006). English vocabulary development in bilingual kindergarteners: What are the best predictors? *Bilingualism: Language and Cognition*, *9*(01), 33–49.

Washburn, K., Thornton, J. F., & Simon, J. I. (1996). *Dumbing down : Essays on the strip mining of American culture* (1st ed.). New York, NY: W.W. Norton.

Waters, H. F. (1977). What TV does to kids. *Newsweek*, February 21, pp. 62–70.

Wierzbicka, A. (1994a). 'Cultural scripts': A new approach to the study of cross-cultural communication. In M. Pütz (Ed.), *Language contact and language conflict* (pp. 69–87). Philadelphia, PA: John Benjamins Publishing Company.

Wierzbicka, A. (1994b). Emotion, language, and cultural scripts. In S. Kitayama & H. R. Markus (Eds.), *Emotion and culture: Empirical studies of mutual influence* (pp. 133–196). Washington, DC: American Psychological Association.

Williams, P. A., Haertel, E. H., Haertel, G. D., & Walberg, H. J. (1982). The impact of leisure-time television on school learning: A research synthesis. *American Educational Research Journal*, *19*(1), 19–50.

Williams, T. M. E., & Siegel, A. E. E. (1986). *The impact of television: A natural experiment in three communities*. Orlando, FL: Academic Press.

Winn, M. (1977). *The plug-in drug*. New York, NY: Viking Press.

Woodard IV, E. H. (1999). The 1999 state of children's television report: Programming for children over broadcast and cable television. Anenberg Public Policy Center, Pennsylvania University, Philadelphia.

Wright, J. C., Huston, A. C., Murphy, K. C., St. Peters, M., Piñon, M., Scantlin, R., & Kotler, J. (2001). The relations of early television viewing to school readiness and vocabulary of children from low-income families: The early window project. *Child Development*, *72*(5), 1347–1366.

Wright, J. C., Huston, A. C., Vandewater, E. A., Bickham, D. S., Scantlin, R. M., Kotler, J. A., ... Finkelstein, J. (2001). American children's use of electronic media in 1997: A national review. *Journal of Applied Developmental Psychology*, *22*(1), 31–47.
Wright, J. C., St. Peters, M., & Huston, A. C. (1990). Family television use and its relation to children's cognitive skills and social behavior. In J. Bryant (Ed.), *Television and the American Family* (pp. 227–252). Hillsdale, NJ: Lawrence Erlbaum.

APPENDIX: OLS STANDARDIZED COEFFICIENTS FOR REGRESSION OF THIRD GRADE READING THETA SCORES ON SELECTED INDEPENDENT MEASURES, TESTING FOR INTERACTION EFFECTS FOR LM STATUS (*T*-VALUES IN PARENTHESES).

	Model 1	Model 2	Model 3
Independent variables			
Reading theta score, kindergarten	.498***	.489***	.489***
	(60.75)	(59.54)	(59.55)
SES	.249***	.243***	.243***
	(28.93)	(27.34)	(27.37)
Gender	.046***	.037***	.037***
	(6.04)	(4.78)	(4.81)
LM status	−.064***	−.087***	−.087***
	(−8.17)	(−5.49)	(−5.48)
Family structure	−.046***	−.044***	−.044***
	(−5.78)	(−5.46)	(−5.48)
TV on weekdays	–	−.009	–
	–	(−1.00)	–
TV on weekdays × LM status	–	*.038**	–
	–	(2.35)	–
TV breakfast – dinner	–	–	−.010
	–	–	(−1.19)
TV breakfast – dinner × LM status	–	–	.008
	–	–	*(0.59)*
TV after dinner	–	–	−.002
	–	–	(−0.23)
TV after dinner × LM status	–	–	*.034***
	–	–	*(2.82)*
TV on weekends	–	.002	.002
	–	(0.27)	(0.27)
TV on weekends × LM status	–	*−.009*	*−.007*
	–	*(−0.55)*	*(−0.42)*
TV rules, composite	–	.002	.002
	–	(0.27)	(0.28)
Homework, days per week	–	−.019*	−.019*
	–	(−2.46)	(−2.42)
Non-school reading, days per week	–	.079***	.079***
	–	(10.00)	(9.98)
Intercept	1.45	1.36	1.36
Adjusted R-squared	.429	.434	.435

Note: Full sample $N = 9{,}777$.
Numbers in parentheses are t-values.
Data Source: ECLS-K 1998–99.
* $p < .05$; ** $p < .01$; *** $p < .001$.

CAN MEDIA CONSUMPTION PREDICT IMMIGRANT ADOLESCENTS' ACCULTURATION-RELATED RISKY HEALTH BEHAVIOR? AN ANALYSIS OF LATINO SAMPLE IN CHIS SURVEY

Yuping Mao and Lu Shi

ABSTRACT

Purpose – *Past studies have shown that acculturation has been linked with media consumption (Shi, 2005; Shohat & Stam, 1996). Some risky health behaviors are associated with immigrants' acculturation into the American society. In this study, we investigate the association between Latino adolescents' recreational use of media with acculturation as related to risky health behaviors such as the consumption of sugar-sweetened beverages, the experience of getting sunburns, smoking, and drinking alcohol.*

Technology and Youth: Growing Up in a Digital World
Sociological Studies of Children and Youth, Volume 19, 421–438

ISSN: 1537-4661/doi:10.1108/S1537-466120150000019015

Methodology/approach – *Regression models were run to analyze the Latino adolescent subsample (aged 12–17) of the 2009 California Health Interview Survey (Ponce et al., 2004).*

Findings – *The regression models show that weekend television and video gaming are significantly associated with the number of sunburns one had in the past year (incident rate ratio = 1.008,* z *= 2.73), and weekend recreational computer use is significantly associated with the number of cans of soda with sugar one drank during the previous day (incident rate ratio = 1.003,* z *= 2.09). The use of English for the interview, age, parents' educational attainment, household size, and gender are also found to be associated with different acculturation-related risky health behaviors investigated in this study.*

Originality/value – *This study is the first to analyze media consumption's association with sunburn among Latino adolescents. Our findings indicate that among Latino adolescents in the United States, a large amount of media consumption can lead to risky health behaviors that were not common in Latin America. Therefore, parents should heed possible behavioral consequences when they decide on the amount of media exposure children have.*

Keywords: Media consumption; Latino adolescents; acculturation; risky health behavior; television watching; video gaming

INTRODUCTION

Extensive literature has shown that immigrants' health behaviors change over time as a result of acculturation, which is also true among Latino immigrants to the United States (Arcia, Skinner, Bailey, & Correa, 2001). Latinos are the largest ethnic minorities in the United States with a population size of around 54 million in 2013, constituting 17 percent of the nation's total population (The United States Census Bureau, 2013). The increasing population size makes it very important to study acculturation and health behaviors among Latinos, both the first-generation immigrants who were born outside the United States, and their children who were born in the United States.

As Berry (1990) argued, acculturation tends to induce more changes in the acculturating group. For Latino Americans, not all changes have been positive. Research has shown that high levels of acculturation are associated with higher rates in Latino American's high risks including poor physical and mental health such as cancer and infant mortality (Clark & Hofsess, 1998; Vega & Amaro, 1994). It is possible that these poor health outcomes are associated with risky health behaviors such as smoking, alcohol use, and high body mass index that Latino Americans increasingly adopt with higher levels of acculturation in the United States (Abraído-Lanza, Chao, & Flórez, 2005; Lara, Gamboa, Kahramanian, Morales, & Bautista, 2005; Pérez-Stable et al., 2001).

Our research contributes to the existing body of literature by examining whether Latino adolescents' use of new media can predict their acculturation-related health behaviors. Media is an important tool that immigrants use regularly to learn the socio-cultural contexts of their hosting countries, which is an important stage of acculturation (Shohat & Stam, 1996; Shi, 2005). Adolescents are active users of mass media and the Internet for entertainment and information seeking (e.g., Blackwell, Lauricella, Conway, & Wartella, 2014; Livingstone & Haddon, 2009). Thus, media may have more impact upon adolescents in comparison with adults, since adolescents are quickly learning and getting various information through media to shape their health attitude and life styles at this stage of their life (Lenhart, Purcell, Smith, & Zickuhr, 2010). Thus, Latino adolescents' time spent on media consumption could be an important indicator of their degree of acculturation. In addition to media usage by adolescents, we also measure the use of the English language as an important indicator of acculturation level as suggested by other researchers (Ceullar, Harris, & Jasso, 1980; Coups et al., 2013; Mendoza, 1989).

Overall, our research investigates how Latino adolescents' time spent on television, video gaming, and recreational computer use, as well as their use of English predict their acculturation-related risky behaviors such as the consumption of sugar-sweetened beverages, the experience of getting sunburns, smoking, and alcohol consumption. This study also considers gender and age factors in examining the relationship between acculturation and health behaviors, which answers the research call for incorporating structural and contextual factors such as gender and age to understand Latino population's acculturation and health behaviors through a more comprehensive theoretical model (Abraído-Lanza, Armbrister, Florez, & Aguirre, 2006). Findings from this study will provide relevant information for practitioners to develop media usage related interventions to prevent/reduce Latino adolescents' acculturation-related risky health behaviors.

LITERATURE REVIEW

Acculturation and Media Use

Acculturation happens when different cultural groups continuously negotiate, and change their original cultural patterns through constant interaction with each other (Redfield, Linton, & Herskovits, 1936). Both unidimensional models and bi-dimensional models have been suggested by scholars who have a different understanding on the information and cultural exchange processes. Scholars taking a unidimensional perspective of acculturation assume that individuals lose their own original cultural paradigms when they acculturate themselves into the new culture in the host country (Cabassa, 2003; Cuellar, Arnold, & Maldonado, 1995). The unidimensional perspective of acculturation emphasizes cultural assimilation. Taking a bi-dimensional perspective to understand acculturation, individuals' adherence to a new dominant culture is independent from maintaining their original culture (Berry, 2005). Acculturation people have two orientations: "(i) a relative preference for maintaining one's heritage culture and identity and (ii) a relative preference for having contact with, and participating in the larger society along with other ethnocultural groups" (Berry, 2009, p. 366). Acculturating groups' preferences to these two orientations lead to different acculturation strategies – integration, assimilation, marginalization, and separation (Berry, 2009). In this study, we adopt the bi-dimensional conceptualization of acculturation since it better captures the complexity of the acculturation process.

Previous research has shown the association between media consumption and acculturation among different acculturating groups in diverse cultural contexts. For example, among Hispanic college male students in the United States, acculturative stress is significantly and positively correlated with endorsement of Western media (Warren & Rios, 2013). Among North Korean refugee young adults in South Korea, the use of South Korean mass media and online media is found to be positively associated with their South Korean identity formation (Han, 2013). Among Chinese students living in the United States, the use of American social networking sites and American micro blog predicts American identification (Zhang, 2012). Among all kinds of acculturation-related behaviors, health behaviors are very important to study since they impact individuals, groups, and societal health in general. Our study investigates how media use is related to acculturation-related health behaviors.

Acculturation and Risky Health Behaviors

Existing literature indicates that soda consumption, smoking, and drinking alcohol are major acculturation-related health behaviors among the Latino population (Lara et al., 2005). In the United States, Latinos tend to rapidly acculturate and develop overweight-related behaviors (diet, smoking, and inactivity) from their U.S.-born relatives (Gordon-Larsen, Harris, Ward, & Popkin, 2003). According to the Centers for Disease Control and Prevention (CDC), Latino adolescents have the highest rates of lifetime use of alcohol and tobacco (Centers for Disease Control and Prevention [CDC], 2008). With the negative health impacts associated with the two substances – alcohol and tobacco use – studying Latino adolescents' substance use is highly relevant. Based on a comprehensive literature review, Lara et al. (2005) and Zemore (2007) drew the same conclusion as each other that the effect of acculturation on drinking patterns is clearer among Latina women than men. The higher the level of acculturation of a Latina woman predicted higher levels of drinking. A more recent study based on 2006 Hispanic Americans Baseline Alcohol Survey confirms that acculturation has a more consistent association with increased drinking and binge drinking among women than among men (Vaeth, Caetano, & Rodriguez, 2012). In particular, high acculturation is associated with a higher volume of drinking among Hispanic women. However, Latino men and non-Latino men share similar drinking behaviors, so the acculturation effect on Latino men's drinking behaviors is ambiguous. Overall, less acculturated Latinos drink more frequently, and in greater volume than those who are more acculturated (Marin & Posner, 1995). Age does not appear to have an effect on the relationship between acculturation and drinking behaviors (Caetano, 1987).

In comparison with drinking, smoking among Latinos is less strongly associated with their level of acculturation (e.g., Coonrod, Balcazar, Brady, Garci, & Van Tine, 1995; Coreil, Ray, & Markides, 1991). Spanish-speaking Latino smokers tend to have a higher level of physical dependence to smoking when their years of residence in the United States increase (Castro et al., 2012). Hispanic youth's smoking initiation increases with acculturation to the U.S. culture, and acculturation has stronger effects on Hispanic girls' smoking than Hispanic boys (Epstein, Botvin, & Diaz, 1998; Lorenzo-Blanco, Unger, Ritt-Olson, Soto, & Baezconde-Garbanati, 2011). Although a higher percentage of Latino men smoke than Latina women, a higher level of acculturation is associated with more smoking among

Latina women (Gorman, Lariscy, & Kaushik, 2014), including smoking during pregnancy (Otero-Sabogal, Sabogal, Pérez-Stable, & Hiatt, 1995). Another study included Latino men and women from different countries of origin, and confirmed the positive association between acculturation and smoking among Latino women, but also found negative association between acculturation and smoking among Latino men (Pérez-Stable et al., 2001). Among Latino youth, acculturation was also found to be positively associated with depressive symptoms and smoking among girls, but the same pattern was not found among boys (Lorenzo-Blanco et al., 2011).

A higher percentage of Mexican-American adolescents are classified as obese (24%) and overweight (20%) than Caucasian adolescents – 16% and 14%, respectively (Ogden, Carroll, Kit, & Flegal, 2012). In the United States, sugar-sweetened beverages (SSB) including carbonated soda significantly contribute to high obesity rates (Malik, Popkin, Bray, Despre´s, & Hu, 2010). From 1988 to 2004, SSB consumption increased significantly among Mexican-American adolescents but remained stable among Caucasian adolescents (Wang, Bleich, & Gortmaker, 2008). If SSB consumption is an indicator of food acculturation in the United States, the increase of SSB consumption among Mexican-Americans could be the result of their increased acculturation levels in the United States, especially since the influence of the Mexican diet is almost lost within one generation in the United States among Mexican-Americans (Batis, Hernandez-Barrera, Barquera, Rivera, & Popkin, 2011).

The increased consumption of carbonated beverages among Mexican-Americans becomes a serious threat to their health, which requires more research on possible intervention methods. Researchers suggest that parents should take the responsibility of educating and controlling SSB consumption of their children (Bogart et al., 2013). Latino children who never or rarely watch television during family meals are found to be less likely to consume soda and chips (Andaya, Arredondo, Alcaraz, Lindsay, & Elder, 2011). Our research will investigate how media consumption including television viewing, video gaming, and recreational computer use over weekends predict soda consumption among Latino adolescents.

Tanning is a trendy behavior among U.S. adolescents, but it also has a high risk of leading to sunburn. Recent statistics show that 34.4% of American adults were sunburned in the past year, and 69% of American adolescents were sunburned in the previous summer (Buller et al., 2011). Research on college students in the Midwestern United States shows that individuals' exposure and attention to health information in mass media – television, magazines, and newspapers – are positively associated with their intentions to avoid unprotected sun exposure (Lovejoy, Riffe, & Lovejoy, 2015). Furthermore, male Caucasian college students' exposure to viewing tanned men's images on television was directly associated with pro-tanning attitudes, and was indirectly associated with pro-tanning intentions, while the same pattern was not found among Caucasian female students (Cho & Choi, 2011). Different cultural groups have different media consumption behaviors. Acculturating group's media usage is likely to be different from the dominant cultural group. The Latino population's behavior pattern in regards to sunburn-related activities and media use requires more research attention. However, research on sunburn-related health behaviors have only been recently conducted on Latino migrant farm workers (Kearney, Phillips, Allen, Hurtado, & Hsia, 2014) and Hispanic adults (Coups et al., 2012). A higher level of acculturation is positively associated with chances of getting sunburn among Hispanic adults in the United States. (Coups et al., 2012). To the best of our knowledge, our study is the first to analyze media consumption's association with sunburn-related health behaviors among Latino adolescents.

During the acculturation process, Marin (1992) argued that changes in the consumption of food, use of media in a multicultural environment, and preference in language use are all important behavioral changes happening at different stages of acculturation. Therefore, our study includes new media consumption and the use of English as independent variables. One serious limitation of the acculturation studies in health research is the general disregard for the impact of material barriers on health behaviors (Hunt, Schneider, & Comer, 2004). To compensate for this limitation, we included material factors such as household income, and social factors such as parent educational level in our analysis. Arcia et al. (2001) found that Latina females reported higher levels of social acceptance than males in the United States, which indicates gender differences in self-reported acculturation levels. In our Latino adolescent sample, we also examine gender differences.

METHOD

Data

We use the Latino adolescent subsample (aged 12–17) of the 2009 California Health Interview Survey (Ponce et al., 2004) – a telephone interview survey that used randomized digit dialing to collect a sample representative of the California adolescent population. For each sampled household, one adolescent aged 12–17 was sampled to answer the telephone interview. Thus, this adolescent dataset was based upon the adolescents' self-report rather than parental report. We chose this dataset partly because California has a sufficient sample of Latino adolescents, and partly because California is one of the few states that have a population-based survey dataset that measures adolescents' media use.

Variables

Two media use continuous variables: (1) the number of hours one spent on the use of television/video gaming on a weekend day and (2) the number of hours spent on recreational computer use on a weekend day were measured in the 2009 California Health Interview Survey. To control for variables that might be related with the adolescent's acculturation, we included in our analysis the following variables: (1) whether the adolescent used English to answer the telephone interview (categorical variable) and (2) how long the adolescent had lived in the United States at the time of the interview (continuous variable). We controlled for the adolescent's outdoor activity with two variables: (1) days the adolescent walked, biked, or skated from school during the past seven days (continuous variable) and (2) whether the adolescent had been to a park or playground during the past 30 days (categorical variable). Finally, we included in our model the adolescent's age, gender, parental educational attainment, household size, and household economic status (as operationalized by "how many times of the federal poverty line," a variable imputed from the household income as reported by either the sampled adult or the most knowledgeable adult in the household if no adult was sampled in the household).

We use four risky behaviors as our dependent variables: (1) whether the adolescent had ever consumed alcohol (categorical variable), (2) whether the adolescent had ever smoked cigarettes (categorical variable), (3) number of cans of soda with sugar the adolescent drank the day before (continuous variable), and (4) number of times the adolescent had been sunburned during the past 12 months (continuous variable).

Four Analysis Models

Given the nature of count variables – the sunburn and the soda variable – we chose to use negative binomial regressions to examine the association between the two media use variables: (1) the number of hours spent on the use of television/video gaming on a weekend day and (2) the number of hours spent on recreational computer use on a weekend day, as well as the first two risky behaviors: the number of sunburns one had in the previous year, and the number of cans of soda with sugar one had during the previous day. For the smoking and alcohol outcomes, we use logistic regression to examine the association between media use variables, and the two substance use variables – alcohol and smoking. STATA 12.0's negative binomial regression and logistic functions were used to conduct the analyses.

We chose not to use the STATA function of multiple imputation to impute the missing values, since we believe that in survey data those who chose not to answer the question could be qualitatively different from any of the defined categories. Assigning these respondents a defined value for the variable they chose not to answer might be conceptually improper and analytically misleading.

RESULTS

Table 1 charts the descriptive statistics of key variables used in the analyses. Among the entire 1,266 Latino adolescents (51.2% male and 48.8% female) in the California Health Interview Survey, an average number of sunburns in the last 12 months was 0.97, and the average number of canned soda with sugar consumed the day before was 0.76. Those who reported having smoked cigarettes constituted 12.81% of the entire Latino adolescent sample, and the prevalence of ever drinking alcohol is 32.14%. An average of 6.13 hours during the weekend was spent on television and/or video games, while the amount of time spent on recreational computer use on the weekend was 10.90 hours. The English language was used by 974 of Latino participants to answer survey questions (79.5% of the Latino subsample used English, and the other 252 used Spanish). The average age was 14.43 (s.d. = 1.69; the mean age was 14.4 among those using English for the interview, and was 14.6 among those using Spanish). On average, those Latino adolescents born outside the United States had spent 3.69 years (s.d. = 0.06; the mean was 3.6 among those using English for the interview and was 2.8 among those using Spanish) in the United States by the time of the 2009 interview. After discarding the observations that had missing values in the variables we included in the models, we were left with a sample size of 1,186 self-identified Latino adolescents in our sample.

Table 1. Descriptive Statistics about the Latino Adolescents in 2009 California Health Interview Survey (N = 1226).

Variables	Mean and Standard Errors		Frequency (for Dummy Variables)
Number of sunburns last 12 months	0.97	(2.67)	
Number of canned soda with sugar the day before	0.76	(1.09)	
Ever smoked cigarettes			12.81%
Ever tried alcohol			32.14%
Weekend TV/video game hours	6.13	(16.30)	
Weekend recreational computer hours	10.90	(27.47)	
Age (years)	14.43	(1.69)	
Number of days walked, biked, or skated from school	2.17	(2.28)	
Years living in the United States	12.07	(4.83)	
Household income as times of federal poverty line	2.08	(2.48)	
Household size	4.64	(1.26)	
English as the interview language			79.45%
Not been to park/playground			27.49%
Female			48.78%

Table 2 presents the analysis results of the four models we ran. Weekend television and video gaming are significantly associated with the number of sunburns one had in the past year (incident rate ratio = 1.008, z = 2.73), and weekend recreational computer use is significantly associated with the number of cans of soda with sugar one had drank during the previous day (incident rate ratio = 1.003, z = 2.09). These two media use variables are not significantly associated with either smoking initiation or alcohol intake experience. Significant and positive predictors of sunburn frequency also include age (incident rate ratio = 1.068, z = 2.05), using English as the interview language (incident rate ratio = 1.570, z = 2.87) and parental educational attainment (incident rate ratio = 1.158, z = 5.93). Cans of soda consumed the day before are negatively associated with household income (incident rate ratio = 0.799, z = −2.18) and the female gender (incident rate ratio = 0.799, z = −2.80).

The adolescent's self-reported age is positively associated with both smoking initiation (odds ratio = 1.766, z = 8.46) and alcohol intake experience (odds ratio = 1.684, z = 11.39), whereas the household size is negatively associated with both smoking initiation (odds ratio = 0.797, z = −2.88) and alcohol intake experience (odds ratio = 0.890, z = −2.07). Number of days walked, biked or skated from school is a significant and positive predictor of alcohol intake experience (odds ratio = 1.061,

Table 2. Four Models about Adolescent Risky Behaviors.

	Times Sunburned Last 12 Months Incident rate ratio	Cans of Soda with Sugar the Previous Day Incident rate ratio	Ever Smoked Cigarettes Odds ratio	Ever Drunk Alcohol Odds ratio
Weekend hours playing TV/Video games	1.008	0.999	0.998	0.991
	(2.73)**	(−0.48)	(−0.29)	(−1.64)
Weekend hours using computer for fun	0.998	1.003	1.001	1.001
	(−1.26)	(2.09)*	(0.37)	(0.41)
Age	1.068	1.031	1.766	1.684
	(2.05)*	(1.20)	(8.46)***	(11.39)***
English as language for interview	1.570	0.954	0.925	1.495
	(2.87)**	(−0.44)	(−0.31)	(2.12)*
# of day walked, biked, or skated from school	0.978	1.025	1.039	1.061
	(−0.96)	(1.43)	(0.93)	(1.96)*
Parents' educational attainment	1.158	0.972	1.031	1.037
	(5.93)***	(−1.40)	(0.69)	(1.13)
Years lived in the United States	1.004	0.998	0.980	0.977
	(0.29)	(−0.19)	(−1.01)	(−1.49)
Times of 100% Federal Poverty Level	1.047	0.954	0.998	0.977
	(1.80)	(−2.18)*	(−0.05)	(−0.75)
Household size	1.043	0.951	0.797	0.890
	(0.99)	(−1.57)	(−2.88)**	(−2.07)*
Not been to park or playground past 30 days	0.890	1.069	0.925	0.965
	(−0.97)	(0.75)	(−0.36)	(−0.23)
Female	1.181	0.799	0.597	0.898
	(1.59)	(−2.80)**	(−2.71)**	(−0.80)
Observations	1186	1186	1186	1186

z statistics in parentheses.
$^{*}p < 0.05$; $^{**}p < 0.01$; $^{***}p < 0.001$.

$z = 1.96$), but positive association with smoking initiation is not statistically significant (odds ratio = 1.039, $z = 0.93$). Similarly, using English for the survey telephone interview is a significant and positive predictor of alcohol intake experience (odds ratio = 1.495, $z = 2.12$) while its association with smoking initiation is not statistically significant (odds ratio = 0.925, $z = -0.031$). Finally, the female gender is a protective factor that is negatively associated with smoking initiation (odds ratio = 0.597, $z = -2.71$) whereas this protective effect is not significant for the outcome of alcohol intake experience (odds ratio = 0.898, $z = -0.80$).

DISCUSSION

The media use-related findings of this study are (1) weekend television and video gaming positively predict Latino adolescents' chances of getting sunburn and (2) weekend recreational computer use positively predict the amount of soda Latino adolescents intake. Therefore, moderating weekend media use, including TV, video gaming, and recreational computer use, can be an effective intervention to reduce the risky health behaviors such as getting sunburn and drinking carbonated beverages. However, available data did not specifically ask the language of the media that the adolescent participants were consuming during weekends. Future studies could further explore how Latino adolescents' use of ethnic media in Spanish and mainstream U.S. media in English predicts those acculturation-related health behaviors.

Our research also reveals some other predictors of acculturation-related risky health behaviors. Age, using English as the interview language, and parental educational attainment positively predict Latino adolescents' sunburn frequency. Household income and female gender are negative predictors of cans of soda that Latino adolescents consumed the day before. Household size and female gender negatively predict Latino adolescents' smoking initiation, while self-reported age positively predicts their smoking initiation. Latino adolescents from a family with a larger household size are less likely to have alcohol intake experience. The following three variables positively predict the chances that Latino adolescents have the experience of drinking alcohol: self-reported age, number of days walked, biked, or skated from school, and using English for the survey telephone interview.

According to Lara et al.'s (2005) review on acculturation and Latino health in the United States, in the theoretical debate on acculturation, some scholars argue that acculturation is a unidimensional process while others believe it should be bi-dimensional. In a unidimensional acculturation model, individuals lose their original cultural paradigms from their home countries while they adopt new cultural paradigms of their host countries. In a bi-dimensional acculturation model, individuals maintain their original culture from the home countries while they acquire and adhere to the new dominant cultures in their host countries. Our findings show that Latinos tend to acculturate in a bi-dimensional process. While Latino adolescents learn and adapt their behaviors to the U.S. culture, they still maintain some important cultural values and norms from their own culture such as appreciation for family (Sabogal, Marin, & Otero-Sabogal, 1987). This might be an explanation as to why household size in this study negatively predicts smoking and alcohol behavior. In traditional Latino culture, large families are preferred. The family size could indicate the appreciation for family that the adolescents in this study have. The traditional Latino culture may be a good protector for adolescents from smoking and drinking alcohol in the United States. In contrast, the preference of using English to participate in this survey is an indicator of a high level of acculturation, which positively predicts alcohol use.

Our study finds that age positively predicts whether Latino adolescents have had smoking and alcohol drinking experience. When we specifically examine the stages of adolescence as to when people start their first time experience of drinking alcohol and smoking, age is an important predictor of whether adolescents have tried those risky health behaviors or not. Moving up along the developmental trajectory of substance use (Chen & Jacobson, 2012), adolescents with older age require more attention in monitoring drinking alcohol and smoking.

Our study confirms West et al.'s (2013) finding that Latina female adolescents are less likely to have ever smoked, but there is an increase of smoking as age increases. West et al. (2013) also found gender, age, and acculturation unrelated to Latino adolescents' alcohol use. Our study has divergent results that two factors positively predict Latino adolescents' alcohol consumption experience: age and the preference of taking the survey in English as an indicator of acculturation level, both hinting that deeper immersion in the American culture might be associated with substance misuse among the immigrant population.

This study is limited by the fact that it is a cross-sectional dataset, and thus no advanced analysis method such as difference-in-difference models can be used to minimize the potential problem of selection bias and omitted variable bias among the observational data. Therefore, the significant predictors of risky behaviors as revealed by our study can only be interpreted as risk factors rather than proven causal determinants. Future studies in this field will need to explore longitudinal or interventional datasets to further explore the relationship between acculturation and adolescent health behavior. Our research shows that Latino adolescents' weekend use of different media formats (weekend television and video gaming, weekend recreational computer use) positively predict their risky health behaviors (getting sunburn and drinking soda); however, this quantitative study does not prevail detailed information on what specific type of media content influences those risky health behaviors. Future research should further examine specific media content's impact on acculturation-related health behaviors.

CONCLUSION

Among Latino adolescents, an increased amount of media consumption could lead to some risky health behaviors that might not be common in Latin America (Andaya et al., 2011). This is consistent with our previous findings about excessive recreational computer use and undesirable food/beverage consumption (Shi & Mao, 2010). As recreational computer use has increased over time among Latino children (van Meijgaard, Shi, & Simon, 2013), and it tends to increase with the adolescent's acculturation process (Shi, van Meijgaard, & Simon, 2012), it is fair to say that recreational computer use has constituted an increasing public health threat to Latino adolescents living in places like California.

For many immigrants, the acculturation process brings significant health benefits like higher health literacy (van Servellen, Brown, Lombardi, & Herrera, 2003), and easier communication with health care providers (Fernandez et al., 2004). However, the acquisition of undesirable health behaviors could also come with the acculturation experience among immigrant adolescents, as evidenced by this paper, as part of the broader literature that documents "the immigrant paradox" (Marks, Ejesi, & García Coll, 2014, p. 59). Parents of Latino adolescents should heed possible behavioral consequences as they decide on the amount of media exposure children should be exposed to.

REFERENCES

Abraído-Lanza, A. F., Armbrister, A. N., Florez, K. R., & Aguirre, A. N. (2006). Toward a theory-driven model of acculturation in public health research. *American Journal of Public Health, 96*(8), 1342–1346. doi:10.2105/AJPH.2005.064980

Abraído-Lanza, A. F., Chao, M. T., & Flórez, K. R. (2005). Do healthy behaviors decline with greater acculturation? Implications for the Latino mortality paradox. *Social Science Medicine, 61*, 1243–1255. doi:10.1016/j.socscimed.2005.01.016

Andaya, A. A., Arredondo, E. M., Alcaraz, J. E., Lindsay, S. P., & Elder, J. P. (2011). The association between family meals, TV viewing during meals, and fruit, vegetables, soda, and chips intake among Latino children. *Journal of Nutrition Education and Behavior, 43*(5), 308–315. doi:10.1016/j.jneb.2009.11.005

Arcia, E., Skinner, M., Bailey, D., & Correa, V. (2001). Models of acculturation and health behaviors among Latino immigrants to the U.S. *Social Science & Medicine, 53*, 41–53. doi:10.1016/s0277-9536(00)00310-5

Batis, C., Hernandez-Barrera, L., Barquera, S., Rivera, J. A., & Popkin, B. M. (2011). Food acculturation drives dietary differences among Mexicans, Mexican Americans, and Non-Hispanic whites. *The Journal of Nutrition, 141*, 1898–1906. doi:10.3945/jn.111.141473

Berry, J. W. (1990). Psychology of acculturation. In J. Berman (Ed.), *Cross-cultural perspectives: Nebraska symposium on motivation* (pp. 201–234). Lincoln, NE: University of Nebraska Press.

Berry, J. W. (2005). Acculturation: Living successfully in two cultures. *International Journal of Intercultural Relations*, *29*, 697–712. doi:10.1016/j.ijintrel.2005.07.013

Berry, J. W. (2009). A critique of critical acculturation. *International Journal of Intercultural Relations*, *33*, 361–371. doi:10.1016/j.ijintrel.2009.06.003

Blackwell, C., Lauricella, A., Conway, A., & Wartella, E. (2014). Youth and the Internet: Developmental implications of website preferences among 8- to 12-year-old children. *Journal of Broadcasting and Electronic Media*, *58*(1), 1–20. doi:10.1080/08838151.2013.875022

Bogart, L. M., Cowgill, B. O., Sharma, A. J., Uyeda, K., Sticklor, L. A., Alijewicz, K. E., & Schuster, M. A. (2013). Parental and home environmental facilitators of sugar-sweetened beverage consumption among overweight and obese Latino youth. *Overweight and Disordered Eating*, *13*(4), 348–355. doi:10.1016/j.acap.2013.02.009

Buller, D. B., Cokkinides, V., Hall, I., Hartman, A. M., Saraiya, M., Miller, E., ... Glanz, K. (2011). Prevalence of sunburn, sun protection, and indoor tanning behaviors among Americans: Review from national surveys and case studies of 3 states. *Journal of the American Academy of Dermatology*, *65*(5), S114.e1–S114.e11. doi:10.1016/j.jaad.2011.05.033

Cabassa, L. J. (2003). Measuring acculturation: Where we are and where we need to go. *Hispanic Journal of Behavioral Science*, *25*(2), 127–146. doi:10.1177/0739986303025002001

Caetano, R. (1987). Acculturation, drinking and social settings among U.S. Hispanics. *Drug and Alcohol Dependence*, *19*(3), 215–216. doi:10.1016/0376-8716(87)90041-x

Castro, Y., Businelle, M. S., Correa-Fernández, V., Kendzor, D. E., Mazas, C. A., Cofta-Woerpel, L., & Wetter, D. W. (2012). Associations between indicators of acculturation and tobacco dependence among Spanish-speaking Latino smokers. *Addictive Behaviors*, *37*, 1101–1108. doi:10.1016/j.addbeh.2012.05.003

Centers for Disease Control and Prevention. (2008). Youth risk behavior surveillance – United States, 2007. *Morbidity and Mortality Weekly Report*, *57*(SS-4).

Ceullar, I., Harris, L. C., & Jasso, R. (1980). An acculturation scale for Mexican American normal and clinical populations. *Hispanic Journal of Behavioral Sciences*, *2*, 199–217.

Chen, P., & Jacobson, K. C. (2012). Developmental trajectories of substance use from early adolescence to young adulthood: Gender and racial/ethnic differences. *Journal of Adolescent Health*, *50*(2), 154–163. doi:10.1016/j.jadohealth.2011.05.013

Cho, H., & Choi, J. (2011). Television, gender norms, and tanning attitudes and intentions of young men and women. *Communication Studies*, *62*(5), 508–530. doi:10.1080/10510974.2011.577500

Clark, L., & Hofsess, L. (1998). Acculturation. In S. Loue (Ed.), *Handbook of immigrant health* (pp. 37–59). New York, NY: Plenum Press.

Coonrod, D. V., Balcazar, H., Brady, J., Garci, S., & Van Tine, M. (1995). Smoking, acculturation and family cohesion in Mexican-American women. *Ethnicity & Disease*, *9*(3), 434–440.

Coreil, J., Ray, L. A., & Markides, K. S. (1991). Predictors of smoking among Mexican-American: Findings from Hispanic HANES. *Preventive Medicine*, *20*(4), 508–517. doi:10.1016/0091-7435(91)90048-9

Coups, E., Stapleton, J. L., Hudson, S. V., Medina-Forrester, A., Natale-Pereira, A., & Goydos, J. (2012). Sun protection and exposure behaviors among Hispanic adults in the United States: Differences according to acculturation and among Hispanic subgroups. *The Brihanmumbai Municipal Corporation Public Health, 12*, 985. Retrieved from http://www.biomedcentral.com/1471-2458/12/985

Coups, E. J., Stapleton, J. L., Hudson, S. V., Medina-Forrester, A., Rosenberg, S. A., Gordon, M. A., Natale-Pereira, A., & Goydos, J. S. (2013). Linguistic acculturation and skin cancer-related behaviors among Hispanics in the Southern and Western United States. *Journal of the American Medical Association Dermatology, 149*(6), 679–686. doi:10.1001/jamadermatol.2013.745

Cuellar, I., Arnold, B., & Maldonado, R. (1995). Acculturation rating scale for Mexican Americans-II: A revision of the original ARSMA scale. *Hispanic Journal of Behavioral Science, 17*(3), 275–304. doi:10.1177/07399863950173001

Epstein, J. A., Botvin, G. J., & Diaz, T. (1998). Linguistic acculturation and gender effects on smoking among Hispanic youth. *Preventive Medicine: An International Journal Devoted to Practice and Theory, 27*(4), 583–589. doi:10.1006/pmed.1998.0329

Fernandez, A., Schillinger, D., Grumbach, K., Rosenthal, A., Stewart, A. L., Wang, F., & Pérez-Stable, E. J. (2004). Physician language ability and cultural competence. *Journal of General Internal Medicine, 19*(2), 167–174. doi:10.1111/j.1525-1497.2004.30266.x

Gordon-Larsen, P., Harris, K. M., Ward, D. S., & Popkin, B. M. (2003). Acculturation and overweight-related behaviors among Hispanic immigrants to the U.S.: The national longitudinal study of adolescent health. *Social Science & Medicine, 57*, 2023–2034. doi:10.1016/s0277-9536(03)00072-8

Gorman, B. K., Lariscy, J. T., & Kaushik, C. (2014). Gender, acculturation, and smoking behavior among U.S. Asian and Latino immigrants. *Social Science & Medicine, 106*, 110–118. doi:10.1016/j.socscimed.2014.02.002

Han, S. M. (2013). *The role of online social media in the acculturation process of North Korean refugee young adults in South Korea*. Wayne State University Dissertations. Paper 767. Retrieved from http://digitalcommons.wayne.edu/oa_dissertations/767

Hunt, L. M., Schneider, S., & Comer, B. (2004). Should "acculturation" be a variable in health research? A critical review of research on U.S. Hispanics. *Social Science & Medicine, 59*, 973–986. doi:10.1016/j.socscimed.2003.12.009

Kearney, G. D., Phillips, C., Allen, D. L., Hurtado, G. A., & Hsia, L. B. (2014). Sun protection behaviors among Latino migrant farmworkers in Eastern North Carolina. *Journal of Occupational and Environmental Medicine, 56*(12), 1325–1331. doi:10.1097/jom.0000000000000275

Lara, M., Gamboa, C., Kahramanian, I., Morales, L., & Bautista, D. (2005). Acculturation and Latino health in the United States: A review of the literature and its sociopolitical context. *Annual Review of Public Health, 26*, 367–397. doi:10.1146/annurev.publhealth.26.021304.144615

Lenhart, A., Purcell, K., Smith, A., & Zickuhr, K. (2010). *Social media and mobile Internet use among teens and young adults*. Washington, DC: Pew Internet & American Life Project.

Livingstone, S., & Haddon, L. (2009). *EU kids online: Final report 2009*. London: EU Kids Online Network.

Lorenzo-Blanco, E., Unger, J. B., Ritt-Olson, A., Soto, D., & Baezconde-Garbanati, L. (2011). Acculturation, gender, depression, and cigarette smoking among U.S. Hispanic youth: The mediating role of perceived discrimination. *Journal of Youth and Adolescence, 40*, 1519–1533. doi:10.1007/s10964-011-9633-y

Lovejoy, J., Riffe, D., & Lovejoy, T. I. (2015). An examination of direct and indirect effects of exposure and attention to health media on intentions to avoid unprotected sun exposure. *Health Communication*, *30*, 261–270. doi:10.1080/10410236.2013.842526

Malik, V. S., Popkin, B. M., Bray, G. A., Despre´s, J.-P., & Hu, F. B. (2010). Sugar-sweetened beverages, obesity, type 2 diabetes mellitus, and cardiovascular disease risk. *Circulation*, *121*, 1356–1364. doi:10.1161/CIRCULATIONAHA.109.876185

Marin, G. (1992). Issues in the measurement of acculturation among Hispanics. In K. F. Geisinger (Ed.), *Psychological testing of Hispanics* (pp. 23–51). Washington, DC: American Psychology Association.

Marin, G., & Posner, S. F. (1995). The role of gender and acculturation on determining the consumption of alcoholic beverages among Mexican-Americans and Central Americans in the United States. *International Journal of the Addictions*, *30*(7), 779–794. doi:10.3109/10826089509067007

Marks, A. K., Ejesi, K., & García Coll, C. (2014). Understanding the U.S. immigrant paradox in childhood and adolescence. *Child Development Perspectives*, *8*(2), 59–64. doi:10.1111/cdep.12071

Mendoza, R. H. (1989). An empirical scale to measure type and degree of acculturation in Mexican-American adolescents and adults. *Journal of Cross Cultural Psychology*, *20*, 372–385. doi:10.1177/0022022189204003

Ogden, C. L., Carroll, M. D., Kit, B. K., & Flegal, K. M. (2012). Prevalence of obesity and trends in body mass index among U.S. children and adolescents, 1999–2010. *Journal of the American Medical Association*, *307*, 483–490. doi:10.1001/jama.2012.40

Otero-Sabogal, R., Sabogal, F., Pérez-Stable, E. J., & Hiatt, R. A. (1995). Dietary practices, alcohol consumption, and smoking behavior: Ethnic, sex, and acculturation differences. *Journal of the National Cancer Institute Monographs*, *18*, 73–82.

Pérez-Stable, E. J., Ramirez, A., Villareal, R., Talavera, G. A., Trapido, E., Suarez, L., … McAlister, A. (2001). Cigarette smoking behavior among U.S. Latino men and women from different countries of origin. *American Journal of Public Health*, *91*, 1424–1430. doi:10.2105/ajph.91.9.1424

Ponce, N. A., Lavarreda, S. A., Yen, W., Brown, E. R., DiSogra, C., & Satter, D. E. (2004). The California Health Interview Survey 2001: Translation of a major survey for California's multiethnic population. *Public Health Reports*, *119*(4), 388. doi:10.1016/j.phr.2004.05.002

Redfield, R., Linton, R., & Herskovits, M. (1936). Memorandum on the study of acculturation. *American Anthropologist*, *38*, 149–152. doi:10.1525/aa.1936.38.1.02a00330

Sabogal, F., Marin, G., & Otero-Sabogal, R. (1987). Hispanic familism and acculturation: What changes and what doesn't? *Hispanic Journal of Behavioral Science*, *9*(4), 397–412. doi:10.1177/07399863870094003

Shi, L., & Mao, Y. (2010). Excessive recreational computer use and food consumption behaviour among adolescents. *Italian Journal of Pediatrics*, *36*(1), 52. doi:10.1186/1824-7288-36-52

Shi, L., van Meijgaard, J., & Simon, P. (2012). The association between acculturation and recreational computer use among Latino adolescents in California. *Pediatric Obesity*, *7*(4), E33–E36. doi:10.1111/j.2047-6310.2012.00057.x

Shi, Y. (2005). Identity construction of the Chinese diaspora, ethnic media use, community formation, and the possibility of social activism. *Continuum: Journal of Media & Cultural Studies*, *19*(1), 55–72. doi:10.1080/1030431052000336298

Shohat, E., & Stam, R. (1996). From the imperial family to the transnational imaginary: Media spectatorship in the age of globalization. In R. Wilson & W. Dissanayake (Eds.), *Global/local: Cultural production and the transnational imaginary* (pp. 145–172). Durham, NC: Duke University Press.

The United States Census Bureau. (2013). *Facts for features: Hispanic heritage month 2014: Sept. 15–Oct. 15*. Retrieved from http://www.census.gov/newsroom/facts-for-features/2014/cb14-ff22.html

Vaeth, P. A. C., Caetano, R., & Rodriguez, L. A. (2012). The Hispanic Americans Baseline Alcohol Survey (HABLAS): The association between acculturation, birthplace and alcohol consumption across Hispanic national groups. *Addictive Behaviors*, *37*, 1029–1037. doi:10.1016/j.addbeh.2012.04.015

van Meijgaard, J., Shi, L., & Simon, P. (2013). Trends in recreational computer use among Latino children in California. *Journal of Immigrant and Minority Health*, *15*(2), 437–441. doi:10.1007/s10903-012-9684-5

van Servellen, G., Brown, J. S., Lombardi, E., & Herrera, G. (2003). Health literacy in low-income Latino men and women receiving antiretroviral therapy in community-based treatment centers. *AIDS Patient Care and STDs*, *17*(6), 283–298. doi:10.1089/108729103322108166

Vega, W. A., & Amaro, H. (1994). Latino outlook: Good health, uncertain prognosis. *Annual Review of Public Health*, *15*, 39–67. doi:10.1146/annurev.pu.15.050194.000351

Wang, Y. C., Bleich, S. N., & Gortmaker, S. L. (2008). Increasing caloric contribution from sugar-sweetened beverages and 100% fruit juices among U.S. children and adolescents, 1988–2004. *Pediatrics*, *121*(6), e1604–e1614. doi:10.1542/peds.2007-2834

Warren, C. S., & Rios, R. M. (2013). The relationships among acculturation, acculturative stress, endorsement of Western media, social comparison, and body image in Hispanic male college students. *Psychology of Men & Masculinity*, *14*(2), 192–201. doi:10.1037/a0028505

West, J. H., Blumbert, E. J., Kelley, N. J., Hill, L., Sipan, C. L., Schmitz, K. E., … Hovell, M. F. (2013). The role of parenting in alcohol and tobacco use among Latino adolescents. *Journal of Child & Adolescent Substance Abuse*, *22*(2), 120–132. doi:10.1080/1067828x.2012.730359

Zemore, S. E. (2007). Acculturation and alcohol among Latino adults in the United States: A comprehensive review. *Alcoholism: Clinical and Experimental Research*, *31*(12), 1968–1990. doi:10.1111/j.1530-0277.2007.00532.x

Zhang, S. (2012). *Babel or Great Wall: Social media use in an acculturation context*. The Pennsylvania State University dissertation. Retrieved from: file:///E:/work%20documents%20November%202014/Erasmus%20University/Research/Sociological%20Studies%20of%20Children%20and%20Youth/literature/revise/Shaoke%20Zhang%2020120726.pdf

UNITED FUTURE LEADERS: A CASE OF THE USE OF TECHNOLOGY IN YOUTH PROGRAMMING AND HIDDEN CURRICULUM

Erin Kostina-Ritchey, Holly E. Follmer-Reece, Sara L. Dodd, Kayla Sherman and Gloria Gonzales

ABSTRACT

Purpose – *The purpose of this paper is to provide a case study of the use of technology as a hidden curriculum in a youth leadership program (United Future Leaders-UFL).*

Methodology/approach – *A description of the UFL program, including theoretical framework and current use of various technology platforms, provide a backdrop to the hidden curriculum implemented by the programming staff. Both intended and unintended outcomes of the use of technology are discussed in the context of UFL values/themes.*

Findings – *A review of technology use in the UFL program resulted in the categorization of realms of influence (Staff ↔ Student Participants,*

Technology and Youth: Growing Up in a Digital World
Sociological Studies of Children and Youth, Volume 19, 439–461

ISSN: 1537-4661/doi:10.1108/S1537-466120150000019016

Staff ↔ Staff, Staff ↔ Parents, Participants ↔ Participants) and five categories of technology use (reminders, communication, sharing of resources, reinforcing learning, increased parental involvement).

Practical/social implications – *Examples of emerging patterns of this hidden curriculum, future directions for technology use within the program, and advice for youth program practitioners are included.*

Originality/value – *This paper adds to the general discussion of types and purpose of technology use, youth programming, and role of technology use as hidden curriculum.*

Keywords: Hidden curriculum; positive youth development (PYD); adolescents/youth; technology

INTRODUCTION

Debates and reports focusing on generational differences between "digital natives" (those born after 1980) and "digital immigrants" (e.g., Jones, 2014; Prensky, 2001) have long been of interest to researchers, youth service providers, and parents. As technology becomes more accessible, media usage for youth has steadily increased. In 2004 young people spent an average of 6.5 hours a day (a total of 8 hours and 33 minutes if including multitasking media use) engaged in technological media; this number increased to 7.38 hours (10 hours and 45 minutes a day if including multitasking activities involving media) by 2009 (Kaiser Family Foundation, 2010). More recent reports from the Kaiser Family Foundation on youth media use have not been released at the time of this publication, but trends are expected to continue to climb.

Online practices of adolescents outside of schoolwork tend to be either friendship or interest-driven (Ito et al., 2013). This finding raises interesting questions as to why youth appear to prefer the use of communication technology over face-to-face or even voice communication to interact with their peers. Madell and Muncer (2007) reported focus group findings that suggested youth may prefer using text and/or Internet-based communication because these media gave them more control over social interactions as well as the ability to be more deliberate about how and what they communicated (asynchronous vs. synchronous communication). Moreover, the use of social networking sites provide valuable training in "critical thinking and argumentation skills ... delivered through peer interaction" (Tynes,

2007, p. 576). However, it is the "virtual" peer interaction itself that could be argued as challenging for some youth.

Given the pervasiveness of technology, parents can be left to wonder what their role is in protecting and promoting technology usage. Families must balance questions of responsible usage with legal implications and best interests for their child (e.g., Boyd, Hargittai, Schultz, & Palfrey, 2011; King, 2010). For example, Riparbelli (2011) reported cases of youth being found guilty of cyberstalking and first degree computer trespassing. Additionally, the use of technology for the purposes of sexting has resulted in cases of minors being charged with harassment, solicitation, and interstate charges of child pornography (e.g., Herman, 2010). Furthermore, legal action continues to be taken against parents and youth who were held responsible for illegal use of peer-to-peer file sharing (Electronic Frontier Foundation, 2008). Finding the right balance of promoting technology use and protecting family members is a difficult endeavor. Given the prevalence of technology use among youth populations and the number of challenges associated with responsible technology usage, parents may turn to youth organizations to help guide character development and responsible technology practices.

Programs for youth audiences should consider their roles in helping to educate and protect participants. In their study involving Finnish teens Leena, Tomi, and Arja (2005) found that adults should bring a more critical eye to the role of communication and information technology in programs designed to promote youth development. As applied to programming, adults need to take active leadership roles in both modeling the use of communication technology and monitoring the use/purpose of technology in youth programming.

One of the overarching goals of youth leadership development is to produce youth who embrace and exemplify healthy living. While it may not be the focus, technology within the program may be an additional method used to reach those overarching goals. For example, the use of technology has been linked to youth empowerment and advocacy (Thackeray & Hunter, 2010). In her 2003–2004 ethnographic study of high school seniors, Brown (2014) identified competence in the use of informational and communication technologies (ICT) as integral to the development of important self-efficacy beliefs such as perceived autonomy, confidence in meeting goals and motivation. She further found that the importance of ICT was associated with student worth when related to self-esteem. Similarly, other research has shown ICT use to be a powerful tool in fostering engagement, motivation, and development of self-efficacy in youth (e.g., Hsieh, Cho, Liu, & Schallert, 2008; Strom, Strom, Wing, & Beckert, 2010). These same self-efficacy beliefs lay a foundation for the development of traits such as empowerment and advocacy – two skills that are emphasized in many youth development programs.

In this paper we share the intended and unintended experiences stemming from the use of technology in a youth leadership program developed for early adolescents. We first provide a brief description of the program and discuss the theoretical framework that guides all facets of the program curriculum to include the virtual, socio-ecological context created by the use of technology. Following an explanation of how technology is used within the program, we explore how implicit rules and expectations that guide technology-driven changes in communication and information-sharing behaviors within the program interact with the face-to-face curriculum to create a hidden curriculum (which some have characterized as a "public pedagogy," e.g., Freishtat & Sandlin, 2010). Four emerging patterns of this hidden curriculum are presented. We conclude the paper with future directions for technology use within the program and advice for youth program practitioners based upon programming reflections.

UNITED FUTURE LEADERS (UFL) PROGRAM

Program Description

Developed as a unique collaboration between local schools, a regional business, a private foundation and a regional public university, the United Future Leaders program (UFL) was established as a means to cultivate skills and qualities of civility, ethics and leadership among 5th and 6th grade student participants as they made the transition between elementary and middle school (junior high). The program utilizes an original curriculum developed by youth professionals with expertise in counseling, education and health. Program facilitators include certified university staff and trained graduate students, student assistants, and student volunteers from the university.

UFL curriculum targets key developmental needs for adolescents, but also employs active learning principles to teach leadership skills which student participants can use to serve and/or influence their peers to choose positive behaviors. The core values of civility, ethics, and leadership are promoted through experiential, developmentally appropriate lessons. These lessons are based upon themes of identity, belonging, physical safety, emotional security, mission and purpose, competence, and service. The themes serve as foundational components that, when experienced frequently and reiterated in a variety of manners, accumulate, combine, and spur the development of the program's core values (civility, ethics, and leadership). Table 1 includes definitions of UFL core values and themes.

UFL student participants self-select into the program. All students enrolled in participating schools in programming grades are eligible to register for programming; however, participants must have parent/guardian consent and commit to active parent/guardian involvement to enroll and participate. In order to keep a small ratio of participants to program facilitators (average 4:1), the group size and the number of groups available at a school may be limited. In cases where the number of student participants is greater than the number of planned groups, either additional groups are added to serve the school or students are accepted into the program on a first-come/first-served basis.

The UFL program began in 2007 as a six-session pilot with almost 30 5th grade students at two elementary schools. Since that time UFL has experienced substantial growth and adaptation to accommodate demand and community needs. Lessons have continued to be developed and expanded. As of the 2014–2015 academic year, UFL curriculum included 18 weekly sessions for 5th grade participants, 12 weekly sessions for 6th grade participants, community service events, a 5th grade team building event, and a 6th grade celebration banquet. The program was offered at 13

Table 1. Definition of UFL Concepts.

UFL Concepts	Definition
Core values	
Civility	Treat others the way you want to be treated
Ethics	Do the right thing when no one is looking
Leadership	Bringing out the best in myself and others
Themes	
Identity	How I see myself and how others see me
Belonging	I belong
Physical safety	I am safe
Emotional security	I know my feelings and I honor them
Mission and purpose	I have direction and purpose
Competence	I have abilities
Service	Helping others while expecting nothing in return

host campuses in three different school districts with an enrollment of 240 students and over 50 program facilitators.

Since the conception of the program, more than 800 student participants have completed the curriculum and "graduated" from UFL. At the request of participants who wanted to remain involved and active within the program beyond 6th grade, the UFL Ambassadors program was developed. The Ambassador program focuses on further mentoring and developing leadership in students 7–12th grades through participation in community outreach/service opportunities held throughout the year. Ambassadors both organize and join in established community outreach opportunities with the goal of promoting the UFL values and themes. Opportunities for Ambassadors to develop their leadership skills are provided in a variety of forums throughout the school year and summer including mentoring younger student participants, individual and group service projects, relationship building, and community outreach.

One of the ultimate programming goals is for UFL student participants is to attend institutions of higher education and to continue to take on leadership roles. For those attending the collaborating university after high school graduation, there are opportunities for active involvement in the UFL program as a volunteer and/or student staff. While not every student who is a graduate from UFL chooses to be an Ambassador, all graduates are informed of and invited to UFL-sponsored outreach/service opportunities through the use of technology.

With almost 1,000 UFL/UFL Ambassador participants, their families, faculty/staff at the 13 host campuses, more than 50 program facilitators and various other stakeholders, communication is essential for coordination of programming components, dissemination of programming information, and maintenance of the interpersonal relationships. UFL staff use traditional methods of communication such as phone calls and letters to personalize communication; however, the use of more modern communication methods such as the use of text messaging, Facebook, Instagram and blogging supplement, reinforce and multiply the information shared. Since 2007, changes to modes of communication have been observed between program stakeholders (e.g., the program directors and student participants, the program directors and parents, and also between the student participants themselves) due to the more widespread use of technology as a communication tool for both students and parents. While the use of technology is very intentional in regards to program support, technology itself is not a program focus. However, we have found technology to be successful means in which to provide positive support for reinforcement of the program's core values and themes which are anchored in positive youth development (PYD). The discussion which follows will investigate how the PYD theoretical framework, the use of technology and a "discovered" hidden curriculum support program processes, outcomes, and goals.

Theoretical Framework

Positive youth development (PYD) serves as the theoretical framework for UFL. Lerner, Phelps, Forman, and Bowers (2009) described PYD as a strength-based model that stems from developmental systems theory and comparative psychology. Development occurs through the reciprocal interplay of the individual and the individual's real-world setting. *Positive* development occurs when personal strengths are aligned with resources for healthy growth found in various socio-ecological contexts such as the home, school, and community (Lerner, 2004).

PYD theorists commonly refer to these strengths and resources as assets. The more assets an individual has, the better equipped s/he is to deal with adversity (e.g., Benson, 2006; Lerner, 2004). The frequent and sustained fusion of assets over time allows the individual to thrive in spite of adversity (Scales, Benson, Leffert, & Blyth, 2000). Thriving is the ultimate outcome of positive development. Indicators of thriving are known as the "Five Cs": competence, confidence, connection, character, and caring/compassion (e.g., Eccles & Gootman, 2002; Lerner, Fisher, & Weinberg, 2000). Table 2 contains definitions for each of the Five Cs and describes how the corresponding UFL values/themes relate to them.

Alignment of personal and ecological assets is not an automatic or passive process; individuals are important and active actors in their own development (Benson, Scales, Hamilton, & Semsa, 2006). Organized youth programs can assist individuals in their self-generated development (Larson et al., 2004) by serving as a safe, yet challenging environment in which they are afforded the opportunity to nurture interests, acquire new skills, and develop a sense of personal and group identity (Roth & Brooks-Gunn, 2003). UFL is an exemplar of this type of program – it is designed to (a) help youth recognize the assets they have within themselves and that are present in their real-world settings, (b) build upon and contribute to those assets, (c) demonstrate how to be an asset and create assets for others, and (d) provide a safe environment in which to experience, practice, and exemplify the Five Cs.

Table 2. Definition of Five Cs and Corresponding UFL Value/Theme.

Five Cs	Definition (Lerner et al., 2000, p. 17)	Corresponding UFL Value/Theme
Competence	"Intellectual ability and social and behavioral skills"	• *Civility* – awareness and application of social competency • *Competence* • *Leadership* – both cognitive and vocational competence[a] • *Mission and purpose* – vocational, setting career goals[a] • *Physical safety* – basic needs must be met (physical, emotional, social …) in order to function[a] (Maslow, 1970)
Connection	"Positive bonds with people and institutions"	• *Belonging* – being a part of and helping others feel they belong (bidirectional) • *Civility* – positive bonds through mutual respect • *Service* – positive relationships fostered by helping others
Character	"Integrity and moral centeredness"	• *Civility* – mutual respect[a] • *Ethics* – integrity • *Identity* – maintain and identify standards based on how I see myself • *Leadership* – internal desire to have good character
Confidence	"Positive self-regard, a sense of self-efficacy and courage"	• *Competence* – confidence in your own skills and the ability to try new things • *Emotional security* – I have the confidence to overcome challenges • *Identity* – positive self-regard • *Leadership* – confidence to lead oneself and others
Caring/ Compassion	"Humane values, empathy and a sense of social justice"	• *Civility* – treat others as you wish to be treated • *Ethics* – social justice • *Service* – helping others in need

[a]See also Lerner et al. (2005).

UFL serves as one of many micro-socializing systems for student participants. As a PYD program, all facets of UFL need to provide an ecological context that connects youth with principled, caring adults (Roth & Brooks-Gunn, 2003). This context exists in the "real world," during in-person program delivery or service events and extends to the "virtual world" that is created through the use of various technology platforms. Real or virtual, the socio-ecological context created by UFL must provide clear and consistent messages about what does and does not promote positive youth development. The atmosphere of both contexts must be one of support (communicating approval, acceptance, value) and empowerment (feeling safe, respected, capable).

Technology Use in UFL

As discussed in the program description (above), UFL's use of technology supplements rather than replaces traditional forms of communication (e.g., phone calls, postal letters). The program directors felt that it was important to continue both traditional and technology-based forms of communication in order not to isolate student participants and families that had limited access to technology outside of publicly accessible sources (school, library) and/or may benefit from special services (e.g., translation from English to Spanish).

Currently, UFL supports the expanded use of technology in the program through e-mail, text messaging, Facebook, Instagram, and a blog. Each technology platform is a separate venue within which specific UFL values/themes are reiterated and demonstrated. While not every value/theme may be incorporated into each technology type, all values/themes are addressed within the collection of technologies.

Civility

The only criteria for UFL program participation is commitment and dedication from students and their families. In return, UFL responds to all communication, questions, and needs of families in a timely manner. This commitment to treating UFL families with equal respect promotes quality program implementation and development. E-mail and text messaging are the technology types used for the reciprocating dialogue between UFL families and UFL directors. Civility is also demonstrated in the more didactic use of Instagram. All pictures shared on the UFL Instagram page demonstrate diverse youth cooperating, collaborating, and learning, while having fun. Followers of the UFL Instagram page are able to see the potential impact of youth who are developing leadership skills and learning how to treat others with respect and acceptance.

Ethics

UFL participants and students do not expect for their successes and accomplishments to be posted and recognized. However, by doing so on the UFL Facebook page, UFL facilitators are able to acknowledge these activities and encourage students to continue their positive behaviors outside of the organization, accomplishing the leadership task of doing the right thing even when no one is watching.

Leadership

UFL program directors send weekly e-mails to volunteer facilitators and staff with announcements, training reminders, and detailed lesson outlines. This communication allows facilitators to be informed of weekly programming needs and encourages active participation in all areas of program implementation, development, and evaluation. Additionally, student assistants, student volunteers, and even student participants often take it upon themselves to seek information about upcoming UFL activities by text messaging program directors. Taking the initiative to ask questions and request support demonstrates leadership competencies.

UFL frequently posts videos and/or pictures of staff and volunteers preparing for delivery of lessons and special events on Facebook. Modeling group communication, teambuilding, dedication to learning, and positive attitudes encourages students to apply similar behaviors in their individual leadership. Encouragement for leadership development also occurs through the UFL blog in which UFL program directors frequently share quotes, resources, and definitions as extensions of the direct UFL curriculum. These additional messages allow student participants, parents, and facilitators to explore different ways to incorporate leadership concepts in their daily lives.

Identity

Facebook provides the unique opportunity to showcase the different talents, gifts, and organizations where students apply their character and leadership skills beyond UFL. Sharing pictures and information about student accomplishments, such as receiving awards and/or recognition for success at school and in the community, creates an audience of support consisting of peers, parents, staff, and educators associated with the UFL program.

Belonging

All UFL participants and parents remain on UFL e-mail distribution lists unless they request to be removed. Consistent and equal communication between UFL and current/past student participants promotes the feelings of belonging to the program and having ongoing support from program directors. A sense of belonging is fostered through the use of Instagram as well. UFL staff post pictures and videos on Instagram to allow students to anticipate and prepare for upcoming lessons and events. Instagram postings also are used to invite parents into program activities. Parents that have been involved in UFL programming for many years often text program directors for information about events and programming. These parents have established a personal connection to UFL and therefore feel comfortable using text messaging as an alternative, less formal means of communication.

Physical Safety

UFL is composed of diverse student participants, which encompass a wide variety of needs and abilities. An example of this includes a student whose parent had concerns regarding the child's ability to remember and safely arrive at the designated program delivery site. For this reason, the parent utilized text messaging to quickly verify the child's attendance. Additionally, UFL program directors utilize Facebook as a means of communicating time sensitive information to parents and students. An example of this would include updates on program cancellations due to inclement weather. Parents and students often receive this information in a more consistent, timely manner because they visit Facebook frequently throughout the day, as opposed to e-mail.

Emotional Security

Instagram is a medium most used by youth because it is intended for the sharing of pictures with minimal text. The UFL Instagram page allows program directors to share fun and interactive pictures of students actively involved in program lessons and events. Student participants enjoy seeing pictures of themselves, which builds their individual self-confidence and self-worth. Parents also feel more confident and connected to the UFL program because they are able to see that students are happy and enjoying their time during program events.

Mission and Purpose
UFL program directors use e-mail to communicate holistic goals and support needs with school district administration and support staff. This method of communication encourages buy-in, commitment, and understanding among UFL host campuses. The UFL blog, on the other hand, is a tool for communicating curriculum objectives and goals to parents and program facilitators. Blog posts provide descriptions of lessons along with questions to promote discussion and reflection of program themes. This thorough information allows key stakeholders to understand the intended purpose behind all UFL themes and lessons.

Competence
Part of the UFL training of undergraduate/graduate student assistants and volunteers includes appropriate and professional methods of communication with program directors. Students are expected to practice professionalism when e-mailing program directors regarding attendance, requests for support, and suggestions for program improvement. Students gain essential competencies to be used in future work environments.

The mission of UFL is to build leadership skills among youth. Program directors frequently post pictures and information on the Facebook page about the competencies gained by student participants through weekly UFL lessons. Parents are then able to include these lessons and skills in the discussions and responsibilities of their child at home. Some specific UFL lessons provide unique opportunities to teach competencies through visual demonstrations. The UFL staff record students completing activities, such as CPR and the Heimlich maneuver, which are then shared on Instagram with students outside of UFL. These visual demonstrations allow program directors to impact youth beyond direct program participants.

Service
E-mail and Facebook are essential technology platforms used to communicate service opportunities to student participants and their families. For example, throughout the two-week UFL Ambassador School Supply Drive parents and participants receive daily e-mail updates of donation progress and remaining needs in order to meet the service goal and upcoming deadlines. UFL partners with local non-profit agencies, such as Ronald McDonald House Charities, to provide additional service opportunities for youth participants. Facebook allows UFL directors to share information posted by these agencies about community needs and service events so that parents and student participants are aware of these opportunities. After the

completion of UFL-hosted service activities, program directors use Facebook to post pictures and information about the impact made by student volunteers. This interactive communication encourages future engagement among program participants and families.

Additional outlets for the use of technology are actively being explored by UFL program directors based on student trends in technology use and how that technology can be used to supplement existing resources (e.g., Blair & Fletcher, 2011; Ito et al., 2013). Student participants, family members, and program facilitators are made aware of each of the types of technology utilized by UFL during recruitment events and through reminders throughout the programming year. If interest is indicated, each stakeholder is encouraged to subscribe/check outlets which contain postings.

Five major categories for the uses of technology have evolved along with the UFL program. These uses include using technology (a) to remind of events and opportunities, (b) as a communication tool, (c) for sharing resources, (d) to expand opportunities to reinforce learning, and (e) to increase parental involvement in curriculum. First and, perhaps, the most obvious use of technology in the UFL program is to provide reminders of upcoming events and opportunities. This use of technology allows for a central location where student participants and their families can access information for upcoming events. The use of technology in this manner has received positive feedback from students and their families and indirectly increased program participation as there are a variety of multiple reminders, that is, text messages, Facebook, Instagram. Second, the use of technology such as Facebook can be used as a communication tool for providing updates of achievements made by student participants and program facilitators. Third, as the UFL program has grown to include an ever increasing number of schools, technology is a resource which allows for a common area where students from different schools who connect through UFL-sponsored events can develop friendships and/or community service partnerships. An unintended byproduct of the increased technology use has been the sharing of resources between UFL participants. Fourth, the use of technology has allowed for UFL staff to post "teasers" of upcoming lessons for student participants. Along with the teasers, technology has served as an outlet for reflection, feedback, reinforcement, and expansion of lessons. Finally, UFL staff realized that special family posts, which are made to keep parents/guardians aware of UFL lessons for the week, allowed for increased family involvement in programming and fostered a deeper shared understanding of the curriculum.

As the five major categories for the uses of technology have evolved, we began to realize that there were unintended "hidden curricula" being taught to the UFL participants and their family members. While UFL program developers were very intentional/mindful of extending PYD principals to the use of technology, hidden curricula that reflected UFL core values of civility, ethics, and leadership were embedded. As we began to "dig deeper" into how our use of technology provided a hidden curriculum, we realized that the effects extended beyond student participants to the program facilitators (university graduate students, student assistants, and student volunteers) who are responsible for program delivery.

THE HIDDEN CURRICULUM IN USE OF TECHNOLOGY

Use of the term "hidden curriculum" was popularized in the United States by Snyder (1971). The original concept was based on two basic assumptions: (a) while some tasks are explicitly outlined, there are parallel expectations that are based on social context, assumptions, and expectations of "students" (program participants) and "teachers" (program directors and facilitators) and (b) "students" will ascertain both what is formally and informally expected, then adapt behavior/actions based on those expectations. Embedded in this idea is the concept of both the social and the relational contexts.

Values that are not made explicit are an example of hidden curriculum (Hopman, de Winter, & Koops, 2014). Chuang et al. (2010) suggested that older students accept values, attitudes, and beliefs they learned from hidden curriculum. Hidden curriculum can take the form of unconscious acts or conscious decisions to be silent on an issue (Manhood, 2011). This said, Chuang et al. noted that the challenge lies in the identification of hidden curriculum – which can be difficult to acknowledge. Thus, the first step is to acknowledge the existence of hidden curriculum (even if it is unintentionally a part of the program) and identify the characteristics that it holds.

Once UFL and center program directors identified the presence of hidden curriculum, they followed suggestions outlined by Chuang et al. (2010) to actively shape the curriculum. Changes were based on the idea that hidden curriculum can be developed. Included is the belief that (a) behavior can be influenced/changed, (b) both structured and unstructured self-reflection influences personal growth and the curriculum, (c) role modeling is the central way individuals can influence hidden curriculum, and (d) any effort to make a positive impact in hidden curriculum is valuable, but changes should ideally occur at both the personal and program levels.

Similar to other work which has sought to understand everyday behavior (Doyle & Arnedillo-Sanchez, 2011), one of the goals of the UFL hidden curriculum became to make an intentional effort to socialize UFL student participants and program facilitators directly, and participant families indirectly, to the "dos and don'ts" of social media use. As UFL is anchored in PYD (e.g., Lerner, 2004) it has not experienced the negative outcomes and/training which have been associated with some other hidden curricula (e.g., Manhood, 2011; Patterson, 2013).

The UFL directors actively modeled the appropriate uses of technology in the types of communications and visual media (photographs and videos) they posted on UFL social media sites (Facebook, Instagram). The postings meet the UFL core values and definitions of civility, ethics, and leadership. For example, only posting positive valence comments concerning experiences, people and programming are modeled. UFL student participants (and parents) began to model their postings and responses to posted materials after the example provided by the program directors on UFL sponsored technology/social media. While the original target audience for this modeling was the UFL student participants, it soon became evident that other UFL staff and volunteers also were benefiting from, and mimicking the style of, posts created by the program directors. The student staff and volunteers began to actively model posts for UFL communication that reflected the standards of UFL core values.

Four patterns of influence in the hidden curriculum emerged which reflected findings by Chuang et al. (2010). These patterns were staff ↔ student participants, staff ↔ staff, staff ↔ parents, and student participants ↔ student participants. Further discussion of communication patterns are discussed below.

Staff ↔ Student Participants

The UFL directors and program facilitators use various technology methods to communicate with student participants. In addition to previously mentioned personal phone calls and print invitations mailed to parents, UFL facilitators post information pertaining to all scheduled events on Facebook and Instagram. The postings provide professional examples not only of layout and design, but of type and style of information which should be communicated for official events. For example, UFL Ambassadors were invited to attend an annual holiday service event benefiting Ronald McDonald House Charities (RMHC). The posted invitation provided event information to include location, date and time, who was invited to attend (Ambassadors grades 6–12, family members, and friends) and event type (service project preparing holiday items to benefit families at RMHC). Using Facebook and Instagram to personally connect with UFL student participants, a medium in which they already actively participate, makes it easy and convenient for the student participants to ask questions and RSVP for events. The option to share information posted on UFL outlets on personal pages allows for student participants to repost the opportunities while sharing the information with their peers.

In addition to information sharing and event notifications, UFL facilitators also utilize social networking sites to reinforce curriculum concepts and build relationships between student participants and program facilitators. Postings are created to show work done at program delivery campuses, UFL staff involved in program/lesson preparation, staff training and reflections to improve programming components. This "behind the scenes" footage is captured through texts, photographs, and videos. Posted footage demonstrates the preparation that goes into successful programming and also how leadership is carried out in all areas of life (e.g., in- and outside of the classroom).

UFL directors/facilitators also communicate with student participants about individual accomplishments. When an active member of the UFL Ambassador program was named "Student of the Month" in a large local school district, UFL posted the information on Facebook along with a link to the local media coverage announcing the student award. This link allowed UFL to recognize the student's achievement and to share the information with the broader UFL and Ambassador audience. In turn, those who knew the student through her volunteer work and Ambassador activities could post comments congratulating her on the honor. This student became a living embodiment of the themes of leadership, ethics, civility, belonging, mission and purpose, competence and service.

Reciprocally, UFL student participants have found the use of Facebook and Instagram as a beneficial way to communicate with program directors and facilitators. After completing the two-year after-school UFL program and moving forward as a UFL Ambassador, student participants do not have scheduled weekly sessions in which they may interact with program directors or facilitators. As a result, UFL Ambassadors often use social networking to share information about their individual activities and remain connected to UFL throughout the year. For example, one UFL

Ambassador established a unique Instagram page specifically for the purpose of sharing information (such as school basketball game schedules and other extracurricular activities in which the UFL Ambassadors participate) with the UFL directors and facilitators. The creation of this Instagram page allowed for UFL staff to know of upcoming events and extend their support virtually and/or by attending posted events. This participant-created page has been frequently updated since its establishment. This example exemplifies how technology (and the hidden curriculum of using technology to communicate) relates to the reinforcement of the curriculum themes of competence, belonging, and leadership.

Staff ↔ Staff

UFL directors realized that while modeling hidden curriculum to student participants, they were in fact also modeling hidden curriculum to program facilitators. UFL social media sites became locations where program facilitators can assess examples of positive posts. In turn, they are encouraged to create their own posts based on the models provided. For example, a picture of four program facilitators who were graduating in the month of December was posted to share with student participants and their families. This picture shows the variety of emotional responses to completing graduation requirements (jumping in the air in celebration, shock, etc.). More importantly, the picture was taken on campus at a known holiday symbol rather than an alternative location which might be popular in other posts not associated with UFL (such as a bar or party). By creating this post, student participants and their families received an update on programming facilitators, some who were remaining with UFL after graduation and others who were not. Indirectly, this post also provided information concerning the program role models and possible future involvement (and thus consistency) in student participant lives. In addition to reinforcing the UFL values and themes (e.g., leadership, belonging, mission and purpose, competence) the post emphasized the importance of pursuing and completing higher education.

Staff ↔ Parents

Establishing an effective parent component in youth programming can be a challenge (see Spencer, Basualdo-Delmonico, & Lewis, 2011; Williams & Sánchez, 2011). The challenge lies in the time and effort required on the part of the programming directors and facilitators to connect with all parents on a personal level. The use of a wide variety of communication methods, including technology, can assist. Utilizing sites such as Facebook or Instagram allows for immediate communication with parents with the additional benefit of seamless response. For example, when an overwhelming RSVP response for an event took place and a change in event venue subsequently occurred, a letter was mailed to families but an announcement was also posted to the Facebook and Instagram sites to remind families of the venue change. Parents/guardians were requested to reply with a click of button and/or short response indicating they received the information. Similar to announcements posted for staff ↔ student participants, the posts can serve to model professionalism and positive reinforcement.

Relationships between UFL directors/facilitators and families have not only increased through the sharing of program logistics, but also through the promotion of program transparency. At times, youth programs may fail to maintain regular communication that informs parents of the lessons and concepts addressed during programming. Although parents should be able to trust program leaders to teach their students positive and beneficial information, they often are unaware of the daily activities, teachings, and interactions. UFL program directors and facilitators frequently post images and reviews of weekly lessons from all host campuses to provide parents with visuals of the activities their students are experiencing. This communication increases the parents' ability to continue discussions of key leadership concepts at home, thus promoting additional student reflection, application, and growth. In addition, UFL-initiated posts allow parents the opportunity to "get to know" the UFL directors and facilitators who work with students each week.

Participants ↔ Participants (and Others)

Technologically savvy student participants may take it upon themselves to take pictures and videos during program delivery sessions for the purpose of sharing these images on their personal social networking sites. While student participants are permitted to make media records of lessons and events, they are expected to follow the same posting practices of UFL directors and program facilitators including civility. These postings are not limited to UFL/Ambassador technology platforms (Facebook, Instagram); student participants may post the information on the social networking sites their peer groups utilize most. For example, one student developed a Flipgram focusing on the UFL lesson related to identity. The posting included pictures, video, music, and text. This student included a written description of the benefits of the lesson and discussion, as well as a reflection of her own identity. The media was created with the specific purpose to share with friends and family. Through sharing programming lessons with friends outside of the program, participants are able to promote key leadership concepts from their own perspective in a fun, age-appropriate manner. Such activities also allow program directors to anecdotally evaluate lessons, thus directly impacting program development, student behavior, and growth.

FUTURE DIRECTIONS

Technology is constantly evolving and changing. The UFL team is dedicated to finding new methods to connect with all stakeholders (program participants, parents, schools, program directors, student assistants, and student volunteers). While these forms of communication are necessary for immediate, short information sharing, they do not allow for a direct connection with program facilitators or long reflections on program experiences. The UFL program is currently exploring new media outlets and recently developed a program blog. This blog is an area where program facilitators are encouraged to write weekly reflections regarding personal experiences with program participants/families and other program facilitators. In addition to allowing for informal curriculum evaluation, individual reflections assist all stakeholders in becoming more familiar with each other. Program facilitators develop their abilities to share various perspectives of the leadership curriculum and philosophy. The use of the blog has allowed for a lengthier outlet where program facilitators can muse about programming while simultaneously employ a technology platform where they are the unilateral providers (developers and responsible-parties for posted content) of information to student participants and families.

ADVICE FOR YOUTH PROGRAMMING PRACTITIONERS

Regardless of whether the use of technology is implicitly or explicitly used in youth development programs, program directors and staff have a responsibility to be mindful of the "do no harm" principal of working with youth (Rhodes & Liang, 2009). Prior to utilizing technology in programming, program directors need signed releases from parents and students to post identifiable photographs of the student participants. This requirement optimizes safety and security while also respecting students' and families' privacy. Students and/or their families could reject photographs and/or videos based on any number of reasons including, but not limited to, religious beliefs (Brougher, 2012), legal concerns (e.g., Berson & Berson, 2006), safety concerns (e.g., Driskell, 2009), and independent nature of the debate concerning child/parent rights (e.g., Lundy, 2007; Schoeman, 1980).

In addition to having photo releases on file for every student participant, security and privacy settings should be set to the most stringent levels available on the site. Those who desire to view the site must first seek permission to view content. This practice allows for media to be available to students and their family members, but helps to protect against unknown or unwanted visitors to the page.

Another important legal issue to consider is the appropriate interaction between program staff and their interaction with minors. By creating digital outlets sponsored by youth programming directors, staff can be held accountable for digital interactions with student participants. Providing platforms which allow for program staff to keep in contact with students/families outside of personal platforms is important as concerns have been raised that younger generations of professionals who have grown up "connected" might face challenges in the blur between personal/professional life and relationship boundaries (Zernike, 2014).

Technology is constantly changing and evolving. Demographics of those who were using one platform of technology today might not be the same people who will be using that platform tomorrow (e.g., Drazenovic, 2014). It is important to track local trends in social network sites used by program participants and their families. Program directors need to be willing to consider new types of technology use in order to reach the greatest number of stakeholders. All forms of communication are necessary to ensure each family receives equivalent information. The implementation of technology (text messages, media sites) should not supersede the use of traditional forms of communication (telephone, mail service).

The use of technology with youth is pervasive. Access to the internet continues to increase. Warschauer and Matuchniak (2010) found that 92% of 12–14 year olds and 96% of 15–17 year olds reported internet use at home or publically accessible locations. Programs that embrace the use of technology can provide active ways to engage students and usher in opportunities to forge greater connections between participants and the program.

However, the use of technology should not be haphazard. Rather, technology use should be seen as a supplement to programming. Done properly, this supplementation requires an active awareness of the hidden curricula being conveyed through technology platforms and includes modeling by program directors and staff.

Given the social changes that have occurred since the concept of hidden curriculum was first developed, questions have arisen about (a) if the concept of hidden curriculum has become less relevant and (b) if there is still a need to consider hidden curriculums (Joughin, 2010). Chuang et al. (2010) noted that hidden curriculum is neutral. It can either reinforce or undermine the goals of the original/formal curriculum depending on how it is used. Thus, the emphasis should be placed not on eliminating hidden curriculum from programming, but rather shaping hidden curriculum to support program goals.

REFERENCES

Benson, P. (2006). *All kids are our kids: What communities must do to raise caring and responsible children and adolescents* (2nd ed.). San Francisco, CA: Jossey-Bass.

Benson, P., Scales, P., Hamilton, S., & Semsa, A., Jr. (2006). Positive youth development: Theory, research, and applications. In R. M. Lerner (Ed.), *Theoretical models of human development* (6th ed., Vol. 1, pp. 894–941). Handbook of Child Psychology (Editors-in-chief: W. Damon & R. M. Lerner.). Hoboken, NJ: Wiley.

Berson, I., & Berson, M. (2006). Children and their digital dossiers: Lessons in privacy rights in the digital age. *International Journal of Social Education, 21*, 135–147.

Blair, B., & Fletcher, A. (2011). "The Only 13-Year-Old on Planet Earth Without a Cell Phone": Meanings of cell phones in early adolescents' everyday lives. *Journal of Adolescent Research, 26*, 155–177.

Boyd, D., Hargittai, E., Schultz, J., & Palfrey, J. (2011). Why parents help their children lie to Facebook about age: Unintended consequences of the 'Children's Online Privacy Act'. *First Monday, 16*, 2–23.

Brougher, C. (2012). *Legal analysis of religious exemptions for photo identification requirements*. Washington, DC: Congressional Research Service.

Brown, T. M. (2014). "I Just Want to Work Hard": Self-efficacy and the social contexts in adolescents' ICT use. *Youth & Society November, 46*, 853–874.

Chuang, A., Nuthalapaty, F., Casey, P., Kaczmarczk, J., Cullimore, A., Dalrymple, J., … Peskin, E. (2010). To the point: Reviewers in medical education taking control of the hidden curriculum. *American Journal of Obstetrics & Gynecology, 203*, 316.e1–316.e6.

Doyle, T., & Arnedillo-Sanchez, I. (2011). Using multimedia to reveal the hidden code of everyday behavior to children with autistic spectrum disorders (ASDs). *Computers & Education, 26*, 357–369.

Drazenovic, M. (2014, May). Elementary school students using Facebook: A case study of Croatian elementary school ban Josip Jelačić. In *Information and Communication*

Technology, Electronics and Microelectronics (MIPRO), 2014 37th International Convention on (pp. 911–913).

Driskell, K. (2009). Identity confidentiality for women fleeing domestic violence. *Hastings Women's Law Journal, 20*, 129–156.

Eccles, J., & Gootman, J. (Eds.). (2002). *Community programs to promote youth. Development/committee on community-level programs for youth.* Washington, DC: National Academy Press.

Electronic Frontier Foundation. (2008). *RIAA v. The People: Five years later.* Retrieved from https://www.eff.org/wp/riaa-v-people-five-years-later

Freishtat, R., & Sandlin, J. (2010). Shaping youth discourse about technology: Technological colonization, manifest destiny, and the frontier myth in Facebook's public pedagogy. *Educational Studies, 46*, 503–523.

Herman, J. (2010). Sexting: It's no joke, it's a crime. *Illinois Bar Journal, 98*, 192–195. Retrieved from http://www.isba.org/ibj/2010/04/sextingitsnojokeitsacrime

Hopman, M., de Winter, M., & Koops, W. (2014). Analyzing the hidden curriculum a method for the analysis of values in youth care interventions. *Methods, 10*, 12–20.

Hsieh, P., Cho, Y., Liu, M., & Schallert, D. (2008). Examining the interplay between middle school students' achievement goals and self-efficacy in a technology-enhanced learning environment. *American Secondary Education, 36*, 33–50.

Ito, M., Baumer, S., Bittanti, M., Boyd, D., Cody, R., Herr-Stephenson, B., … Tripp, L. (2013). *Hanging out, messing around and geeking out: Living and learning with new media.* Cambridge, MA: MIT Press.

Jones, A. (2014, January 22). Sex and the single tween. *Newsweek*. Retrieved from http://www.newsweek.com/sex-and-single-tween-245090

Joughin, G. (2010). The hidden curriculum revisited: A critical review of research into the influence of summative assessment on learning. *Assessment & Evaluation in Higher Education, 35*, 335–345.

Kaiser Family Foundation. (2010). *Generation M2 media in the lives of 8 to 18 year olds.* Menlo Park, CA: Kaiser Family Foundation.

King, A. (2010). Constitutionality of cyberbullying laws: Keeping the online playground safe for both teens and free speech. *Vanderbilt Law Review, 63*, 845–884.

Larson, R., Jarrett, R., Hansen, D., Pearce, N., Sullivan, P., Walker, K., … Wood, D. (2004). Organized youth activities as contexts of positive development. In P. A. Linley & S. Joseph (Eds.), *Positive psychology in practice* (pp. 540–560). Hoboken, NJ: Wiley.

Leena, K., Tomi, L., & Arja, R. (2005). Intensity of mobile phone use and health compromising behaviors – How is information and communication technology connected to health-related lifestyle in adolescence? *Journal of Adolescence, 28*, 35–47.

Lerner, J., Phelps, E., Forman, Y., & Bowers, E. (2009). Positive Youth Development. In R. M. Lerner & L. Steinberg (Eds.), *Handbook of adolescent psychology* (3rd ed., pp. 524–558). Hoboken, NJ: Wiley.

Lerner, R. (2004). Diversity in individual ↔ context relations as the basis for positive development across the life span: A developmental systems perspective for theory, research, and application. *Research in Human Development, 1*, 327–346.

Lerner, R., Lerner, J., Almerigi, J., Theokas, C., Phelps, E., Gestsdottir, S., … von Eye, A. (2005). Positive youth development, participation in community youth development programs, and community contributions of fifth grade adolescents: Findings from the first wave of the 4-H Study of Positive Youth Development. *Journal of Early Adolescence, 25*, 17–71.

Lerner, R. M., Fisher, C. B., & Weinberg, R. A. (2000). Toward a science for and of the people: Promoting civil society through the application of developmental science. *Child Development*, *71*, 11–20.

Lundy, L. (2007). 'Voice' is not enough: Conceptualizing article 12 of the United Nations convention on the rights of the child. *British Educational Research Journal*, *33*, 927–942.

Madell, D., & Muncer, S. (2007). Control over social interactions: An important reason for young people's use of the Internet and mobile phones for communication? *CyberPsychology & Behavior*, *10*, 137–140.

Manhood, S. (2011). Medical education beware the hidden curriculum. *Canadian Family Physician-Le Medecin de Famille Canadien*, *57*, 983–985.

Maslow, A. (1970). *Motivation and personality* (2nd ed.). New York, NY: Harper & Row.

Patterson, J. (2013). Constructions of violence and masculinity in the digital age. In C. Fowley, C. English, & S. Thouesny (Eds.), *Internet research, theory, and practice: Perspectives from Ireland* (pp. 113–134). Dublin: Research-publishing.net.

Prensky, M. (2001). Digital natives, digital immigrants. *On the horizon*, *9*, 1–6.

Rhodes, J., & Liang, B. (2009). First do no harm: Ethnical principals for youth mentoring relationships. *Professional Psychology: Research and Practice*, *5*, 452–485.

Riparbelli, L. (2011). 12-year-old sentenced for cyberstalking classmate. *ABC News*. Retrieved from http://abcnews.go.com/Technology/12-year-sentenced-washington-cyberstalking-case/story?id=14072315

Roth, J. L., & Brooks-Gunn, J. (2003). What is a youth development program? Identification and defining principles. In F. Jacobs, D. Wertlieb, & R. M. Lerner (Eds.), *Enhancing the life chances of youth and families: Public service systems and public policy perspectives: Vol. 2. Handbook of applied developmental science: Promoting positive child, adolescent, and family development through research, policies, and programs* (pp. 197–223). Thousand Oaks, CA: Sage.

Scales, P., Benson, P., Leffert, N., & Blyth, D. (2000). Contribution of developmental assets to the prediction of thriving among adolescents. *Applied Developmental Science*, *4*, 27–46.

Schoeman, F. (1980). Right of children, rights of parents and the moral basis of the family. *Ethics*, *91*, 6–19.

Snyder, B. (1971). *The hidden curriculum*. New York, NY: Knopf.

Spencer, R., Basualdo-Delmonico, A., & Lewis, T. O. (2011). Working to make it work: The role of parents in the youth mentoring process. *Journal of Community Psychology*, *39*, 51–59.

Strom, P., Strom, R., Wing, C., & Beckert, T. (2010). Adolescent learning and the internet. *Education Digest*, *75*, 10–16.

Thackeray, R., & Hunter, M. (2010). Empowering youth: Use of technology in advocacy to affect social change. *Journal of computer-mediated communication*, *15*, 575–591.

Tynes, B. M. (2007). Internet safety gone wild? Sacrificing the education and psychosocial benefits of online social environments. *Journal of Adolescent Research*, *22*, 575–584.

Warschauer, M., & Matuchniak, T. (2010). New technology and digital worlds: Analyzing evidence of equality in access, use and outcomes. *Review of Research in Education*, *34*, 179–225.

Williams, T., & Sánchez, B. (2011). Identifying and decreasing barriers to parent involvement for inner-city parents. *Youth & Society*, *45*, 54–74.

Zernike, K. (2014, October 3). Abuse cases at 2 schools, with technology at the root. *The New York Times*. Retrieved from http://www.nytimes.com

ABOUT THE AUTHORS

Åsa Andersson has a Ph.D. in Ethnology and is a Senior Lecturer in Cultural Studies at University West, Sweden. Her research and teaching focus mainly on youth culture, gender, and ethnicity. She has led research projects on identification processes among young women in multicultural neighborhoods. She has also been involved in an action research project on empowerment in Roma communities in western Sweden. Her most recent research project is about young men and women's experiences of unemployment.

Troy E. Beckert is an Associate Professor of Family and Human Development at Utah State University. His interests in adolescents and their families shape a research focus on intergenerational learning, autonomy, and psychosocial development. His current direction of inquiry allows for culturally diverse assessments of cognitive autonomy and psychosocial development as they relate to technology use in adolescent populations.

Margareta Bohlin has a Ph.D. in Psychology and is a Senior Lecturer at University West, Sweden. Her research focuses on risk-taking among adolescents from existential, social, and gender perspectives. She leads projects concerning adolescents' use of social media and on risk discourses in the media, both online and offline. She also participates in projects about sexual development in relation to the Internet and social media as well as in research about children's and young people's experiences of type 1 diabetes, risks, and treatment. She mainly teaches Developmental Psychology and courses in intersectional and gender perspectives, research methods, and risk research.

Antonella Brighi, <antonella.brighi@unibo.it> Ph.D., is Professor of Developmental Psychology at Bologna University, Italy. She has coordinated European and National research projects on bullying and cyberbullying and has been part of an international network of researchers on cyberbullying. She has been International Research Fellow at Flinders University, Australia. Her research focuses on risk factors, incidences, and intervention programs for the prevention of bullying and cyberbullying in childhood and in adolescence, exploring the emotional impact of social exclusion and aggression on mental health as well as the link between processes of socialization in the offline and online contexts. She is consultant for schools for the implementation of programs for socioemotional competences and promotion of psychological well-being.

Ana Campos-Holland, Ph.D., is an Assistant Professor of Sociology at Connecticut College who specializes in the study of childhood and adolescence. She prioritizes methodological approaches that create platforms for youth's voices to be heard.

Shelia R. Cotten is a Professor in the Department of Media and Information and the Director of the Trifecta Initiative at Michigan State University. Her research examines the impacts of information and communication technologies across the life course, and the health, educational, and social impacts of that use. She is a Past Chair of the Communication and Information Technologies section of the American Sociological Association. Her work has been published in *Computers & Education*, *Computers & Human Behavior*, *Journal of Gerontology*, *Social Science & Medicine*, and the *Journal of Family Issues*, among others.

Elmir de Almeida, Ph.D., is Professor at the *Universidade de São Paulo*, linked to the Department of Education, Information and Communication, as well as to the Postgraduate Program in Education of the *Faculdade de Filosofia, Ciências e Letras de Ribeirão Preto* (Philosophy, Sciences and Literature College). Researcher of themes related to youth and youngsters and their interactions with the schooling and non-schooling educational spheres.

Brooke Dinsmore is a M.A. candidate at the University of Virginia dedicated to the study of childhood and adolescence at the intersectionalities of inequality.

Sara L. Dodd, Ph.D., became the Director of the Center for Adolescent Resiliency as of July 1, 2011. She is an Assistant Professor in the College of Human Sciences and is jointly appointed to Texas AgriLife Extension of the Texas A&M University as a state specialist in 4-H and Youth Development. Prior to her faculty appointment, she worked at Texas Tech as a research associate and project director for an NIH grant studying the effects of frequent work travel upon families (Dr. Anisa Zvonkovic, principal investigator). Prior to graduate studies, Dr. Dodd worked as an analyst in the IT industry and as a legislative staff member for a Texas state senator. She earned her bachelor's degree (government) from Abilene Christian University and her MBA (management) and Ph.D. (family/consumer Sciences) from Texas Tech. Dr. Dodd's research interests include leadership theory and development, adolescent health and wellness, and consumer/behavioral economics.

Harry T. Dyer is an Associate Tutor and Ph.D. candidate at the University of East Anglia. He achieved a scholarship to complete his Masters in Social Science Research Methods at the University of East Anglia at the age of 23, before continuing to receive scholarship funding for his ongoing Ph.D., under the supervision of Professor Victoria Carrington. At the University of East Anglia, Harry leads seminars and lectures on a wide range of social theories, methodologies, and practices at undergraduate level. He is a Digital Sociologist, whose research focuses on online identity formation. Using Actor-Network Theory, his research aims to explore how technology and site design can affect the creation and maintenance of identities online across a wide range of technologies and Social Networking Sites. He hopes to continue this work after his Ph.D., and to carry on researching and exploring the ever-changing face of Digital Sociology. Harry would like to thank his loving wife, Vanessa, for altruistically putting up with the long hours, the heavy-handed use of the third-person tense, and the distant frame of mind that so often goes hand-in-hand with academia.

Holly E. Follmer-Reece, M.S., Research Associate, Center for Adolescent Resiliency, Texas Tech University. Ms. Follmer-Reece is a doctoral candidate in Curriculum and Instruction at Texas Tech University. She joined the Center for Adolescent Resiliency in August 2011 and is responsible for designing and maintaining the research platform of United Future Leaders, Transforming through Transition, Food Challenge, and other special projects. Prior to working with the center, Ms. Follmer-Reece worked as a researcher on a NIH grant and for the Department of Human Development & Family Studies. She holds an M.S. in Human Development & Family Studies from Texas Tech University and a B.A. in Child & Family Studies and French from Baylor University.

Giovanna Gianesini, Ph.D., is Post-Doc Researcher at Bologna University, Italy. She is a Social Psychologist with extensive experience as Marriage and Family Counselor and Human Resources consultant. Her research focuses on resilience, mobbing, and psychological well-being in the workplace, the dynamics of violence in intimate relationships, bullying, family policies, aging, and gender identity. She has developed a Relational and

Emotion Regulatory Model (RERR) of resilience and is validating its measure in clinical (RERR-S) and organizational (RERR-HRM) settings. She has been member of the Interdisciplinary Center for Family Studies (CIRF) at University of Padua, editor of the column "Family Crucial Matters" for the Interdisciplinary Journal of Family Studies, member of the executive board of the European Society on Family Relation (ESFR) and co-editor of their Newsletter, member of the Editorial Board of CPFR (Contemporary Perspective on Family Research).

Philip Gibson is a Medical Sociologist interested in organizational dynamics and individual outcomes. He has previously studied the effects of technology use in education on student achievement and life chances. He is currently studying the effects of social structures within religious denominations on clergy burnout.

Gloria Gonzales is the Director of the United Future Leaders (UFL) program at the Center for Adolescent Resiliency. Since 1990, she has worked with children and adolescents as a teacher, counselor, and administrator. While serving as a school counselor in Lubbock ISD, she developed an elementary to middle school transition program which was adopted district-wide. In addition to overseeing the continued growth and development of UFL, Gloria serves as a presenter and speaker for staff development, conferences, and professional organizations and writes curriculum in the areas of leadership and wellness. In 2006, Gloria received the Texas Tech University's Excellence in the Field of Education *Tribute to Teachers Award*; in 2010 and 2012 the UFL program was recognized by the National Middle School Association as a *Team That Makes a Difference* and in 2011, the program was honored as the recipient of the Lubbock Volunteer Center's *Get Involved Award.* Gloria earned her Bachelor of Science in Education from Lubbock Christian University, her Master of Education in Counseling from Sul Ross State University and Mid-Management Certification in Administration from Lubbock Christian University.

Timothy M. Hale, Ph.D., is a Research Scientist at Partners' Center for Connected Health and an Instructor in Dermatology at Harvard Medical School. He received his doctorate in medical sociology from the University of Alabama Birmingham, in 2011. His research examines the use of new information and communication technologies in healthcare, emerging digital health lifestyles, and digital inequalities. His work has been published in the *Journal of Gerontology*, *Journal of Health Communication*, *Journal of*

Medical Internet Research, and *Information, Communication & Society*. He is a co-editor of two forthcoming volumes, *Digital Distinctions & Inequalities*, and *New Media Cultures*, in the *Emerald Studies in Media and Communications: Communication and Information Technologies Annual*.

Jodie L. Hertzog, Ph.D., is an Associate Professor and is currently Chair of the Sociology Department at Wichita State University where she also serves as undergraduate coordinator. She received her doctorate in Child Development and Family Studies from Purdue University. Her research and teaching specialties revolve around intimate relationships, gender inequalities, violence prevention, and qualitative research methods.

Erin Kostina-Ritchey, Ph.D., joined the Center for Adolescent Resiliency (CAR) team in Fall 2013. Prior to her experience with CAR, Erin worked as a researcher on a NIH grant, for the Department of Human Development & Family Studies and for the Department of Bilingual Education & Diversity Studies. She holds a Ph.D. in Human Development & Family Studies, M.Ed. in Bilingual Education & Diversity Studies, and an M.A. in Interdisciplinary Studies (International Development and Public Policy) all from Texas Tech University. In addition, Erin is certified in Family Life Education. Prior to her graduate studies, Erin was a social services and education consultant for a non-profit organization in Central Asia. Her research interests include adolescent/emerging adult development, curriculum development, and community development.

Linda Lundin has a Ph.D. in Psychology and is a senior lecturer at University West, Sweden. She mainly teaches Developmental Psychology and courses in research methodology. Her research interests revolve around how people deal with challenges in life and the consequences of facing early challenges, such as growing up in a dysfunctional family. She has previously been involved in research projects on eating disorders, social anxiety, and parents' insights into their teenagers' Internet activities. Linda is currently involved in an international research project on how experiences of living with tinnitus interact with previous life history.

Yuping Mao (Ph.D., Ohio University) is Assistant Professor in the Department of Media and Communication at Erasmus University Rotterdam, the Netherlands. Her research focuses on intercultural, organizational, and health communication. Yuping's research interests in health communication focus on the following areas: cross-cultural interpersonal communication between patients and health professionals, health information diffusion among different cultural groups, and media effects on health behaviors such as substance use. Yuping is particularly interested in health communication issues among migrants. Yuping has published a few book chapters, and her work has also been published in peer-reviewed journals such as *Communication Research*, *Canadian Journal of Communication*, *China Media Research*, *International Journal of Health Planning and Management*, *Journal of Substance Use*, and *Italian Journal of Pediatrics*.

Vanderlei Mariano, M.S., is Professor at the *Centro Universitário Fundação Santo André*, linked to the *Faculdade de Filosofia, Ciências e Letras* (Philosophy, Sciences and Literature College), Mathematics Department. Researcher of themes related to (a) youth and youngsters and their interactions with the schooling and non-schooling educational spheres and (b) educational practices in the field of mathematics.

Christine McCauley Ohannessian is an Associate Professor of Pediatrics and Psychiatry at the University of Connecticut School of Medicine. She is also the Director of the Children's Center for Community Research at Connecticut Children's Medical Center. Dr. Ohannessian's research interests focus on adolescent psychological problems, substance use, and technology use. She is especially interested in risk and protective factors involved in adolescent problem behaviors.

Gregory J. Mills is currently a Ph.D. student in the Department of Sociology at the University of Connecticut, and received his M.A. in Sociology from the University of South Florida in Tampa. As a Master's student, his thesis analyzed the experiences of first-generation immigrants. The work in this chapter aligns with these themes. More broadly, his current research interests also include the causes of residential segregation by both race and income, as well as how segregation works to structure everyday experiences and opportunity structures. He has published on American homeownership trends, and how homeownership, segregation, and mobility have historically been intimately tied together in the 20th and 21st century American contexts. His dissertation will encompass these themes by looking at the causes and consequences of growing class segregation, particularly in terms of educational achievement and family formation patterns.

Marilena Nakano, Ph.D., is Professor at the *Centro Universitário Fundação Santo André*, linked to the *Faculdade de Filosofia, Ciências e Letras* (Philosophy, Sciences and Literature College), Pedagogy Department. Researcher of themes related to (a) youth and youngsters and their interactions with the schooling and non-schooling educational spheres; (b) educational practices; and (c) social movements.

Jessica Niblack graduated from Wichita State University in 2010 with a Bachelor's in Women's Studies and in 2013 with a Master's in Sociology. She has taught Sociology courses at Wichita State University, Newman University, and is currently employed as an instructor in Sociology at Hutchinson Community College. She also teaches for Fort Hays State University's virtual college. Her courses include Introduction to Sociology, Social Inequality, Diversity, and Gender Roles. Her interests primarily center on gender and inequality.

LaToya O'Neal Coleman is a postdoctoral research associate in the Department of Media and Information at Michigan State University. She recently completed her doctorate in medical sociology at the University of Alabama at Birmingham. Her research examines the role of risk and protective factors in developmental psychopathology among children and adolescents. She is also interested in the social reproduction of health and educational inequality.

Gina Pol is a B.A. candidate at Connecticut College, double majoring in Sociology and Economics, who is interested in the study of consumerism, adolescence, and inequality.

Matthew H. Rafalow is a Ph.D. Candidate in Sociology at University of California-Irvine with research interests in education and technology. His earlier work on technology use in schools was published in the *International Journal of Sociology of Education* and *Revue Internationale d'Éducation Sèvres*. He is also co-authoring a book on learning online from work with the Connected Learning Research Network, a MacArthur Foundation-funded research initiative. Prior to graduate work in Irvine, he received his Master's in Education from Columbia University Teachers College.

Kayla Sherman became a staff member with the Center for Adolescent Resiliency in February 2012 and currently serves as the Assistant Director of the United Future Leaders program. Her fascination with the multiple facets of media and communication studies, as well as a passion for working with youth, greatly influenced her education and career goals. Kayla graduated from Texas Tech University (TTU) in 2011 with a dual degree Bachelor of Arts in Communication Studies and Public Relations. She also completed her teaching certification program in the area of secondary speech education during her undergraduate studies. Kayla is currently pursuing a Master of Science in Family and Consumer Sciences Education at TTU and hopes to complete her graduate studies in 2016. Through her work with CAR, Kayla has been actively involved in the development and implementation of program curriculum, served as a presenter for various local, state, and national conferences and organizations, and assisted with the development of many research initiatives.

Lu Shi received his Ph.D. in Public Policy Analysis from F.S. Pardee RAND Graduate School and an M.A in media studies from Syracuse University. He is currently an Assistant Professor in Department of Public Health Sciences, Clemson University. Previously a research scientist at UCLA School of Public Health, Lu Shi has been developing the microsimulation model to forecast future health trends for the national and state populations. He is also designing the dissemination strategy of research results, using his graduate training and professional experience as a communication expert. Before coming to the United States of America, he worked for three years as a journalist in China. He has published many peer-reviewed journal articles in disease prevention and early detection, including substance abuse, obesity prevention, health communication, and cancer screening. He recently finished a randomized controlled trial of mindfulness-based meditation to explore its cost-effectiveness as a health intervention.

Brian Simpson is a Professor of Law in the School of Law at University of New England in Armidale, Australia. He has previously held appointments at various Australian universities and at Keele University in the United Kingdom. His principal research interests are in children's rights, social justice and law, and the law's regulation of urban problems. His work is interdisciplinary in nature. In the area of children and the law, his work seeks to understand how the law, read as a narrative, helps to explain constructions of childhood and family life. Within that context, he is particularly concerned with the manner in which law accepts or rejects the degree to which children are afforded autonomy, and how this connects with different conceptions of childhood.

Jennifer E. Simpson is a regional academic manager working for The Open University, on its undergraduate social work program in the south of England. As a social work practitioner, she specialized in children and family social work, in particular foster care. She has also worked extensively with educational establishments in the area of child protection and building resilience in fostered children. Jennifer is currently undertaking a Ph.D. with Edinburgh University where the focus of her research is fostered children and issues related to contact. Her study of this area provides an opportunity, unlike existing research, to capture how contact between fostered children and their birth families has changed as a result of 3G telecommunications and social media, and whether or not there is a tangible impact in terms of key areas of local authority care, for example, placement stability, reunification, and safeguarding. Jennifer has also published in peer-reviewed journals such as Child and Family Social Work; Journal of Adoption and Fostering and Every Child Journal.

Radhamany Sooryamoorthy is Professor of Sociology and an Academic Leader in the School of Social Sciences at the University of KwaZulu-Natal in South Africa. He has taught at the Acharya Nagarjuna University, and Loyola College of Social Sciences (both in India), the University of Calgary (Canada), and the Luleå University of Technology (Sweden). His current research interests lie in the areas of communication, scientific collaboration, and productivity. His previous works include *Science in Participatory Development* (co-author) (Zed Books, 1994) *NGOs in India* (co-author) (Greenwood Press, 2001), and *Transforming Science in South Africa: Development, Collaboration and Productivity* (Palgrave Macmillan, 2015).

Emma Sorbring is Associate Professor in Psychology and is the Research Director for Child and Youth Studies at University West, Sweden. Her research and teaching interests lie in the area of children, adolescents, and families. She is the site project manager for a worldwide parenting project, Parenting Across Cultures, which spans 13 cultures, has 1,300 participating families, and will continue for nine years. Other projects focus on teenagers' Internet use and parental strategies, sexual development in traditional and new settings (Internet), dating violence, parental behavior and children's adjustment, and young people's decision-making. Her most recent project "Particip@tion on the Internet?" is about pupils with intellectual disabilities and identification processes on the Internet.

Sarah Tulane is a Clinical Assistant Professor of Family and Human Development at Utah State University. Her research interests include family life education, adolescent psychosocial development, and the combination of the two. Her current research focuses on adolescent social development in relation to interactive technology such as text messaging and social media. Dr. Tulane's research includes an examination of the social rules and contexts of text messaging, parent involvement in social media, and social media use in marital relationships.

J. Mitchell Vaterlaus, Ph.D., LMFT, is an Assistant Professor in the Department of Health and Human Development at Montana State University. Dr. Vaterlaus studies the role of new media and technology on family and human development. His published research has focused on motivations for adolescent and young adult technology use in different contexts and the connection between technology and health. Dr. Vaterlaus is particularly interested in the role technology plays in the parent-child relationship during adolescence and young adulthood. His studies in this area have focused on parental mediation, youth culture, and generational differences in technology knowledge.

Maria Elena Villar e Villar, Ph.D., is Professor at the *Centro Universitário Fundação Santo André*, linked to the *Faculdade de Filosofia, Ciências e Letras* (Philosophy, Sciences sand Literature College), Pedagogy Department. Researcher of themes related to (a) youth and youngsters and their interactions with the schooling and non-schooling educational spheres; (b) infancy; and (c) educational practices.

Kevin Zevallos is a B.A. candidate at Connecticut College, majoring in Sociology, who is interested in the study of politics and masculinity within youth's lived realities.